What is the Electron?

Edited by **Volodimir Simulik**

Apeiron
Montreal

Published by C. Roy Keys Inc.
4405, rue St-Dominique
Montreal, Quebec H2W 2B2 Canada
http://redshift.vif.com

First Published 2005

Library and Archives Canada Cataloguing in Publication

What is the electron? / edited by Volodimir Simulik.

Includes bibliographical references.
ISBN 0-9732911-2-5

1. Electrons. I. Simulik, Volodimir, 1957-

QC793.5.E62W48 2005 539.7'2112 C2005-902252-3

Cover design by François Reeves.
Cover graphic by Gordon Willliams.

Table of Contents

Preface

The electron is the first elementary particle, from both the physical and the historical point of view. It is the door to the microworld, to the physics of elementary particles and phenomena. This book is about electron models.

The year 1997 marked the centenary of the discovery of the electron as a particle by J.J. Thomson. We have already passed the centenary of Planck's great discovery and the beginning of quantum physics; 2001 marked the 75th anniversary of Schrödinger's equation and the beginning of quantum mechanics, while the year 2003 was the 75th anniversary of the Dirac equation and Dirac's model of the electron.

Today the most widely used theoretical approaches to the physics of the electron and atom are quantum mechanical and field theoretical models based on the non-relativistic Schrödinger and the relativistic Dirac equations and their probabilistic interpretation. This is the basis of modern quantum field theory. More than 75 years is a long time for a physical theory! This theory is the basis for all contemporary calculations of physical phenomena.

After 75 years most physical theories tend to be supplanted by new theories, or to be modified. The theory's successes, as well as its difficulties, are now evident to specialists. There is no proof of the uniqueness of the quantum field theory approach to the model of the electron and atom. Are other approaches possible? Quantum field theory may be sufficient to describe the electron, but is it necessary? This theory and its mathematics are very complicated; can we now propose a simpler construction? Is the electron an extended structure, a compound object made up of sub-particles, or is it a point-like elementary particle, which does not consist of any sub-particles? What is the limit of application of modern classical physics (based either on the corpuscular or wave model) in the description of the electron? These and many other questions remain without definitive answers, while experiments on quantum entanglement have given rise to new discussion and debate. New high-precision experimental data, *e.g.*, on the electric and magnetic dipole moments of the electron, may prove decisive.

This book, *What is the electron?*, brings together papers by a number of authors. The main purpose of the book is to present original papers containing new ideas about the electron. *What is the electron?* presents different points of view on the electron, both within the framework of quantum theory and from competing approaches. Original modern models and hypotheses, based on new principles, are well represented. A comparison of different viewpoints (sometimes orthogonal) will aid further development of the physics of the electron.

More than ten different models of the electron are presented here. More than twenty models are discussed briefly. Thus, the book gives a complete picture of contemporary theoretical thinking (traditional and new) about the physics of the electron.

It must be stressed that the vast majority of the authors do not appeal to quantum field theory, quantum mechanics or the probabilistic Copenhagen interpretation. The approaches adopted by these authors consist in using "lighter" mathematics and a "lighter" interpretation than in quantum theory. Some of them are sound approaches from the methodological point of view.

The editor will not presume to judge the models or the authors. We will not venture to say which model is better, and why. The reasons are simple. (i) Readers can reach their own conclusions themselves. (ii) Investigation of the electron is by no means finished. (iii) My own point of view is presented in my contribution to the book. So I want my paper to be on an equal footing with other new models of the electron presented here.

The general analysis of the electron models presented here shows that they can be classified as follows: corpuscular and wave, classical and quantum, point and extended, structureless and with structure. The reader can compare and ponder all these approaches! I would like to thank the authors for their contributions.

It is my hope that this volume will prove worthwhile for readers, and encourage them to pursue further investigation of electron models.

Volodimir Simulik
Senior Research Associate
Institute of Electron Physics
Ukrainian National Academy of Science.
Uzhgorod, Ukraine

A Comprehensive Theory of the Electron from START

Jaime Keller
Departamento de Física y Química Teórica
Facultad de Química
Universidad Nacional Autónoma de México
AP 70-528, 04510, México D.F., MEXICO
E-mail: keller@servidor.unam.mx
and keller@cms.tuwien.ac.at

Space and Time are primitive concepts in science, used to describe material objects in relation with other material objects and the evolution of those relations. Mathematical description of those relations results in an observer's geometric frame of reference. To describe the object's behaviour, we add one more geometric element: the Action attributed to the system of objects. A fundamental concept is that of action carriers. The resulting Theory has a deductive character. A comprehensive (mass, charge, weak charge, spin, magnetic moment) theory of the electron is presented from this point of view. The main emphasis is given to the mathematical structures needed and the epistemological issues of the theory.

PACS number(s): 01.55.+b, 31.15.Ew, 71.10.-w, 71.15.Mb

Keywords: Space-Time-Action, Electron, Neutrino, Magnetic Interaction, Weak Interaction, START.

1. Introduction: space, time and material objects – mathematical structures

This paper contributes to the construction of a deductive theory of matter, starting from first principles and using a single mathematical tool, geometric analysis. We present a comprehensive theory, where the analysis is centered in the theory of the electron.

It represents a logical continuation of the material presented in the volume *The Theory of the Electron, A theory of matter from START* and a series of publications [1-5].

We recast here our fundamental philosophical and methodological remark. The theory of the electron developed in the above mentioned book is based on two main theoretical considerations: the nature of a scientific theory and the elements used to describe nature. The basic purpose of the theory presented here is a description of what can be observed, inferred, related and predicted within the fundamental limitations of experimental and theoretical science. We do not go beyond these limitations in any sense, nor seek to derive fundamental concepts from model structures which might be supposed to be more fundamental.

We use three basic elements of physical objects and phenomena: time, space, and action density. The first element, a one-dimensional manifold *time*, an evolution parameter, (a primitive concept, universally accepted) is defined by its mathematical properties. The concept of *space*, frame of reference, is defined, using the same considerations, through its mathematical description; this requires a three dimensional manifold in agreement with our anthropological apprehension of nature. The third element of physical nature considered here is given the unfortunate name of density of *action*, and describes the existence of physical objects, assuming action is a one dimensional manifold joined to the previous four in a geometrical unity. We have refrained from giving this concept a new name because we want to emphasize that we are presenting new conceptual and mathematical structures (*Principia Geometrica Physicae*). In our presentation an action density field is introduced into the space-time frame of reference to describe matter through the properties of this action density distribution. The space-time-matter concept is tautological: it is a set of non-separable concepts in nature.

Geometry is introduced through the use of a quadratic form to give a quadratic space structure to the variables:

Quadratic Forms from Pythagoras to the XXI Century

Quadratic form	Dim/diff. Op	Group
$l^2 = x^2 + y^2 + z^2 \qquad (\Delta t)$	$3-D \quad \nabla, \nabla^2$	Statics Galileo
$s^2 = (ct)^2 - (x^2 + y^2 + z^2)$	$4-D \quad D, D^2$	Kinetics Poincaré
$S^2 = (ct)^2 - \left(x^2 + y^2 + z^2\right) - w^2$	$5-D \quad K, K^2$	Dynamics START

$$w = \kappa_{(0)}\overline{a}, \qquad \kappa_{(0)} = \frac{d_{(0)}}{h} = \frac{c}{E_{(0)}}, \cdots \left(\overline{a}\right)^2 = \sum_\mu a_\mu^2$$

When distributions of action in space-time are made to correspond to physical objects, we conclude that, as time evolves, the permanence of these objects is related to a set of symmetry constraints on that action distribution. The presentation used is then both a mathematical and an epistemological approach to the study of matter, and of physics itself.

1.1   Epistemological approach

The procedure followed in this article is:

1.   To define a frame of reference to describe physical objects as a distribution field (carrier), the geometric space-time-action frame allows the definition of velocities and of energy-momentum as derivatives.

2.   To define fields of carriers through a set of properties.

3. To describe interaction as the possibility of exchange of energy-momentum among the carriers (of sets of properties).
4. To find the equations for the interaction fields.
5. To find the sources of the interaction fields.
6. To determine the physical properties of a field of sources in such a form that those fields can be used as carriers.
7. To find the equations for the (source) carriers.
8. To find the observable properties of those carriers and identify them as the observed electron and its elementary particle partner, the neutrino.

1.2   Position and localization

Once the sub-frame space-time is defined there is a fundamental difference between position and localization. *Position* refers to a mathematical point x in space (which in general can be described by an anchored vector). The fixing point is called *Coordinate Origin* and a Poincaré coordinate transformation includes a change of this reference point. *Localization* refers to the possibility of assigning a restricted, continuous, set of position points $\rho(\mathbf{x}) \neq 0$ to a physical object (or phenomena). Localized objects are those for which a domain of position points can be assigned, the size of the domain being defined as the *size* of the physical object (or phenomenon). *Non-localized* objects correspond to those for which the domain of explicitly considered position points is larger than some assumed size of the object (or phenomenon).

1.3   Mass, charge, action, space and time

In our theory action, as a fundamental variable, is distributed among a set of *carrier of action fields*. An action density $w(\mathbf{x}, t)$, action w per unit space-time hypervolume $\Delta x_0 \Delta x_1 \Delta x_2 \Delta x_3$ at point $(\mathbf{x}, t)$ with $x_0 = ct$, is the fundamental concept defining space (parameterized by x), time (parameterized by t), and action density (parameterized by a scalar analytical function $w(\mathbf{x}, t)$, as *primitive* concepts from which all other physical quantities will be derived or at least related directly or indirectly. The different forms of distributing the action among these carriers define the carriers themselves. This is fundamental in the practical use of the four principles below. For an elementary carrier n we will define $w_n(x, t) = f_n \rho_n(x, t)$. With constant in space f_n.

Within our fundamental formulation we will have to define properties of the fields we call carriers. A carrier will have physical significance through its set of properties. The density ρ of an elementary carrier field can be defined through a set of scalar constants, such that the integral of the product of these constants, and the density gives the experimentally attributed value of a property for that carrier. We will use an example: a carrier field identified with an electron will have a density $\rho(\mathbf{x}, t)$, and if the property is Q we will define $Q = \int q(\mathbf{x}, t) d\mathbf{x} = \int q\rho(\mathbf{x}, t) d\mathbf{x}$ for all t, which determines that Q is a constant property (in space and time) for that field. The set of properties $\{Q\}$ characterizes a carrier field and in turn establishes the conditions for a density field to correspond to an acceptable carrier.

The concept of *charge* appears in the theory first of all from the necessity to define the objects which exchange action (charges are always relative properties) in order to give a formal meaning to the principle that action will be exchanged in integer units of the Planck constant. In this context for an electron-like carrier both mass and electric charge belong to the generic name of 'charges'. This program can obviously not be achieved if the formulation is not suitable to deduce of the theory of elementary particles, giving a geometric meaning to this theory.

The definition of w is as a finite analytical action density and, to agree with standard formulations, the energy density $\tilde{E} = \partial w / \partial t$ and the momentum density $p_i = \partial w / \partial x^i$ are the fundamental rates of change of the primitive concept of action (considering a unit time-like interval $\Delta x_0 = 1$).

In our full geometrization scheme a vectorial representation $X = x_\mu e^\mu$ for $\{ct, x, y, z; x_\mu, \mu = 0,1,2,3\}$ is used, and from the space-time gradient of w we recover the positive semi-definite energy-momentum expression

$$E^2 / c^2 - p_x^2 - p_y^2 - p_z^2 = P^2, \qquad P = p_\mu e^\mu, \qquad (1)$$

as well as the space-time $ds^2 = c^2 dt^2 - dx^2 - dy^2 - dz^2$. Action change $dK = P \cdot dX$ is introduced through quadratic terms $|dK|^2$ (see appendix)

$$dS^2 - ds^2 = -|dK|^2 = -\kappa_0^2 \left\{ (E^2 / c^2) c^2 dt^2 - p_x^2 dx^2 - p_y^2 dy^2 - p_z^2 dz^2 \right\}, \qquad (2)$$

creating a unified geometrical quadratic form dS^2. The dK vector, the directional in space-time change of action, is a new theoretical quantity formally defined by (2). Notice that the generalization $ds^2 \Rightarrow dS^2$ also corresponds to a (generally curved) generalization of the space-time metric

$$dS^2 = \left(1 - \kappa_0^2 E^2 / c^2\right) c^2 dt^2 - \left(1 - \kappa_0^2 p_x^2\right) dx^2 - \left(1 - \kappa_0^2 p_y^2\right) dy^2 - \left(1 - \kappa_0^2 p_z^2\right) dz^2$$

1.4   Hypotheses and principles of START

The set of hypotheses and principles which are explicitly included in our theory are called START [3]:

> Physics is the science which describes the basic phenomena of Nature within the procedures of the Scientific Method.

> We consider that the mathematization of the anthropocentric primary concepts of space, time and the existence of physical objects (action carriers), is a suitable point of departure for creating intellectual structures which describe Nature.

> We introduce a set of principles: *Relativity, Existence, Quantization and Choice* as the operational procedure, and a set of 3 mathematical postulates to give these principles a formal, useful, structure.

> We have derived in this and previous papers some of the fundamental structures of Physics: General Relativity, Density Functional Theory, Newtonian Gravitation and the Maxwell formulation of Electromagnetism. A fundamental common concept is the definition of energy (action) carriers. Most of the relations presented here are known, our procedure derives these structures and theories from START.

1.5 Energy, momentum and interaction fields

There are in the theory two different forms of studying contribution to energy and momentum: the quantities defined in the paragraphs above and, second, quantities that will be called relative energy or relative momentum.

Principle of Space-Time-Action Relativity. In a space-time-action manifold an unstructured observer cannot determine his own state of motion; he can only determine the relative motion of other bodies in relation to himself and among the other bodies themselves. Light in the space-time-action manifold is assigned the "speed" c.

An observer of a "system of bodies" will describe first each body as in motion relative to the observer with the concept: *motion originated momentum* (**p**), and, second, the motion of that body in relation to the rest of the system with the concept: *interaction originated momentum* ($\Delta \mathbf{p} = e\mathbf{A}$). The interaction originated momentum is the result of a non-unique description procedure, this freedom of definition will mathematically appear as a "gauge freedom" in the formulations below. The *total momentum* to enter in the descriptions for bodies in interaction is $\mathbf{p} + \Delta \mathbf{p}$.

1.6 Action carriers in START

Consider a set of scalar field "carriers" in such a form that the total action density in space-time is the sum of the action attributed to the carriers. Some properties arise from the START geometry itself, others from the description of a physical system as a time evolving energy distribution. In stationary systems, for a given observer, an elementary carrier field c *is defined* to have an energy density $\varepsilon_c \rho_c(\mathbf{x})$ with ε_c being a constant in space, and an integer number of carriers N_c of type c. The density $\rho_c(\mathbf{x},t)$ obeys $\int_V \rho_c(\mathbf{x},t)\, dx = N_c$ in the system's volume V.

We make a sharp distinction between action density and Lagrangian density. The Lagrangian contains, in general, prescriptions (and Lagrange multipliers) for the description of the system.

Both the action density function $w(X)$ and the splitting among carrier fields will be considered analytically well-behaved functions. A *description* is introduced when we treat the energy $\tilde{E}(t)$ of a system as a sum of the different carrier types $\{c\}$ such that $\tilde{E}(t) = \sum_c \tilde{E}_c(t)$, a sum of constants $\tilde{E}_c(t)$ in space for a given observer.

1.7 Carriers and physical bodies

The charges are to be defined in our theory from a geometrical analysis of the distribution $w(X)$ when momentum is described in two ways: the amount which is related to the rate of change of action with respect to relative position, and the amount, per unit charge, which is pairwise shared, adding to zero, among the carriers.

The rates of change of relative energy and momentum are called forces. A carrier for which a current of charges can be defined is by definition a body. A body corresponds to our hitherto undefined concept of matter.

Our study below will show that we cannot define an elementary body unless other properties, in addition to charge, are given to the carrier.

> In our presentation the word "particle" is systematically avoided as for many authors it refers to a "point" body, with no spatial dimensions. The word body, on the other hand, conveys the idea of spatial distribution. Point-like distributions can only be introduced as a practical tool for handling a distribution confined to a region of space small in relation to the total system's volume.

1.7.1 Maxwell equations from START

In our formalism [3,4] the Maxwell equations in their standard textbook form are analytical properties of the third derivatives of the action density attributed to a test carrier (with 'electric' charge) as induced by a collection of interacting carriers. The energy per carrier can be considered the derivative of a scalar field, but the momentum for interacting carriers cannot be solely considered the gradient of a scalar field. In this particular case, assume that we describe a set of carriers as interacting by partitioning an amount of energy (the interaction energy $\tilde{E}_e(X)$) among them, allowing the partitioning to be described as the sum of the overall momentum $(\partial w_e(X)/\partial x^i)e^i$ plus the momentum $\Delta_R p_{e,i} e^i$ induced by interactions among the carriers. These interaction moment fields might then have a non-null rotational part.

Consider, in the reference frame of a given observer, the induced action density (arising from the interaction), denoted by $\partial w_e(X)$, per unit charge ($\Rightarrow$ p.u.ch) of a test carrier at space-time point $X = x^\mu e_\mu$. Here the Greek indices $\mu = 0,1,2,3$ and $x^0 = ct$ whereas the space vectors $\mathbf{q} = q^i \mathbf{e}_i = q_i \mathbf{e}^i$, $\mathbf{e}_i = e_0 e_i$, $i = 1,2,3$ are written in bold face letters, and we use the standard definitions of "dot" and "cross" products. From it define the related energy density $\mathbf{E}_e(X)$ and the total (external plus induced) momentum density $\mathbf{p}_e$, *per unit charge of the test carrier*, as

$$\mathbf{E}_e(X) = \frac{\partial w_e(X)}{\partial t}, \quad \mathbf{p}_e = p_{e,i} e^i = \left(\frac{\partial w_e(X)}{\partial x^i} + \Delta_R p_{e,i} \right) e^i, \ \{\text{def. 1}\} \qquad (3)$$

also, by definition, the electric field strength $\mathbf{E}$ as the force (p.u.ch) corresponding to these terms

$$\mathbf{E} = \left(\frac{\partial \mathbf{E}_e(X)}{\partial x^i} + \frac{\partial p_{e,i}}{\partial t} \right) e^i = \nabla \mathbf{E}_e(X) + \frac{\partial \mathbf{p}_e}{\partial t}, \qquad (4)$$

with time dependence

$$\frac{\partial \mathbf{E}}{\partial t} = \left(\frac{\partial^2 \mathbf{E}_e(X)}{\partial t \partial x^i} + \frac{\partial^2 p_{e,i}}{\partial t \partial t} \right) e^i = 2 \frac{\partial^3 a_e(X)}{\partial t \partial x^i \partial t} e^i + \frac{\partial^2 (\Delta_R p_{e,i})}{(\partial t)^2} e^i.$$

By definition of interacting carriers, we have added in {def. 1} the term $\Delta_R p_{e,i} e^i$ as the effect of the conservation of *interaction transverse moment* between the fields representing the rest of the carriers with that sort of charges. This is by definition the origin, in START, of a magnetic field intensity

$B = B_k e^k$ that will appear as the curl of the momentum (p.u.ch) of an interaction field acting on a carrier of type b. The axial vector

$$B = \left(\frac{\partial p_{e,l}(X)}{\partial x^i}\right) e^j \times e^i = \nabla \times p_e,$$

with time dependence

$$\frac{\partial B}{\partial t} = \left(\frac{\partial^2 p_{e,l}(X)}{\partial t \partial x^i}\right) e^j \times e^i.$$

Otherwise the space variation of E, including the *interaction transverse moment*, $\nabla E = \nabla \cdot E + \nabla \times E$, will also include a transversal (rotational) term

$$\nabla \times E = \left(\frac{\partial^2 p_{e,j}(X)}{\partial x^i \partial t}\right) e^i \times e^j = -\frac{\partial B}{\partial t} \quad \text{\{2nd Maxwell Equation\}} \tag{5}$$

relation which is the direct derivation in START of this well known Maxwell equation. The scalar term $\nabla \cdot E$ being a divergence of a vector field should be defined to be proportional to a source density

$$\nabla \cdot E = \frac{1}{\varepsilon_0} \rho = \sum_i \left(\frac{\partial^2 w_e(X)}{\partial x^i \partial x^i \partial t}\right) = \frac{\partial}{\partial t} \nabla^2 w_e(X), \quad \text{\{1st Maxwell Equation\}}$$

and will be given full physical meaning below.

For the space variation of B we have

$$\nabla B = \nabla \cdot B + \nabla \times B.$$

The first term vanishes identically in our theory because it corresponds to the divergence of the curl of a vector field

$$\nabla \cdot B = 0, \quad \text{\{3rd Maxwell Equation\}}$$

while the last term, using $U \times V \times W = V(U \cdot W) - (U \cdot V)W$

$$\nabla \times B = \nabla(\nabla^2 w_e(X)) - \nabla^2 p_e = \mu_0\left(J + \varepsilon_0 \frac{\partial E}{\partial t}\right), \quad \text{\{4th Maxwell Equation\}}$$

The additional dimensional constant μ_0 is needed to transform from time units (used in the conceptual definition of a current $J = \nabla(\nabla^2 a_e(X))/\mu_0$) into distance units. The units of $\varepsilon_0\mu_0$ are of T^2/D^2 or inverse velocity squared, in fact (see below) $\varepsilon_0\mu_0 = c^{-2}$ corresponding to have used above twice the derivative with respect to t and not to $x_0 = ct$.

The (4th Maxwell Equation), defining J, is related to the analog of the (1st Maxwell Equation) and the analog of the (2nd Maxwell Equation), also to a Lorentz transformation of the (1st Maxwell Equation).

The Maxwell equations can be formulated in 4-D form ($\square = e^\mu \partial_\mu$)

$$e_0 \square = \frac{1}{c}\partial_t + \nabla = \frac{1}{c}\partial_t + e_i\partial_i, \quad \tilde{J} = e^\mu J_\mu, \quad J_0 = \rho, \quad \mu = 0,1,2,3, \quad x_0 = ct):$$

$$F = E + cB, \quad \nabla F = \frac{1}{\varepsilon_0}\left(\rho + \frac{1}{c}J\right) - \frac{\partial}{c\partial t}F, \quad \square F = \frac{1}{\varepsilon_0}\tilde{J}.$$

The here derived Maxwell equations are formally equivalent to the original Maxwell equations, then they are: first local equations and second linear in the sources (ρ and J).

Both the (4th Maxwell Equation), defining J, related to a Lorentz transformation of the (1st Maxwell Equation) defining ρ, can immediately be integrated using geometric analysis techniques, the standard approach being of fundamental conceptual consequences in START. The space divergence of a non-solenoidal vector field like E is immediately interpreted as its 'source' using the standard geometric theorem that the volume integral of a divergence $\nabla \cdot E$ equals the surface integral of the normal (to the surface) component of the vector field $\mathbf{n} \cdot \mathbf{E}$. Consider:

$$\int_V (\nabla \cdot E)dV = \int_V \frac{4\pi}{\varepsilon_0} \rho(r')r'^2 dr' = \frac{1}{\varepsilon_0}Q = \int_S E(r)\frac{(\mathbf{r} \cdot \mathbf{n})}{r}dS = 4\pi r^2 E(r),$$

$$E = E(r)\frac{\mathbf{r}}{r} = \frac{Q}{4\pi\varepsilon_0 r^2}\frac{\mathbf{r}}{r}, \qquad \mathbf{r} \cdot \mathbf{n} = r.$$

That is: the inverse square law of the Newtonian and Coulomb forces are geometrical consequences of the definition of interaction among charged carriers. Nevertheless this is not a derivation of the value(s) of the (Newtonian and) Coulomb constant(s) G and ε_0.

For a small ($l \leq r$) current source at the origin of coordinates: (in the sphere $\mathbf{r}'(\theta,\phi) \cdot \mathbf{r}^{ct} = 0$, $(\mathbf{r}')^2 = (\mathbf{r}^{ct})^2 = 1$)

$$\int_V (\nabla \times B)dV = \int_S B(r)(\mathbf{r}'(\theta,\phi) \times \mathbf{n})dS = 4\pi r^2 fB(r)\mathbf{r}^{ct},$$

$$\int_V 4\pi\mu_0 J\delta(r')r'^2 dr' = \mu_0 M\mathbf{r}^{ct} = 4\pi r^2 fB(r)\mathbf{r}^{ct} \Rightarrow B = B(r)\mathbf{r}' = \frac{\mu_0 M}{4\pi r^2 f}\mathbf{r}',$$

and its Amperian inverse square law is also a geometrical consequence of the definition of transverse interaction among charged carriers.

1.7.2 Beyond Newtonian gravity

The analysis above depends only on the assumption of the decomposition of the action and of the energy momentum into contributions per carrier. The analysis above can applied to gravitation considering the mass $M = E/c^2$. The Newtonian gravitational potential equation per unit test mass m

$$V(r) = -G\frac{M}{r}, \quad \text{that is} \quad E = -G\frac{M}{r^2},$$

the usual relations in the textbook formulation of Newtonian gravity. The constant $G = 1/4\pi\varepsilon_0^{(g)}$. If we define $c^2\mu_0^{(g)}\varepsilon_0^{(g)} = 1$ then $\mu_0^{(g)} = 4\pi G/c^2$.

In this approach to gravitation there is no quantization properly, there being no exchange of action, only a description of the sharing of energy between a source carrier and a test carrier. We include the transverse momentum in the interaction between sources of the gravitational field:

$$\nabla \times E_g = \frac{\partial^2 p_j^{(g)}(X)}{\partial x^i \partial t} e^i \times e^j = -\frac{\partial B_g}{\partial t},$$ (6)

$$\nabla \times B_g = \nabla\left(\nabla^2 w_e^{(g)}(X)\right) - \nabla^2 p_e^{(g)} = \mu_0^{(g)}\left(J_g + \varepsilon_0^{(g)} \frac{\partial E_g}{\partial t}\right) \frac{4\pi G}{c^2} J_g + \frac{1}{c^2} \frac{\partial E_g}{\partial t}$$

1.8 Formal definition of carrier fields

We follow our presentation in [4] (Keller and Weinberger).

A carrier-domain **B** is a connected open set whose elements can be put into bijective correspondence with the points of a region (domain in some instances) *B* of an Euclidian point space **E**. *B* is referred to as a configuration of **B**; the point in *B* to which a given element of **B** corresponds is said to be "*occupied*" by that *element*. If **X** denotes a representative *element* of **B** and **x** the position relative to an origin **0** of the point **x** occupied by **X** in *B*, the preceding statement implies the existence of a function ϑ: **B** $\rightarrow$ B$_0$, (B$_0$, stands for the totality of the positions relative to **0** of the points of *B*) and its inverse Θ: B$_0$ $\rightarrow$ **B** such that

$$x = \vartheta(X), \quad X = \Theta(x)$$ (7)

In a motion of a carrier-domain the configuration changes with time

$$x = \phi(X,t), \quad X = \Phi(x,t)$$ (8)

In a motion of **B** a typical element **X** occupies a succession of points which together form a curve in **E**. This curve is called the path of **X** and is given parametrically by equation (8). The rate of change v of **x** in relation to t is called the velocity of the element **X**, (our definitions run parallel to those of an extended body in continuum mechanics; see for example Spencer 1980 [8]). The velocity and the acceleration of **X** can be defined as the rates of change with time of position and velocity respectively as **X** traverses its path. "Kinematics" is this study of motion *per se*, regardless of the description in terms of physical forces causing it. In space-time a body is a bundle of paths.

Equations (8) depict a motion of a carrier-domain as a sequence of correspondences between elements of **B** and points identified by their positions relative to a selected origin **0**. At each **X** a scalar quantity is given, called carrier density $\rho(X)$, such that if $x = \phi(X,t)$ then $\rho(X) \rightarrow \rho(x,t)$ defines a scalar field called local carrier density.

As already mentioned, a carrier will have physical significance through its set of properties. We used charges as example. The set of properties $\{Q\}$ characterizes a carrier field and in turn establishes the conditions for a density field to correspond to an acceptable carrier.

1.9 Carriers in interaction

In **B** the carrier has existence only, whereas in *B* the carrier c has a distribution characterized by the density $\rho_c(x,t)$. There is no restriction in defining a reference space B$_R$ where the carrier exists in the points **x** with constant density ρ_0 occupying a volume V_0 such that $\rho^{(0)}V_0 = 1$. These two quantities are unobservable as far as any "observation" requires an "interaction," only then the distribution acquires meaningful space dependence as a function, by definition, of an *external interaction* $V(\mathbf{x},t)$, which will be defined below. Here it is important to state that as a result of this interaction, and of the properties attributed to the carrier, the density evolves into a current: $\rho^{(0)} \Rightarrow \vec{j}_c^V(\mathbf{x},t)$. The density is characterized by the properties of the carrier and the self-organization of the carrier, which adapts to the external interactions.

1.10 Composite, decomposable, elementary, average and average description of carriers

There are several ways to analyze the density. Each allows a physical interpretation. For example:

- A *composite* carrier is defined as one for which the density

$$\rho_C(\mathbf{x},t) = \sum_c A_C^c \rho_c(\mathbf{x},t), \tag{9}$$

with the definition of each of the $\rho_c(\mathbf{x},t)$ being also meaningful as a description of a carrier.

- Similarly a *non-decomposable* carrier is defined as one for which (9) applies but for which the meaning of each of the $\rho_c(\mathbf{x},t)$ cannot be defined without reference to the global $\rho_C(\mathbf{x},t)$.

- An (*non-decomposable*) *elementary* carrier is one for which a single $\rho_c(\mathbf{x},t)$ is all it is needed; in this case we emphasize the discrete nature of an elementary carrier, but we do not assume a point-like or any internal structure for them.

- An *average* carrier is defined as one for which its density can be described as ($W = \Sigma_{c=1,n} A_A^c$)

$$\rho_A(\mathbf{x},t) = \frac{1}{W} \sum_{c=1,n} A_A^c \rho_c(\mathbf{x},t), \tag{10}$$

with the definition of each of the $\rho_c(\mathbf{x},t)$ being meaningful as a description of a carrier itself.

- Similarly an *average description of a carrier* can be defined either as a space average over carrier descriptions as in (10) or as a time average of a description, or sum of descriptions ($\bar{W} = \sum_{c=1,n} \frac{1}{r} \int_{t=t_0}^{t=t_0+\tau} A^c(t)dt$, the choice $\bar{W} = 1$ presents less manipulation difficulty)

$$\overline{\rho(\mathbf{x})} = \frac{1}{\bar{W}} \sum_{c=1,n} \frac{1}{r} \int_{t=t_0}^{t=t_0+\tau} A^c(t)\rho(\mathbf{x},t)dt. \tag{11}$$

This paper is centered on the definition of the elementary carriers and their correspondence with the fields describing the elementary particles, in particular the electron and its partner particle, the neutrino.

1.11 The density

For a physically acceptable carrier density:

D1. $\rho_c(\mathbf{x},t)$ is a real function $\rho_c(\mathbf{x},t) \subset \mathbb{R}$.

D2. The density $0 \le \rho_c(\mathbf{x},t) < \infty$ in order to represent a finite amount of charges and of action.

D3. The derivatives of the density $-\infty < \partial_\mu \rho_c(\mathbf{x},t) < +\infty$ in order to represent a finite amount of energy-momentum.

Theorem 1 If $\Psi(\mathbf{x},t)$ is an analytical quadratic integrable complex or multivector function, conditions D1, D2 and D3 are fulfilled identically if $\rho_c(\mathbf{x},t) = |\Psi_c(\mathbf{x},t)|^2$. Here $|f|^2$ means the real quadratic form of any more general function f, even if f itself is not necessarily a real function and we define: if $|f|^2 = f^+ f$ then $\partial_\mu |f|^2 = (\partial_\mu f^+)f + f^+(\partial_\mu f)$.

Condition D1 is fulfilled by the definition $\rho_c(\mathbf{x},t) = |\Psi_c(\mathbf{x},t)|^2$, D2 by the requirement of quadratic integrability, D3 by the definition $\partial_\mu |f|^2 = (\partial_\mu f^+)f + f^+(\partial_\mu f)$ and the analytical properties of $\Psi(\mathbf{x},t)$. It is seen that the conditions D1, D2, D3 and $\int \rho_c(\mathbf{x},t)d\mathbf{x} = N_c$ correspond to the $\Psi(\mathbf{x},t)$ being quadratic integrable Hilbert functions.

1.12 Wave function quantum mechanics and density functional theory from START

We proceed now to establish the basic theoretical aspects of the study of carriers, which result in a stationary state Wave Function Quantum Mechanics and Density Functional Theory of the carriers.

- The total energy of the system is a functional of the density, which can be defined in two steps. The first is to establish that there is a ground, least action, minimum energy, state of the system, which defines the carriers themselves:

$$E_0^{(N)} = \int E_0^{(N)}(\mathbf{x})d\mathbf{x} = \int \rho_0^{(N)}(\mathbf{x})\varepsilon d\mathbf{x} = N\varepsilon, \tag{12}$$

$$\int \rho_0^{(N)}(\mathbf{x})d\mathbf{x} = N \qquad (N = \text{number of carriers}), \tag{13}$$

where the density of energy $E_0^{(N)}(\mathbf{x})$ at a given space point $\mathbf{x}$ has been factorized as the product of the energy ε per carrier and the carrier density $\rho_0^{(N)}$. This by itself is the definition of *elementary carriers* of a given type: they *are indistinguishable, equivalent, and the energy of the carrier is a constant in space*, for all points of the distribution and, in a given system, the same for all elementary carriers of the given type.

- The constant defining the energy per carrier is a real functional of the carrier density and of the auxiliary function $\Psi(\mathbf{x})$.

$$\varepsilon = \varepsilon\left[\rho_0^{(N)}(\mathbf{x}), \Psi(\mathbf{x})\right]. \tag{14}$$

Because the reference energy has to be freely defined, this constant may be positive, negative, or null. The functional may, in some cases, become a *local density functional* (LDF).

The energy density, assuming *indistinguishable (independent or interacting) carriers* of a given type is now subject to the needs or desires of the observer describing the system. This defines independent carriers from interacting carriers, in that this energy appears as a property of the *carrier in the system* (a pseudo-carrier in condensed matter physics language), different from an isolated carrier.

Physics studies both the system in itself and, mainly, its response to external excitations. In the simplest approximation the necessary description is that of the possible stationary states of the system.

- The study of different excitation energies of the system $h\nu$ is now equivalent to the Heisenberg approach to studying a physical system through its excitation spectra, which was properly termed quantum mechanics due to the direct use of Planck's constant h.

- Density functional theory describes the self-organization of the carrier system with density $\rho(\mathbf{x})$ in the presence of some external potential.

1.12.1 The density as the basic variable

It is convenient to define the action in a form that distinguishes the part corresponding to the self-organization of the distribution and the part that corresponds to the 'external' influences on the distribution.

The volume (in space) of integration is considered large enough for the 'kinetic' energy to be internal; there should be no need to change the integration domain as a function of time. If the external influence is represented by the external potential $V(X)$ we can write for the total (invariant) action

$$A = \int dt \left[\tilde{E}_I[\rho(X)] + \int dx V(X)\rho(X) \right], \tag{15}$$

where the functional $\tilde{E}_I[\rho(X)]$ corresponds to the energy of the distribution of carriers $\rho(X)$. This functional $\tilde{E}_I$ has the interesting property that at a given time

$$\frac{\delta \tilde{E}_I}{\delta \rho(X)} = -V(X). \tag{16}$$

This is a basic relation in Action-DFT as far as there is an intrinsic definition of the external potential. This shows the tautological nature of the concept of carriers, once they are defined, by $\tilde{E}_I[\rho(X)]$, the external potential is defined through the density of the carriers themselves. The tautological cycle is closed when given $V(X)$ and $\rho(X)$ the kinetic energy and the interaction terms define $\tilde{E}_I[\rho(X)]$. Reminder: in practice more general forms of $V(X)$ should also be acceptable.

From the definitions above we can extend the description to consider a set $\{b\}$ of types of carriers, each carrier type with density ρ_b. In this case for each b the 'external potential' depends in all types $b' \neq b$.

1.12.2 Introducing gauge freedom for the description of the action

The density $\rho(X)$ at space-time point X is required to be gauge invariant, whereas the description of the energy (action) is gauge dependent. This is achieved by constructing the energy density as the product of an average energy per carrier ε with the two conjugated quantities $\Psi(X)$ and $\Psi^\dagger(X)$ such that $\rho(X) = \Psi^\dagger(X)\Psi(X)$ is gauge invariant. Here we define an auxiliary quantity: a gauge phase $\phi(X)$, similar to that proposed by Klein and by Fock as early as 1926 [10]:

$$\Psi(X) = A(X) \; ^{geom}\!\sqrt{\rho(X)} \; e^{-ia_0(X)+i\phi(X)} \; P_{+\uparrow}, \tag{17}$$

where we are restricted (even if $\phi(X)$ can be very general [3] and can represent electroweak, color and gravitational interactions), by definition, to

$$\hbar \frac{\partial(a_0(X) - \phi(X))}{\partial t} = \varepsilon, \tag{18}$$

showing the gauge freedom of the description of the energy (action) associated with the carrier. We have then recovered the equivalent to the Hohenberg-Kohn Theorems [11] and, with our definition of $^{geom}\!\sqrt{\rho(X)}$ below, the Hartree-Fock or the Kohn-Sham minimization procedures [12] from

$$\delta\left(\tilde{E}[p] - \varepsilon\left\{\int\rho(\mathrm{x})dx - N\right\}\right) = 0, \tag{19}$$

allowing the direct self-consistent determination of $\rho(\mathrm{x})$ and ε (see [5]).

1.13 Least action amplitude functions in START

We can now follow the START definitions and the Schrödinger procedure to obtain the stationary action states of the elementary carriers system.

1. Let the Schrödinger (1926) definition of action $W(\mathrm{x},t)$ in terms of an auxiliary function $\Psi(\mathrm{x},t)$ be

   $$W(\mathrm{x},t) = K \ln \Psi(\mathrm{x},t) = -K \ln \Psi^\dagger(\mathrm{x},t), \tag{20}$$

   that is: action is considered a sum of terms. The action $W(\mathrm{x},t)$ is required to correspond to the stationary states of the system to be described, if ensured through a variational optimization procedure.

2. Let the carrier density ρ be the real quantity defined above

   $$\rho(\mathrm{x},t) = \Psi^\dagger(\mathrm{x},t)\Psi(\mathrm{x},t), \tag{21}$$

   where , $\rho(\mathrm{x},t)$, Ψ and $\Psi^\dagger$ are: unique-valued, continuous and twice-differentiable and obey the additional condition $\rho(\mathrm{x},t)\big|_{space\ boundary} = 0$.

3. Let the canonically conjugated variables be $X = (\mathrm{x},t)$ and $\Box W = iK\Box\ln\Psi = -iK\Box\ln\Psi^\dagger$, with $\Box = e^\mu \partial_\mu$ the space-time gradient operator.

4. Let the local energy description be ($\tilde{E}_0$ is not a density)

   $$K^2 \frac{(\Box\Psi^\dagger)(\Box\Psi)}{\Psi^\dagger\Psi}c^2 = \tilde{E}^2 - (Pc)^2 = (\tilde{E}_0)^2 = (m_0c^2)^2, \tag{22}$$

(in the case where an interaction, through a gauge, is assumed to exist $(E-V)^2 - (Pc - eA)^2 = (m_0c^2)^2$) with the Euler-Lagrange (density of energy and constrain) function

$$J = K^2(\Box\Psi^\dagger)\cdot(\Box\Psi)c^2 - (m_0c^2)^2\Psi^\dagger\Psi, \tag{23}$$

and perform the variational search for the *extremum* energy $\tilde{E}$ (minimum of action for a stationary state system) $\delta J = 0$ to obtain from the standard variational approach the condition ($K^2 = \hbar^2$)

$$K^2[\Psi^\dagger(\Box^2\Psi) + (\Box^2\Psi^\dagger)\Psi] = m_0c^2\Psi^\dagger\Psi, \tag{24}$$

and then the equation for the auxiliary function Ψ (the Schrödinger-Klein-Gordon-like Equation (SKG)) is

$$\left[\hbar^2\left(\frac{\partial^2}{\partial t^2} - \left(\frac{\partial^2}{\partial x^2} + \frac{\partial^2}{\partial y^2} + \frac{\partial^2}{\partial z^2}\right)c^2\right) - (m_0c^2)^2\right]\Psi = 0. \tag{25}$$

We must emphasize that in the relativistic (and in the non-relativistic) case we obtain, through the Schrödinger optimization procedure, the Ψ (or $\Psi^\dagger$) *function which minimizes the action of the system*. A geometric factorization of the operator in the SKG equation transforms it into a Dirac-like equation. The gauge potentials are to be added.

1.13.1   General case of the auxiliary amplitude function Ψ

The auxiliary amplitude function Ψ describing the (set of) carrier(s) is constructed from sums of per carrier c, contributions ψ_c (sets of sums also). The space-time distribution ψ_c of a carrier and its intrinsic properties is given by the (four factors) functions:

$$\psi_c = {}^{geom}\!\sqrt{\rho_{(1)}(x,t)}R(x,t)e^{-i(\omega t - kx)}P_{+\uparrow} \tag{26}$$

the first factor, the geometric square root ${}^{geom}\!\sqrt{\rho}$, describes the per carrier local density, the second, the multivector transformation $R(x,t)$, the carrier local properties, the third $e^{-i(\omega t - kx)}$ the observer-relative carrier local motion and the last, the $P_{+\uparrow}$, is a projector describing the reference sign of the mass and the reference direction of the spin.

1.13.2   First order equation as a factorization-projection

Consider (here again $k^2 = (m_0c/\hbar)^2$, $i^2 = -1$)

$$\frac{1}{c^2}\frac{\partial^2}{\partial t^2}\psi - \nabla^2\psi + k^2\psi = 0, \tag{27}$$

and propose the factorization of the operator in the Dirac sense ($\Box = e^\mu\partial_\mu$)

$$(\Box + ik)(\Box - ik)\psi = (\gamma^\mu\partial_\mu + ik)(\gamma^\mu\partial_\mu - ik)\psi, \tag{28}$$

defining the projected function

$$\Psi = (\gamma^\mu\partial_\mu - ik)\psi, \quad (\gamma^\mu\partial_\mu + ik)\Psi = 0 \tag{29}$$

which obeys, by construction, the well known Dirac equation, showing that the auxiliary function Ψ, *optimized to obtain the least action*, is a geometric function (using a representation $\gamma(e^\mu) = \gamma^\mu$ of the geometry).

For a massless carrier field $k = 0$, this factorization is not unique, as

$$(\partial^\mu \partial_\mu + m^2) = (D^\dagger + mi)(D - mi),\tag{30}$$

requires that

$$-D^\dagger m + mD = 0 \quad \text{and} \quad D^\dagger D = \partial^\mu \partial_\mu = \Box^2,\tag{31}$$

therefore we can have a set of choices, either:
1. any value of m and $D^\dagger = D$ (the standard Dirac operator $D_0 = \Box$);
2. or when $m = 0$ the possibility $D^\dagger \neq D$ also becomes acceptable.

The basic requirement $D^\dagger D = DD^\dagger = \partial^\mu \partial_\mu$ limits the choices of D. Here they will be written in the Lorentz invariant form. The $\Gamma^\mu_{(f)}$ are a generalization (ir-reducible or reducible representation) of the Dirac γ^μ matrices. The limitation is so strong that the only possible choice, within the algebra, is when the chirality generator $i\gamma^5$, which has the same action on all γ^μ, that is $i\gamma^5 \gamma^\mu = -\gamma^\mu i\gamma^5$, is used (see Keller [3]). We define the differential

$$\partial^{(d)}_\mu = \left\{ 1\cos(n + t^d_\mu)\frac{\pi}{2} + i\gamma^5 \sin(n + t^d_\mu)\frac{\pi}{2} \right\} \partial_\mu,\tag{32}$$

with n and t^d_μ integers, a choice which results in the simplest multi-vector. Here, to take the electron as a reference, we use $n = 1$.

Then, in a particular frame we have the 'diagonal' structure:

$$\partial^{(d)}_\mu = \begin{cases} \partial_\mu & \text{if } n + t^d_\mu \text{ are even,} \\ i\gamma^5 \partial_\mu & \text{if } n + t^d_\mu \text{ are odd.} \end{cases}\tag{33}$$

The vectors, which can be represented by the standard γ^μ matrices, correspond to an irreducible representation of the Clifford algebra $C_{1,3}$ useful for writing the wave equations of the fundamental family of leptons and quarks $(e^-_R, e^-_L, \nu_L, \{u_L, d_l; \text{ color}\}\})$ of elementary particles. The electron requires a combination of two massless fields $e^- = (e^-_R, e^-_L)$ for the standard phenomenology of electroweak-color interactions. The case of the neutrino presented here is the simplest of these structures.

1.14 Interaction fields and charges

Consider the particular case of an initial situation without electromagnetic phenomena being present $E = 0$, $J = 0$, $B = 0$ and $\rho = 0$, and that in the process of creating a pair of interacting carriers with electric charges Q, an initial pulse of current $J(r, t) = Qv(r)\delta(t_0)$ is assumed to have been generated. This induces an electric field for $t > t_0$ from the Maxwell Equations derived above:

$$\partial_t E = -J(r, t)/\varepsilon_0 + \nabla \times B/\varepsilon_0 \mu_0 = Qv(r)\delta(t_0)/\varepsilon_0,\tag{34}$$

$$E(r, t; t > t_0) = -Qv(r)/\varepsilon_0 \quad \text{then} \quad \rho(r, t) = -Q\nabla \cdot v(r),\tag{35}$$

and

$$\partial_t B(r, t; t > t_0) = -\nabla \times E(r, t; t > t_0) = Q\nabla \times v(r)/\varepsilon_0,\tag{36}$$

showing that this virtual mechanism (in our process to establish a partitioning of energy and momentum among charged carriers) requires the actual allocation of physical properties, to the collection of created carriers, since the divergence of the current pulse creates a charge and the rotational of the current pulse of velocity field v(r) creates a magnetic dipole. We can see that the definitions are a circular procedure: sources create fields or fields generate the concept of sources.

In START the charge Q corresponds to the rotations (a new type of "spin") in the planes generated by the (space axis)-(action axis) basis vectors. Notice that a charged source with a circular current generates an electric and a magnetic field, the case of the electron, and the carrier also shows the presence of the spin associated to the solenoidal current. Within the postulates above the action of circular currents will have to be quantized in terms of Planck's $\hbar$, a pair of currents in terms of $\hbar/2$ each. The emitted, excess energy-momentum-angular momentum (electromagnetic wave), is itself (from the Maxwell Equations) an action carrier traveling at the speed $c^{-2} = (\varepsilon_0 \mu_0)$. We see that the concept of electromagnetic (light) wave is basic to the study of action density and its distribution in space-time, and the quantization condition is also fundamental in this case.

2.   The experimental electron

The matter fields enter into the theory as charge-current distribution densities $j_\mu \rightarrow (\rho, J)$. The currents J (for example those generated by the electron field) can, in general, be decomposed into their solenoidal j_{sol} and irrotational j parts. In Dirac's theory the solenoidal parts analyzed *via* the Gordon decomposition contain two components: one which is intrinsically solenoidal; and a second which is solenoidal only with reference to the boundary conditions and the observer's frame of reference. Then the electron sources of the electromagnetic fields, in units of the electron charge e, are described in fact by a set of seven basic quantities:

$$\rho, j_i, j_j, j_k, j_{i,sol}, j_{j,sol} \text{ and } j_{k,sol} \tag{37}$$

We have already reminded the reader that an electron cannot exist without its electromagnetic fields, that is, it exists with an electrostatic field generated by the electron's charge, an *intrinsic magnetic field* generated by its intrinsic *solenoidal current* and an additional electromagnetic field generated by the, extrinsic, electric current. A satisfactory theory considers physical entities as constituted by whatever is *observable*.

The intrinsic solenoidal current of the electron implicates not only a magnetic moment but also an angular momentum

$$S = \int S(x) = \frac{1}{2}\hbar, \tag{38}$$

then in (39) above j_{sol} could also be replaced by an angular moment field $\vec{S}(x)$. Dirac's theory shows that the magnitude of $\vec{S}(x)$ is

$$S(x) = S\rho(x), \tag{39}$$

then only the direction of $\vec{S}(x)$ is independent of $\rho(x)$ but not its magnitude. This is one of the most important features of the geometrical content of the electron theory. It says that, even if the analysis of an electron distribution shows some solenoidal current, there is a curl of the distribution at every point and, as is well known in vector analysis, the overall intrinsic solenoidal current is the result of the application of Gauss' theorem to the ensemble: *every point of the distribution contains the same amount of intrinsic angular momentum per unit density*.

There is no indication whatsoever of a structure giving rise to spin, and in fact a spin field $S = \psi\gamma_{12}\psi^* = \rho\psi\gamma_{12}\psi^{-1}$ is one of the most fundamental quantities of the standard theory.

In all experiments performed up to date an electron appears as a distribution of charge, currents and electromagnetic (electroweak, in fact) fields. Problems arise from the attempt to rationalize the experimental facts starting from a point particle idea as the basis for the interpretation of experiment or for the interpretation of the results of the now standard quantum mechanical calculations. Experiment shows that there is no internal structure of the electron, but the experiment does not disagree with the existence of distribution. The 'interpretation' of the distribution is a not a question of quantum mechanics, nor of the electron theory. That is, there is no experiment resolving the electron 'cloud' into instantaneous positions of a 'point' particle, nor at the same time is there any evidence at all of a possible excitation of internal structures of an electron.

We could speak in terms of electromagnetic quantities alone. The densities, which we commonly refer to the sources, can be substituted by electromagnetic quantities through the integral form of the Maxwell equations. For example, to relate E and $\nabla \cdot$ E

$$E(\vec{r}_2) = \frac{1}{4\pi\varepsilon_0} \int \frac{\varepsilon_0 \nabla \cdot E(\vec{r}_1)}{r_{12}^2} \frac{\vec{r}_{12}}{r_{12}} dV_1, \tag{40}$$

or to relate $\nabla \times H$ and $\nabla \cdot E$ for time independent E,

$$\nabla \times H = (\nabla \cdot E)v, \tag{41}$$

and we can even think of the electromagnetic potentials A^μ as quantities related to the sources in special forms

$$\nabla^2 A^0 + \nabla \cdot \frac{\partial \vec{A}}{\partial t} = \frac{\nabla \cdot D}{\varepsilon_0}. \tag{42}$$

We can then assume that besides the field intensities E and H we have a vector distribution (reminder $\tilde{J} = \varepsilon_0(\rho + J)$)

$$\nabla \cdot E \rightarrow \rho, \tag{43}$$

$$(\nabla \cdot E)v \rightarrow J, \tag{44}$$

the energy-momentum related to this vector being

$$E = \gamma m_0 \varepsilon_0 \nabla \cdot E/e, \tag{45}$$

$$P = \gamma m_0 \varepsilon_0 (\nabla \cdot E) v'/e. \tag{46}$$

Here m_0 appears as a parameter providing the correct dimensions and v' corresponds to the relative velocity between the inertial system where $\nabla \cdot E$ has been computed and that of the observer. Remember that relativistically E and H cannot be separated, nor do they have a unique formulation; in fact, they can always be expressed as Lorentz transformations and duality rotation of a reference bi-vector $H = \psi \gamma_{12} \psi *$.

2.1   Spin, magnetic moment and mass

We now give a meaning to the proposed amplitude function ψ_c in the free particle approximation. The rest mass parameter m_0 of the carrier will be directly related to the amplitude term of the non-dispersive wave packet. This is a consequence of the fact that a non-dispersive wave packet, ψ, is a solution of the equation

$$\Box \psi = 0, \quad \text{where } \Box^2 = \nabla^2 = \frac{1}{c^2} \frac{\partial^2}{\partial t^2}. \tag{47}$$

Then a non-dispersive wave for a carrier of mass m travelling in the $+x$ direction with velocity v relative to the observer takes the form [13, 14, 15])

$$\psi = \sqrt[geom]{\rho} (\sin k_0 r / k_0 r) \exp[i(\omega t - kx) P_{+\uparrow}, \tag{48}$$

where

$$k_0 = m_0 c / \hbar, \quad r = \left\{ \frac{(x - vt)^2}{1 - (v^2/c^2)} + y^2 + x^2 \right\}^{\frac{1}{2}}, \tag{49}$$

$$\omega = mc / \hbar, \quad k = mv / \hbar, \quad k_0^2 = (w/c)^2 - k^2,$$

with $\rho(x,t) = $ constant representing the time average of a steady state. That (48) is a solution of (47) follows by simple substitution. It is also one form of the standard spherically symmetrical solution of (47) after it has been subjected to a Lorentz transformation.

Then a solution of (47) takes the form (we leave out the reference projector $P_{+\uparrow}$ in this part of the discussion; note $R > 0$)

$$\psi = \sqrt[geom]{\rho} R \exp[iS]. \tag{50}$$

Inserting this ψ into (47) and then separating real and imaginary parts, the following two equations are obtained:

$$\Box^2 R - R \left\{ (\nabla S)^2 - \frac{1}{c^2} \left(\frac{\partial S}{\partial t} \right)^2 \right\} = 0, \tag{51}$$

$$R \Box^2 S + 2 \left\{ \nabla R \nabla S - \frac{1}{c^2} \left(\frac{\partial R}{\partial t} \right) \left(\frac{\partial S}{\partial t} \right) \right\} = 0. \tag{52}$$

If one takes the $\exp[iS]$ to be the de Broglie wave, so that $S = \omega t - kx$, (52) now leads directly to the result that

$$\frac{\Box^2 R}{R} = \frac{m_0^2 c^2}{\hbar^2}. \tag{53}$$

It is well known that equation (48) represents the superposition of two spherically symmetrical waves, one converging and one diverging, both having phase velocity c [15], and if the waves are electromagnetic waves, this combination constitutes a phase-locked cavity similar to that proposed by Jennison [16], who has also shown that such cavities have many of the inertial properties of particles. The structure of the field can be associated with the electron. We now compute the spin of the distribution (48). The momentum $\mathbf{P}$ of the field is

$$\mathbf{P} = \frac{\hbar}{4i}[\psi^\dagger \nabla \psi + \psi^\dagger \alpha(\alpha^\mu \nabla_\mu)\psi] + \text{h.c.}, \tag{54}$$

$$\mathbf{P} = \frac{\hbar}{2i}[\psi^\dagger \nabla \psi - (\nabla \psi^\dagger)\psi + \frac{\hbar}{4}\nabla \times \psi^\dagger \sigma\psi], \tag{55}$$

and, for $k = 0$ corresponding to a particle at rest,

$$\mathbf{P} = \rho(x,t)\frac{\hbar}{4}\left(\frac{k^2}{2\pi^2}\right)\frac{\partial}{\partial r}\left(\frac{\sin^2 kr}{k^2 r^2}\right)(-2y\hat{x} + 2x\hat{y}), \tag{56}$$

which represents a circular flow of the field in the plane $\hat{x} \wedge \hat{y}$. The angular momentum is given by

$$\mathbf{J} = \frac{\hbar}{2i}\int x \times [\psi^\dagger \nabla \psi - (\nabla \psi^\dagger)\psi]d^3 x + \frac{\hbar}{2}\int \psi^\dagger \sigma\psi d^3 x, \tag{57}$$

and again the second term, spin, will be the relevant quantity. If we assume R to be normalized, then the integral of the spin part would be trivially of magnitude $\hbar/2$. As for a de Broglie wave packet $k_0 = m_0 c/\hbar$, then the same prefactor $R(r)$ that generates the mass generates the spin of the total wave. This appears to be the real origin of the structural parts discussed above. Notice that $4\pi r^2 R(r) = 0$ as $r \to 0$.

The prefactor $R(r)$ provides, additionally, a connection with the standard model of elementary particles given that

$$R(r) = \frac{\sin kr}{kr} = \frac{e^{ikr}}{ikr} - \frac{e^{-ikr}}{ikr}, \tag{58}$$

and it corresponds to a standing spherical wave: e^{ikr}/kr is an outgoing spherical wave and e^{-ikr}/kr an incoming spherical wave. Given a spin direction they will have opposite helicities, and the standing spherical wave will be the realization of the well-known sum of a left-handed and a right handed wave.

2.1.1 Conserved electromagnetic quantities

The integrated quantities

$$\mathrm{E} = \int \mathrm{E}(x,t)(dx)^3, \quad \mathrm{P} = \int P(x,t)(dx)^3, \quad \text{and} \quad \mathrm{M} = \int (P \times x)(dx)^3 \tag{59}$$

electromagnetic energy-momentum and angular momentum for a steady current $\tilde{\mathrm{J}}$ are time independent: $d\mathrm{E}/dx^0 = d\mathrm{P}/dx^0 = d\mathrm{M}/dx^0 = 0$.

2.2 The basic structural relationship between charge, magnetic moment, spin and mass

A crucial argument of the present paper is that once we have defined a field of sources for an electromagnetic field, which contains a static electric and a static magnetic part, and we have shown that this field carries a spin $1/2\hbar$, the field configurations correspond to a charged "particle" with spin. The charge of this particle is e and its spin is $\left|M_z\right| = \hbar/2$. In this case we have (see Appendix)

$$\frac{ge}{2} = \frac{\hbar}{2} \quad \text{or} \quad g = e\frac{\hbar}{e^2} = \frac{e}{\bar{\alpha}},\tag{60}$$

where $\bar{\alpha} = e^2/\hbar c = 7.29735308 \times 10^{-3} \approx 1/137$ is the fine structure constant (above in units where $c = 1$). The relation e/g is a fundamental dimensionless structural constant of the formulation of the theory. We now argue that the usual relationship between a magnetic moment μ and a spin s is

$$\mu = \frac{e}{m}s\tag{61}$$

and, as for the carrier field, we have defined that the magnetic moment is μ_0 the carrier should be attributed a mass m_0. This being a structural relationship, which should be obeyed at each point of the distribution with carrier density ρ. The space integral of ρ for one carrier is 1, and then the relation (61) is obeyed as a relation between physical constants at each and every point of the carrier distribution, its space integral being μ_0, e, m_0 and s. The radius r_0 is implicitly contained in the definitions.

2.3 Action and energy involved in the interaction

The logical cycle of the interaction structure would close when the energy and action related to these logical and mathematical structures are determined. From the definition of the divergence of the interaction fields as the sources

$$\partial^\beta F_{\alpha\beta} = -\frac{1}{\varepsilon_0} J_\alpha,\tag{62}$$

which allows the calculations of the energy given off by the source itself, the energy of the interaction field.

For this purpose, consider a variation of the four-vector potential A^μ and the scalar product of these δA^μ with the source carrier current to obtain, from (62) after integration in a volume Ω of four-dimensional space

$$\int\left(\partial^\beta F_{\alpha\beta} + \frac{1}{\varepsilon_0} J_\alpha\right)\delta A^\alpha d\Omega = 0,\tag{63}$$

this quantity refers to the action related to the source and also to the field. Notice that where $J_\alpha \neq 0$ the integrand vanishes by definition. An integration by parts, using a boundary condition $\delta A^\mu(\text{boundary}) = 0$ and the antisymmetry of the $F_{\mu\nu} = -F_{\nu\mu}$ gives

$$\int\partial^\beta(F_{\alpha\beta})\delta A^\alpha d\Omega = \int[\partial^\beta(F_{\alpha\beta}\delta A^\alpha) - F_{\alpha\beta}\partial^\beta(\delta A^\alpha)]d\Omega = \int -F_{\alpha\beta}\partial^\beta(\delta A^\alpha)d\Omega..\tag{64}$$

because $\delta A^\alpha = 0$ on the boundary.

$$\int -F_{\alpha\beta}\partial^{\beta}\left(\delta A^{\alpha}\right)d\Omega = \frac{1}{2}\int\left[-F_{\alpha\beta}\partial^{\beta}\left(\delta A^{\alpha}\right) - F_{\beta\alpha}\partial^{\alpha}\left(\delta A^{\beta}\right)\right]d\Omega$$

$$= \frac{1}{2}\int\left[-F_{\alpha\beta}\delta\left(\partial^{\beta}A^{\alpha} - \partial^{\alpha}A^{\beta}\right)\right]d\Omega = \frac{1}{2}\int F_{\alpha\beta}\delta\left(F^{\alpha\beta}\right)d\Omega = \frac{1}{4}\int\delta\left(F_{\alpha\beta}F^{\alpha\beta}\right)d\Omega,$$

(65)

to obtain finally

$$\int\delta\left(\frac{1}{4}F_{\alpha\beta}F^{\alpha\beta} + \frac{1}{\varepsilon_0}J_{\alpha}A^{\alpha}\right)d\Omega = 0.$$

(66)

The field energy density is

$$\tilde{E}(X) = \frac{1}{4}F_{\alpha\beta}F^{\alpha\beta} + \frac{1}{\varepsilon_0}J_{\alpha}A^{\alpha}.$$

(67)

We have followed the Huang and Lin [9] analysis in his equivalent work to obtain the Lagrangian of the electromagnetic field. We should remember that in our approach *carriers do not interact with themselves*, and the quantity in (67) should be taken to be zero if $J_{\alpha}(x,t) \neq 0$.

3.   The many-electron problem

The N electron problem (fermions) should solve the set of equations

$$\rho = N\rho_{one\ electron\ in\ the\ N\ electrons\ system} = \Psi^{\dagger}\Psi.$$

The statistics are the Fermi-Dirac statistics and require:

- The density for the N equivalent fermion carriers system is to be constructed as a sum of M independent alternative contributions $\rho(\mathbf{x},t) = \Sigma_{(i=1,M\geq N)}\rho_i(\mathbf{x},t,s_i)$.
- There should be at least one linearly independent function (pseudo-carrier amplitude) contributing to the density for each of the N equivalent carriers in the system. A linear transformation would then give one different spin-orbital (SO or "state") per electron, the usual argument.

$$\psi_c(\mathbf{x},t) = \frac{1}{\sqrt{N}}\sum_{i=1}^{M\geq N}a_c^i\phi_i(\mathbf{x},t,s_i)\qquad a_c^i = b_c\alpha_i$$

$$a_c^ia_c^{i'} = -a_c^{i'}a_c^i\quad \left|a_c^i\right|^2 \leq 1;\ \sum_{i=1}^{M}\left|a_c^i\right|^2 = N$$

The total amplitude function should be a sum of single (pseudo-) carrier amplitude functions $\psi_c(x,t,s)$, such that the exchange among two carriers of the space–time–spin (x,t,s) descriptions.

$$\Psi = \sum_c \varpi^c\psi_c(\mathbf{x},t,s)\ \text{and}\ \overline{\Psi} = \sum_c \varpi^c\psi_c^{\dagger}(\mathbf{x},t,s)$$

This defines Ψ as a vector, linear form, expressed in the basis $\{\varpi^c\}$.

The Ψ are defined and the products ordered to obtain

$$\int\|\Psi\|^2 d\mathbf{x} = \int\overline{\Psi}\Psi d\mathbf{x} = N$$

$$\int \phi_i(\mathbf{x},t,s_i)\phi_i(\mathbf{x},t,s_i)d\mathbf{x} = \delta_{ii}; \quad \omega^c\omega^{c'} = -\omega^{c'}\omega^c; also \ \varpi^c\omega^{c'} = \delta^{cc'}$$

The 1st condition is double: first the orthonormality among the φ_i functions (requiring them to be eigenfunctions of the same differential equation operator) to fulfill the condition of making linear independent combinations ψ_c and second the Grassmann character of the a_c^i coefficients, to make the local density per carrier corresponds to the sum of the squares

$$\left| a_c^i \phi_i(\mathbf{x},t,s_i) \right|^2,$$

and the 2nd condition, equivalent to the Pauli principle, defines the ω^c as Grassmann variables and the ϖ^c as their Grassmann conjugates.

The (stationary state) Hamiltonian for a many-electron atom (or molecule or solid) may be written in the form

$$\hat{H} = \sum_{c=1}^{N} \left[\omega^c \hat{H}^{core}(x)\varpi^c + \sum_{c'=c+1}^{N} \int dx_j \omega^{c'} \omega^c \left(e^2/r_{ij}^{cc'} \right) \varpi^c \varpi^{c'} \right],$$

$$\hat{H} = \sum_{c=1}^{N} \omega^c \left[\hat{H}^{core}(x) + \hat{H}^{interaction}(x) \right] \varpi^c, \quad E = \int \Psi^* \hat{H}\Psi dx$$

where the core Hamiltonian for the electron c, consists of the kinetic-energy and nuclear attraction terms for electron c with coordinates i. Hamiltonian $H_{core}(x) + H_{interaction}(x)$ is a one-electron operator, even if $1/r_{ij}^{cc'}$, being dependent on the inter-electron distance, is a two-electron operator; ω^c and ϖ^c act here as projection operators. The energy of the N-electron system is given by E and we should determine both Ψ and E

There is a total density function $\rho_N(x)$ which should be integrable in a final volume, and everywhere in that volume should be a finite and non negative *function*, corresponding to a many electron function $\Psi_N(\{x_n\}; n = 1,...,N)$ where $\rho_N(x) = |\Psi_N|^2 = \sum_{c=1}^{N} \rho_c(x) = \sum_{a=1}^{M \geq N} \rho_a(x)$. This Hermitian square can be described as both a sum of $\rho_c(x) = |\psi_c|^2$ or as a sum of SO contributions $\rho_a(x) = |b_a\phi_a|^2$.

Third, in the case of the many electron (fermion) system we are studying *all N electrons (fermions) are equivalent*. This requires that the density itself is a sum $\rho_N(x) = |\Psi_N|^2 = N\rho_{electron}(x)$, and each $\rho_{electron}(x)$ should be generated by equivalent contributions. That is $\rho_{electron}(x) = 1/N \sum_a \rho_a(x)$.

As density appears as a sum of densities, then the wave function should both be the square root of the total density and also provide the square root of each one of the contributions to the total density. For this we require the use of geometric (multivector analysis) techniques. In fact the problem is similar to that of finding the linear form (geometric square root) $\overline{d} = ae_1 + be_2 + ce_3 + ...$ which corresponds to the quadratic form $d^2 = a^2 + b^2 + c^2 + ...$.

3.1   Configuration space and real space

A basic concept in the study of a many-electron system (N interacting fermions) is, from the considerations above, the simultaneous, repeated, use of real space (the space of the observer) for each one of the fermions of the system: configuration space. Then, if $\mathbf{x}$ represents a point in real space, it is customary to represent by $\mathbf{X} = \{\mathbf{x}_a ; a = 1, \ldots, N\}$ the set of points in the configuration space $\mathbf{X}$ for N fermions.

Here and in the rest of our presentation we use a geometric notation $\mathbf{X} = \sum \omega_a \mathbf{x}; \{a = 1, \ldots, N; \omega_n \omega_m = -\omega_m \omega_n\}$ and the projection operators ω_a such that $\omega_a \omega_b = \delta_{ab}$ selecting the part of the configuration space which corresponds to electron $a: \omega_a \mathbf{X} = \mathbf{x}_a$. This allows a clear formal definition of the electrons involved in each part of the calculation. Our geometric procedure introduces the statistics of the fermion system from the beginning because the interchange of two electrons in a given expression will change the sign of the corresponding terms.

3.2   The energy calculation

In correspondence with our formal definition of configuration space the total electronic energy operator or Hamiltonian is

$$\hat{H}_{KKW} = \sum_n \omega_n \left(-\frac{\hbar^2 \nabla^2}{2m_e} - \frac{Ze^2}{|x_n|} + \frac{1}{2} \sum_{m \neq n} \omega_m \frac{e^2}{|x_{nm}|} \omega_m \right) \omega_n, \text{(H)} \qquad (68)$$

and the wave function Ψ_N^{KKW} is , with $\alpha_i \alpha_j = -\alpha_j \alpha_i$ and $\overline{\alpha}_i \alpha_j \delta_{ij}$

$$\Psi_N^{KKW} = \sum_{n=1}^N \psi \omega_n \text{ where } \psi = \sum_{i=1}^{M \geq N} b_i \alpha_i \phi_i (x_i) \text{ (WF)} \qquad (69)$$

Here the electron, or pair of electrons, under consideration is explicitly selected. Note that a double set of Grassmann numbers $\{\omega_n ; \alpha_i\}$ has been introduced; this has an analytical analogue in the HF method, where in a determinant the exchange of columns or of rows changes the sign of the determinant. The exchange terms arise from the definition of the wave function (WF) when used in (H).

In (H) the core Hamiltonian for the electron n with coordinates i consists of the kinetic-energy term and the nuclear attraction local potential. In the calculation of ψ the effective Hamiltonian is $H_{core}(x) + H_{interaction}(x)$ where the second term is a one-electron operator, even if the electron repulsion, being dependent on the inter-electron distance, is a two-electron (i for n, j for m) operator. The resulting exchange and correlation potential is the same for all components of ψ. Orthonormality and equivalence are used

$$E_{core} = \frac{1}{N} \int \overline{\Psi} \left\{ \sum_{c=1}^N \omega^c \hat{H}^{core}(c) \omega^c \right\} \Psi dx_c$$

$$= \frac{1}{N} \int \sum_d \omega^d \psi_d^\dagger (x_d) \left\{ \sum_{c=1}^N \omega^c \hat{H}^{core}(c) \omega^c \right\} \sum_d \omega^d \psi_d (x_d) dx_c$$

$$E_{core} = \frac{N}{N} \int \psi_1^\dagger (x_1) \hat{H}^{core}(1) \psi_1 (x_1) dx_1 = \int \left[\sum_i \varepsilon_i^{core} (x_1) \frac{|a_i|^2 \overline{\rho_i (x_1)}}{\rho_i (x_1)} \right] \rho_i (x_1) dx_1$$

$$= \int \varepsilon^{core} (x_1) \rho_1 (x_1) dx_1 \text{ (formal definition)}$$

For the electron-electron interaction (e-e), because the equivalence of the N electrons and using the expansion of the ψ, we obtain

$$E_{int} = \frac{N}{2} \sum_i \varphi_i^*(1)\alpha_i^* \sum_j \varphi_j^*(2)\alpha_j^* \frac{e^2}{|x_{12}|} \sum_k \varphi_k(2)\alpha_k \sum_l \varphi_l(1)\alpha_l dx_1 dx_2 \; .$$

Considering the property $\alpha_i^* \alpha_j = \delta_{ij}$, there are 3 types of e-e terms:

I) $j = k$ and $i = l$ which gives

$$\int \left\{ \frac{1}{2} \int \left[\sum_i \sum_{j \neq i} |a_i^j|^2 \overline{\rho}_j(2) \frac{e^2}{|x_{12}|} \right] dx_2 \right\} |a_i^l|^2 \overline{\rho}_l(1) dx_1 = \int V_I(\mathbf{x}_1) \rho_1(\mathbf{x}_1) dx_1$$

II) $j = l \neq i$ and $i = k$ (one interchange $a_i a_j = -a_j a_i$ is needed!)

$$\int \left\{ \frac{1}{2} \int \left[-\sum_{i,j\neq i} \delta_i' |a'|^2 |a'|^2 \varphi_i^*(2)\varphi_j(2) \frac{e^2}{|x_{12}|} \varphi_j^*(1)\varphi_i(1)/\rho_1(\mathbf{x}_1) \right] dx_2 \right\} \rho_1(\mathbf{x}_1) dx_1$$

$$= \int V_{II}(\mathbf{x}_1) \rho_1(\mathbf{x}_1) dx_1 \text{ (the } \delta_i' \text{ arises from } i \neq j \text{, spins orthonormal)}$$

III) Null terms, where ($i \neq l$ and $i \neq k$) or ($j \neq l$ and $j \neq k$). The *total electron-electron interaction energy*:

$$E^{inter}[\rho, \Psi] = \frac{N}{2} \int (V_I(\mathbf{x}_1) + V_{II}(\mathbf{x}_1)) \rho_1(\mathbf{x}_1) dx_1 = N \int \varepsilon^{inter}(\mathbf{x}_1) \rho_1(\mathbf{x}_1) dx_1$$

Then we have two different contributions which will also contribute to the formal interpretation of the Pauli Exclusion Principle: a given electron is not interacting with itself and there is an "exchange" term for fermions, where from $a_i a_j = -a_j a_i$ a negative sign appears. Those terms are related, and similar in structure, to the integrals related to "exchange-correlation" in the HF+CI sense. Finally the total energy of N equivalent electrons is

$$E[\rho] = N \int \{ \varepsilon^{core}(\mathbf{x}_1) + \varepsilon^{inter}(\mathbf{x}_1) \} \rho_1(\mathbf{x}_1) dx_1$$

$$\varepsilon[\rho] = \varepsilon^{core}(\mathbf{x}_1) + \varepsilon^{inter}(\mathbf{x}_1) = \text{constant} , E[\rho] = N\varepsilon[\rho] \int \rho_1(\mathbf{x}_1) dx_1 = N\varepsilon$$

In principle it should be written $E = E[\rho, \Psi]$. The variational procedure is to be carried with respect to the ψ's. The basic set of equations for our KKW method, presented in comparison with HF and HF+CI, is as follows:

KKW	HF	(HF)+CI
$\widehat{H}\Psi = N\mathcal{E}\Psi$	$\widehat{H}\Psi_0^{SC} = \varepsilon_{HF}\Psi_0^{SC}$	$\widehat{H}\Psi_{CI}^{SC} = N\mathcal{E}\Psi_{CI}^{SC}$
2 $\;\Psi = \sum_{n=1}^{N}\omega_n\psi$ $\omega_n\omega_m = -\omega_m\omega_n$ $\psi = \sum_{i=1}^{M\geq N} a_i\phi_i$	$\Psi_0^{SC} = \frac{1}{\sqrt{N!}}[\varphi_i]$ $[\varphi_i]\equiv\det\lvert\varphi_i\rvert$	$\Psi_{CI}^{SC} = N\left(\Psi_0 + \sum_B c_B\Psi_B\right)$ $B=\{i\}\to\{i'\}_B$
$H^{\widehat{KKW}}\psi = \varepsilon\psi$	$H_{(i)}^{\widehat{HF}}\varphi_i^{SC} = \varepsilon_i\varphi_i^{SC}$	$\left[\left\langle\Psi_B^{SC}\widehat{H}\Psi_D^{SC}\right\rangle - \varepsilon\delta_{BD}\right]=0$
4 $\;\begin{aligned}&\rho(x)=\sum_{i=1}^{M\geq N}\rho_i(x)\\&\rho_i=\lvert b_i\rvert^2\lvert\phi_i\rvert^2\\&\int\rho_i dx=\lvert b_i\rvert^2\\&H^{\widehat{KKW}}\phi_i=\varepsilon_i\phi_i\\&\sum_i\varepsilon_i(\rho_i/\rho)=\varepsilon\\&a_i=b_i\alpha_i\\&\overline{\alpha_i}\alpha_j=\delta_{ij}\\&\alpha_i\alpha_j=-\alpha_j\alpha_i\end{aligned}$	$\rho_{HF}=\sum_i\rho_i^{HF}$ $\rho_i^{HF}=\lvert\varphi_i^{SC}\rvert^2$	$\rho = D\left(\rho_{HF}+\sum_B\lvert c_B\rvert^2\rho_B\right)$ $\left\langle\Psi_B^{SC}\Psi_D^{SC}\right\rangle=\delta_{BD}$ $\rho_B=\lvert\Psi_B^{SC}\rvert^2$ $N=1/\sqrt{1+\sum_B\lvert c_B\rvert^2}$ $D=N^2$

Row 1 presents the basic equation, row 2 the structure of the wave function, row 3 the resulting equation after the variation, energy minimization procedure, row 4 the definitions for the total density in each method and some auxiliary conditions.

Appendix: Definitions and Notation

We use the term *space* to denote the 3-D space of our perception of the distribution of physical objects in Nature and for its mathematical representation as an $\mathbb{R}^3$ manifold with a quadratic form. Its points are denoted by the letter $\mathbf{x}$ and represented as a vectorial quantity $\mathbf{x} = x^i e_i$. We use the traditional indices $i = 1,2,3$.

We use the term *time* to denote the 1-D space of our perception of the evolution of physical phenomena in Nature and for its mathematical representation as an $\mathbb{R}^1$ manifold with a quadratic form. The normal-face letter t denotes its points.

We use the term *space-time* to denote the 4-D Minkowski space of our perception of the physical world in the sense of relativity theory, and for its mathematical representation as an $\mathbb{R}^4$ manifold with a quadratic form: $ds^2 = g_{\mu\nu}dx^\mu dx^\nu$, $(\mu,\nu = 0,1,2,3)$. Its points are denoted by the Normal-face letter X and represented as a vectorial quantity $X = X^\mu e_\mu$. We use the traditional indices $\mu = 0,1,2,3$. The vectors e_μ in the geometry of space-time generate the G_{ST} 16 dimensional space-time geometry of multivectors. The basis vectors $\{e_0, e_1, e_2, e_3\}$, with $e_0^2 = -e_1^2 = -e_2^2 = -e_3^2 = 1$ and the definition property $e_\mu e_\nu = -e_\nu e_\mu$ generate a Clifford group $Cl_{1,3}$. We also use the notation $e_{0j} = e_0 e_j = \mathbf{e}_j$ $(j = 1,2,3)$ and $e_5 = e_0 e_1 e_2 e_3 = e_{0123}$. A special property of the pseudo-scalar (and also hypervolume and inverse hypervolume) in space-time e_5 is that $e_5 e_\mu = -e_\mu e_5$ (from $e_\mu e_\nu = -e_\nu e_\mu$, $\mu \neq \nu$) and then it has the same commuting properties with the generating vectors of G_{ST} as generating vectors have among themselves.

A vector e_4 can be used to introduce an additional basis vector, giving one more dimension (*action*). We thereby obtain the five dimensional carrier space spanned by the basic vectors e_u, $u = 0,1,2,3,4$ (identified as $e_u \Rightarrow e_\mu$, $u = \mu$ and e_4) with metric $g_{uv} = \mathrm{diag}(+1,-1,-1,-1,-1)$. This is used to construct a geometrical framework for the description of physical processes: a unified space-time-action geometry G_{STA}, mathematically a vector space with a quadratic form. An auxiliary element j anti-commutes with all e_μ : $e_\mu \mathrm{j} = -\mathrm{j}e_\mu$ and $\mathrm{j}^2 = +1$.

Multivector Representation. The base space $\mathbb{R}^5$ corresponds to the real variables set $\{ct, x, y, z, K_0\alpha\} \leftrightarrow \{x^u; u = 0,1,2,3,4\}$, that is, time, 3-D space and action (in units of distance introducing the universal speed of light in vacuum c and the system under observation dependent, using the Compton wavelength λ for a system with energy mc^2 : $K_0 = \lambda/h = 1/mc$). Time is usually an independent evolution coordinate. Action is distributed in space, then we consider the functions $x(t), y(t), z(t)$ and $w(t,x,y,z) = K_0\alpha(t,x,y,z)$. The nested vectors

$$dS = \sum_\mu dx^u e_u \; ; u = 0,1,2,3,4 \qquad 5-D$$

$$ds = \sum_\mu dx^\mu e_\mu \; ; \mu = 0,1,2,3 \qquad 4-D$$

$$d\mathbf{x} = \sum_i dx^i \mathbf{e}_i \; ; i = 1,2,3 \; ; \mathbf{e}_i = e_0 e_i \qquad 3-D$$

are members of a Clifford algebra generated by the definition of a quadratic form

$$dS^2 \equiv (dS)^2 = \left(\sum_\mu dx^u e_u \right)^2 = \sum_{\mu\nu} g_{uv}^{START} dx^u dx^v \, ,$$

$$g_{uv}^{START} = diag(1,-1,-1,-1,-1), e_u e_v = -e_v e_u$$

$$e = e_0 e_1 e_2 e_3 e_4 = -e^\dagger, \quad e_u e = e e_u$$

This 5-D geometry has two types of rotations: *space rotations* associated with *angular momentum* (in particular *spin $\frac{1}{2}\hbar$ and intrinsic magnetic moment*) and *"rotations" in the action-space planes*, with degeneracy 3 and intrinsic value $\wp\hbar/2mc$, associated with *the electric charge of the field*.

Observable objects are extended in space described by an action density α in space-time. Then a) defining $m(\mathbf{x},t)c^2 = \varepsilon_{total}(\mathbf{x},t)$, b) the inverse of the space-time volume $e_0 e_1 e_2 e_3/\Delta x \Delta y \Delta z \Delta t$, c) the space-time d'Alembertian operator $\Box = \Sigma_\mu e^\mu \partial_\mu$ (for a given observer with time vector e_0 the operator $\Box$ has the property $e_0\Box = -\partial_t + \nabla = -\partial_t + e_i\partial_i$), d) along $b = \Sigma_\mu b^u e_\mu$ the directional change operator is $\partial b = \Sigma_\mu db^\mu \partial_\mu$ (apply for $b = cte_0, xe_1, ye_2, ze_3$ to obtain the sum of *directed changes of w*) to obtain:

$$\mathbf{a}(\mathbf{x},t)e_4 = K_0 \alpha(\mathbf{x},t)e_4 = K_0 \frac{m(\mathbf{x},t)c^2 \Delta t}{\Delta x \Delta y \Delta z \Delta t} e = \frac{1}{m_0 c} \frac{m(\mathbf{x},t)c^2 \Delta t}{\Delta x \Delta y \Delta z \Delta t} e = \frac{(m(\mathbf{x},t)/m_0)c \Delta t}{\Delta x \Delta y \Delta z \Delta t} e$$

$$\mathbf{a}(\mathbf{x},t)e_4 = \frac{\left(m(\mathbf{x},t)/m_0\right)c_\Delta t}{\Delta x_\Delta y_\Delta z_\Delta t}e = \frac{w(\mathbf{x},t)}{\Delta x_\Delta y_\Delta z_\Delta t}e = w(\mathbf{x},t)e$$

$$edw = \sum_\mu \left[\left(\partial_\mu w(\mathbf{x},t)\right)dx^\mu\right]e_\mu e$$

$$(dS)^2 = (dS)(dS)^\dagger$$

$$= \left(1-\left(\kappa_0 p_0\right)^2\right)(cdt)^2 - \left(\left(1-\left(\kappa_0 p_1\right)^2\right)(dx)^2 + \left(1-\left(\kappa_0 p_2\right)^2\right)(dy)^2 + \left(1-\left(\kappa_0 p_3\right)^2\right)(dz)^2\right)$$

here $p_\mu = \partial_\mu \alpha(\mathbf{x},t)$ is a momentum density. Notice that $w(\mathbf{x},t)$ is the distance equivalent to a *reduced action* density, this makes the approach universal for all systems.

We use the term *action a* to denote the 1-D space of our perception of the objects of physical phenomena in Nature and for its mathematical representation as an $\mathbb{R}^1$ manifold with a quadratic form da^2.

e) We use the term *space-time-action* to denote the 5-D space of our perception of physical phenomena in Nature and for its mathematical representation as an $\mathbb{R}^5$ manifold with a quadratic form

$$dS^2 = ds^2 - \kappa_0^2 da^2 = g_{\mu\nu}dx^\mu dx^\nu - \kappa_0^2 da^2 = g_{AB}dx^A dx^B,$$

$$(A,B = 0,1,2,3,4), \quad (\mu,\nu = 0,1,2,3).$$

(70)

Its points are represented by the set $(X,\kappa_0 a)$, $\kappa_0 = 1/m_0 c$.

f) We use the term *description* to denote the partitioning of the total action (or energy-momentum) into carriers c. We use the term *theoretical structure* for a set of defining mathematical considerations.

Acknowledgement

J.K. is a member of the SNI (CoNaCyT, México). The author is grateful to Mrs. Irma Vigil de Aragón for technical support.

References

[1] Keller J. *Advances in Applied Clifford Algebra*, **9**(2), 309-395 (1999).

[2] Keller J. *Rev. Soc. Quim. Mex.*, **44**(1), 22-28 (2000).

[3] Keller J. *The Theory of the Electron; A Theory of Matter from START*, Foundations of Physics Series 117. Dordrecht: Kluwer Academic Publishers (2001).

[4] Keller J. Unification of Electrodynamics and Gravity from START, *Annales de la Fondation Louis de Broglie*, **27**(S), 359-410 (2002); Keller J. *Advances in Applied Clifford Algebras*, **11**(S2), 183-204 (2001). Keller J. A Theory of the Neutrino from START, *Electromagnetic Phenomena*, **3**,1(9) 122-139 (2003). Keller J. and Weinberger P. A Formal definition of Carriers, *Advances in Applied Clifford Algebras*. **12**(1), 39-62 (2002).

[5] Keller J., Keller A., Flores J.A. La Búsqueda de una Ecuación para la (Raíz Cuadrada de la) Densidad Electrónica, *Acta Chimica Theoretica Latina*, **XVIII** (4), 175-186 (1990). Flores, J.A. and Keller, J., Differential Equations for the Square Root of the Electronic Density in Symmetry Constrained Density Functional Theory, *Phys. Rev. A*, **45** (9), 6259--6262 (1992).

[6] de Broglie L. *Annales de Physique*, **3**, 22, (1925).

[7] Schrödinger E. *Annalen der Physik*, **79**, 361, 489, 734; **81**, 109 (1926).

[8] Spencer A. J. M. *Continuum Mechanics*, New York: Longman (1980).

[9] Huang L.S. and. Ling C. L. *Am. J. Phys.*, **70** (7), 471-743 (2002).

[10] Fock V. A. *Z. Phys.*, **39**: 226 (1926).

[11] Hohenberg P. and Kohn W. *Phys. Rev. B*, **136**, 864-867 (1964).

[12] Kohn W. and Sham, L.J. *Phys. Rev. A.*, **140**, 1133-1138 (1965).

[13] Mackinnon L. *Lett. Nuovo Cimento*, **31**, 37 (1981).

[14] Mackinnon L. *Found. Phys.*, **8**, 157 (1978).

[15] de Broglie, L. *C. R. Acad, Sci.*, **180**, 498 (1925); *Théorie Générale des Particules à Spin.* Paris: Gauthier-Villars (1943); *Ondes Électromagnetiques et Photons.* Paris: Gauthier-Villars (1968).

[16] Jennison, R. C. *J. Phys. A: Math, Nucl. Gen.*, 11, 1525 (1978).

The Electron in the Unified Composite Model of All Fundamental Particles and Forces

Hidezumi Terazawa
Institute of Particle and Nuclear Studies,
High Energy Accelerator Research Organization,
1-1 Oho, Tsukuba, Ibaraki, 305-0801, Japan
E-mail address: terazawa@post.kek.jp

The electron, one of the most fundamental particles in nature, is described in detail in the unified composite model of all fundamental particles and forces, a candidate for the most fundamental theory in physics.

I.   Introduction

In 1897, J.J. Thomson discovered the electron, one of the most fundamental particles in nature. For more than a century since then, the electron has played a key role in physics as well as science and technology. What is the electron? From the remarkable progress in experimental and theoretical physics in the twentieth century, it has become well known that matter consists of atoms, an atom consists of a nucleus and electrons, a nucleus consists of nucleons (protons or neutrons) and a nucleon consists of quarks. There exist at least twenty-four fundamental fermions, the six flavours of leptons including the electron and the eighteen (six flavours and three colors) of quarks. In addition, there exist at least twelve gauge bosons including the photon, the three weak bosons, and the color-octet of gluons. The quarks have the strong interaction with the gluons while both the quarks and leptons have the electroweak interactions with the photon or the weak bosons. In addition, all these fundamental particles have the gravitational interaction with themselves (or through the graviton). Furthermore, it has also become clear that the strong and electroweak forces of these fundamental particles fit the standard model in which the strong interaction can be described by quantum chromodynamics, the Yang-Mills gauge theory of color $SU(3)$, while the electroweak interactions can be described by the unified gauge theory of weak-isospin $SU(2) \times$ hypercharge $U(1)$. The latter theory assumes the existence of additional fundamental particles, the Higgs scalars, which should be found in the near future. Since there exist so many fundamental particles in nature and so many parameters in the standard model of fundamental forces, it is now hard to believe that all these particles are fundamental, and it is rather natural to assume that the standard model is not the most fundamental theory, but what can be derived as an effective theory at low energies from a more (and probably the most) fundamental theory in physics.

In this paper, I shall describe the electron in detail in the unified composite model of all fundamental particles and forces, a candidate for the most fundamental theory in physics. This paper is organized as follows: In Section II, I will introduce the unified composite model of all fundamental particles and forces, in which not only all quarks and leptons, including the electron, but also all gauge bosons including the photon, the weak bosons and even the graviton as well as the Higgs scalars are taken as composite states of subquarks, the more (and most) fundamental particles. In Section III, I will explain all properties of the electron such as the electric charge, the intrinsic spin angular momentum and the mass in the unified composite model. In Section IV, I will describe all interactions of the electron such as the electroweak and gravitational interactions as effective interactions at low energies (or at long distances) in the unified composite model. Finally, the last Section will be devoted to conclusions and further discussion. Throughout this paper, the natural unit system of $\hbar(\equiv h/2\pi) = c = 1$ where h [$\cong 6.62606876 \times 10^{-34}$ Js] is the Planck constant and c (= 299792458 m/s) is the speed of light in vacuum should be understood for simplicity unless otherwise stated. Also, note that electric charges should be understood as in units of electron charge e [$\cong 1.602176462 \times 10^{-19}$ C] unless otherwise stated.

II.   Unified composite model of all fundamental particles and forces

The unified composite model of all fundamental particles and forces consists of an iso-doublet of spinor subquarks with charges $\pm\frac{1}{2}$, w_1 and w_2 (called "wakems" standing for weak and electromagnetic) [1] and a Pati-Salam color-quartet of scalar subquarks with charges $+\frac{1}{2}$ and $-\frac{1}{6}$, C_0 and C_i ($i = 1,2,3$) (called "chroms" standing for colors) [2]. The spinor and scalar subquarks with the same charge $+\frac{1}{2}$, w_1 and C_0, may form a fundamental multiplet of $N = 1$ supersymmetry [3]. Also, all the six subquarks, w_i ($i = 1,2$) and C_α ($\alpha = 0,1,2,3$), may have "sub colors," the additional degrees of freedom [4], and belong to a fundamental representation of sub color symmetry. Although the sub color symmetry is unknown, a simplest and most likely candidate for it is SU(4). Therefore, for simplicity, all the subquarks are assumed to be quartet in sub color SU(4). Also, although the confining force is unknown, a simplest and most likely candidate for it is the one described by quantum subchromodynamics (QSCD), the Yang-Mills gauge theory of sub color SU(4) [4]. Note that the subquark charges satisfy not only the Nishijima-Gell-Mann rule of $Q = I_w + (B - L)/2$ but also the "anomaly-free condition" of $\Sigma Q_w = \Sigma Q_C = 0$.

In the unified composite model, we expect at least 36 (= 6 × 6) composite states of a subquark (a) and an antisubquark (a^* or $\mathbf{a}$) which are sub color-

singlet. They include: 1) $16 = (4 \times 2 \times 2)$ spinor states corresponding to one
generation of quarks and leptons and their antiparticles of

$$v_e = C_0^* w_1, \quad e = C_0^* w_2, \quad u_i = C_i^* w_1, \quad d_i = C_i^* w_2,$$

and their hermitian conjugates $(i = 1, 2, 3)$; 2) $4 (= 2 \times 2)$ vector states corresponding to the photon and weak bosons of

$$W^+ = \underline{w_2 w_1};$$

$$\gamma, \ Z = \underline{w_1 w_1}, \ \underline{w_2 w_2}, \ \underline{C_0 C_0}, \ \underline{C_i C_i};$$

$$W^- = \underline{w_1 w_2};$$

or $4 (= 2 \times 2)$ scalar states corresponding to the Higgs scalars of

$$\varphi_{ij} = \left[(\underline{w_1 w_1}) \ (\underline{w_2 w_1}) \ / \ (\underline{w_1 w_2}) \ (\underline{w_2 w_2}) \right] \ (i, j = 1, 2);$$

and 3) $16 = 4 \times 4$ vector states corresponding to a) the gluons, "leptogluon" and
"barygluon" of

$$G_a = \underline{C_i} \left(\lambda_a / 2 \right)_{ij} C_j; \quad G_0 = \underline{C_0 C_0}; \quad G_9 = \underline{C_i C_i} \ (i, j = 1, 2, 3),$$

where $\lambda_a \ (a = 1, 2, 3, ..., 8)$ is the Gell-Mann's matrix of $SU(3)$ and b) the "vector leptoquarks" of

$$X_i = \underline{C_0 C_i}$$

and the hermitian conjugates $(i = 1, 2, 3)$ or $16 = 4 \times 4$ scalar states corresponding to the "scalar gluons," "scalar leptogluon," "scalar barygluon" and "scalar
leptoquarks" of

$$\Phi_{\alpha\beta} = \underline{C_\alpha C_\beta} \ (\alpha, \beta = 0, 1, 2, 3).$$

Quarks and leptons with the same quantum numbers but in different generations can be taken as dynamically different composite states of the same constituents. In addition to these "meson-like composite states" of a subquark and
an antisubquark, there may also exist "baryon-like composite states" of 4 subquarks, which are sub color-singlet.

III.   Quantum numbers and electron mass

In the unified composite model the electron of charge -1 and spin ½ is taken as
a composite S-wave ground state of the spinor subquark w_2 of charge $-½$ and
spin ½ and the scalar antisubquark $\overline{C_0}$ of charge $-½$ and spin 0. The quantum
numbers of the electron come from those of subquarks, the constituents of the
electron. In order to explain the mass of the electron, we must consider all the
masses of quarks and leptons together, since the electron is not the only isolated member but one of the at least twenty-four fundamental fermions, the
quarks and leptons. By taking the first generation of quarks and leptons as almost Nambu-Goldstone fermions [5] due to spontaneous breakdown of approximate supersymmetry between a wakem and a chrom, and the second generation of them as quasi Nambu-Goldstone fermions [6], the superpartners of
the Nambu-Goldstone bosons due to spontaneous breakdown of approximate

global symmetry, we have not only explained the hierarchy of quark and lepton masses, $m_e \ll m_\mu \ll m_\tau$, $m_u \ll m_c \ll m_t$, $m_d \ll m_s \ll m_b$, but also obtained the square-root sum rules for quark and lepton masses [7], $m_e^{1/2} = m_d^{1/2} - m_u^{1/2}$ and $m_\mu^{1/2} - m_e^{1/2} = m_s^{1/2} - m_d^{1/2}$, and the simple relations among quark and lepton masses [8], $m_e m_\tau^2 = m_\mu^3$ and $m_u m_s^3 m_t^2 = m_d m_c^3 m_b^2$, all of which are remarkably well satisfied by the experimental values and estimates. By solving a set of these two sum rules and two relations [9], given the inputs of $m_e = 0.511$ MeV, $m_\mu = 105.7$ MeV, $m_u = 4.5 \pm 1.4$ MeV, $m_c = 1.35 \pm 0.05$ GeV and $m_b = 5.3 \pm 0.1$ GeV [10], we can obtain the following predictions:

$$m_\tau = 1520 \text{ MeV } (1776.99 + 0.29/- 0.26 \text{ MeV}),$$

$$m_d = 8.0 \pm 1.9 \text{ MeV } (5 \text{ to } 8.5 \text{ MeV}),$$

$$m_s = 154 \pm 8 \text{ Mev } (80 \text{ to } 155 \text{ Mev}),$$

$$m_t = 187 \pm 78 \text{ GeV } (174.3 \pm 5.1 \text{ or } 178.1 + 10.4/-8.3 \text{ GeV}),$$

where the values in the parentheses denote either the experimental data or the phenomenological estimates [10], to which our predicted values should be compared. Furthermore, if we solve a set of these two sum rules and these two relations, and the other two sum rules for the W boson mass m_W and the Higgs scalar mass (m_H) derived in the unified composite model of the Nambu-Jona-Lasinio type [1],

$$m_W = \left(3\langle m_{q,l}^2\rangle\right)^{1/2},$$

$$m_H = 2\left(\Sigma m_{q,l}^4 / \Sigma m_{q,l}^2\right)^{1/2},$$

where m_q, l's are the quark and lepton masses and $<>$ denotes the average value for all the quarks and leptons, we can predict not only the four quark and/or lepton masses such as m_d, m_s, m_t, and m_τ as above but also the Higgs scalar and weak boson masses as

$$m_H \cong 2m_t = 366 \pm 156 \text{GeV},$$

$$m_W \cong (3/8)^{1/2} m_t = 112 \pm 24 \text{GeV},$$

which should be compared to the experimental value of $m_W = 80.423 \pm 0.039$ GeV [10].

What is left for future theoretical investigations is to try to complete the ambitious program for explaining all the quark and lepton masses by deriving more sum rules and/or relations among them and by solving a complete set of the sum rules and relations. To this end, my private concern is to see whether one can take the remarkable agreement between my prediction of $m_t = \left(m_d m_c^2 m_b^2 / m_u m_s^3\right)^{1/2} \cong 180$ GeV and the experimental data as an evidence for the unified composite model. Recently, I have been more puzzled by the "new Nambu empirical quark-mass formula" of

$$M = 2^n M_0$$

with his assignment of $n = 0,1,5,8,10,15$ for u, d, s, c, b, t [11], which makes my relation of $m_u m_s^3 m_t^2 = m_t m_c^3 m_b^2$ hold exactly. More recently, I have been even more puzzled by the relations of

$$m_u m_b \cong m_s^2 \text{ and } m_d m_t \cong m_c^2$$

suggested by Davidson, Schwartz and Wali (D-S-W) [12], which can coexist with my relation and which are exactly satisfied by the Nambu assignment. If we add the D-S-W relations to a set of our two sum rules, our two relations and our sum rules for m_W and if we solve a set of these seven equations by taking the experimental values of $m_e = 0.511 \text{ MeV}$, $m_\mu = 105.7 \text{ MeV}$ and $m_W = 80.4 \text{ GeV}$ as inputs, we can find the quark and lepton mass spectrum of

$$m_\tau = 1520 \text{ MeV } (1776.99 + 0.29/ - 0.26 \text{ MeV}),$$

$$m_u = 3.8 \text{MeV } (1.5 \text{ to } 4.5 \text{MeV}),$$

$$m_d = 7.2 \text{MeV } (5 \text{ to } 8.5 \text{MeV}),$$

$$m_s = 150 \text{MeV } (80 \text{ to } 155 \text{MeV})$$

$$m_c = 0.97 \text{GeV } (1.0 \text{ to } 1.4 \text{GeV}),$$

$$m_b = 5.9 \text{GeV } (4.0 \text{ to } 4.5 \text{GeV}),$$

$$m_t = 131 \text{GeV } (174.3 \pm 5.1 \text{GeV or } 178.1 + 10.4/ - 8.3 \text{GeV}),$$

where an agreement between the calculated values and the experimental data or the phenomenological estimates seems reasonable. This result may be taken as one of the most elaborate theoretical works in elementary particle physics.

IV.   Interactions and coupling constants of the electron

In the unified composite model, the unified gauge theory of Glashow-Salam-Weinberg for electroweak interactions of the composite quarks and leptons [13] is not taken as the most fundamental theory, but as an effective theory at low energies which can be derived from the more (and, probably, most) fundamental theory of quantum subchromodynamics for confining forces of elementary subquarks [4]. It is an elementary exercise to derive the Georgi-Glashow relations [14],

$$(\sin \theta_w)^2 = \Sigma (I_3)^2 / \Sigma Q^2 = 3/8 \text{ and}$$

$$(f/g)^2 = \Sigma (I_3)^2 / \Sigma (\lambda_a / 2)^2 = 1,$$

for the weak-mixing angle θ_w, the gluon and weak-boson coupling constants (f and g), the third component of the isospin (I), the charge (Q) and the color-spin $(\lambda_a / 2)$ of subquarks without depending on the assumption of grand unification of strong and electroweak interactions. The experimental value [10] is $[\sin \theta_w (M_z)]^2 = 0.23113 \pm 0.00015$. The disagreement between the value of $3/8$ predicted in the subquark model and the experimental value might be excused by insisting that the predicted value is viable as the running value renormalized à la Georgi, Quinn and Weinberg [15] at extremely high energies (as high as 10^{15}GeV, given the "desert hypothesis."

The CKM quark-mixing matrix V [16] is given by the expectation value of the subquark current between the up and down quark states as [17]

$$V_{us} = \langle u | \underline{w_1 w_2} | d \rangle .$$

By using the algebra of subquark currents [18], the unitarity of quark-mixing matrix $V^\dagger V = V V^\dagger = 1$ has been demonstrated although the superficial non-unitarity of V as a possible evidence for the substructure of quarks has also been discussed by myself [19]. In the first order perturbation of isospin breaking, we have derived the relations of $V_{us} = -V_{cd}$, $V_{cb} = -V_{ts}$,..., which agree well with the experimental values of $|V_{us}| = 0.219 - 0.226$ and $|V_{cd}| = 0.219 - 0.226$ [10] and some other relations such as $|V_{cb}| = (|V_{ts}|) \cong (m_s / m_b) |V_{us}| \cong 0.021$, which roughly agree with the latest experimental value of $|V_{cb}| = 0.038 - 0.044$ [10]. In the second-order perturbation, the relations of $|V_{ub}| \cong (m_s / m_c) |V_{us}||V_{cb}| \cong 0.0017$ and $|V_{td}| \cong |V_{us}||V_{cb}| \cong 0.0046$ have been predicted. The former relation agrees remarkably well with the latest experimental data of $|V_{ub}| \cong 0.0025 - 0.0048$ [10]. The predictions for V_{ts} and V_{td} also agree fairly well with the experimental estimates from the assumed unitarity of V, $|V_{ts}| \cong 0.037 - 0.044$ and $|V_{td}| \cong 0.004 - 0.014$ [10]. In short, we have succeeded in predicting all the magnitudes of the CKM matrix elements except for a single element, say, V_{us}. On the contrary, the lepton-mixing has a different feature.

In 1998, the Super-Kamiokande Collaboration [20] found an evidence for the neutrino oscillation [21] due to neutrino-mixing among three generations of neutrinos $(\nu_e, \nu_\mu, \nu_\tau)$ in the atmospheric neutrinos. More recently, neutrino-mixing has been confirmed not only by the K2K Collaboration [22] for long-base-line neutrino oscillation by neutrino beams from KEK to Super-Kamiokande, but also by the SNO Collaboration [23] for solar neutrinos at the Sudbury Neutrino Observatory. They have concluded that the data are consistent with two-flavour $\nu_\mu \leftrightarrow \nu_\tau$ oscillations with $(\sin 2\theta_{\mu\tau})^2 \geq 0.88$ and $\Delta m_{\mu\tau}^2 = 2 \times 10^{-3}$ to 5×10^{-3} (eV)^2 [20]. The neutrino oscillation indicates not only the non-vanishing mass of neutrinos but also the breakdown of lepton number conservation [24]. I have found a simple model of neutrino masses and mixings [25], whose predictions are consistent not only with such a large mixing and such a small mass-squared difference between ν_μ and ν_τ suggested by the Super-Kamiokande data but also with a small mixing $((\sin \theta_{e\mu})^2 = 2 \times 10^{-3}$ to 4×10^{-2} and a large mass-squared difference $\Delta m_{e\mu}^2 = 0.3$ to 2.2(eV)^2 between ν_e and ν_μ suggested by the LSND data [26] but not with the solar neutrino deficit [27]. However, the LSND data has not been confirmed by any other experiments [26, 28] but seems to contradict the latest result from the KamLAND Collaboration [29], which has excluded all oscillation solutions but the 'Large Mixing Angle' solution to the solar neutrino problem with a large mixing $[(\sin 2\theta_{e\mu})^2 \cong 0.86$ to $1.00]$ and a small mass-squared difference $(\Delta m_{e\mu}^2 \cong 6.9 \times 10^{-5}$ (eV)$^2)$. Also note that the CHOOZ experiment [28] has given the constraints of $(\sin \theta_{e\tau})^2 < 0.15$ and

$\Delta m_{e\mu,e\tau}^2 < 1 \times 10^{-3}$ (eV)^2. Furthermore, the Heidelberg-Moscow group has recently reported the first evidence for neutrinoless double beta decay deducing the effective neutrino mass of 0.11 to 0.56 eV with a best value of 0.39 eV [30]. On the other hand, very lately, the determination of absolute neutrino masses from Z-bursts caused by ultrahigh energy neutrinos scattering on relic neutrinos has predicted the heaviest neutrino mass to be $2.75 + 1.28/-0.98$ eV for galactic halo and $0.26 + 0.20/-0.14$ eV for extragalactic origin [31]. More lately, by comparing the power spectrum of fluctuations derived from the Two Degree Field Galaxy Redshift Survey with power spectra for models with four components: baryons, cold dark matter, massive neutrinos and a cosmological constant, an upper limit on the total neutrino mass of 1.8 eV has been obtained [32]. As it stands now, it seems difficult to make a simple model of neutrino masses and mixings which is consistent with all the experimental results since some experimental results contradict others.

In the unified pregauge and pregeometric theory of all fundamental forces, the gauge-coupling and gravitational constants are related to each other through the most fundamental length scale of nature. A pregauge theory is a theory in which a gauge theory appears as an effective and approximate theory at low energies (lower than a cut-off λ_1) from a more fundamental theory [33], while a pregeometric theory (or pregeometry) is a theory in which Einstein theory of general relativity for gravity appears as an effective and approximate theory at low energies (lower than a cut-off λ_2) from a more fundamental theory [34]. Let us suppose that the cut-off in electrodynamics $\lambda_{e.m}$ and the cut-off in geometrodynamics λ_{grav} are the same or at least related to each other as $\lambda_{e.m} \approx \lambda_{grav}$. In most pregeometric theories of gravity in which Einstein-Hilbert action is induced as an effective and approximate action at long distances by quantum effects of matter fields, the Newtonian gravitational constant is naturally related to the ultra-violet cut-off as $G \approx \lambda_{grav}^{-2}$. If this is the case, these two equations lead to the relation

$$\lambda_{e.m} \approx G^{1/2}.$$

This is the famous conjecture by Landau in 1955 [35]: there must be a natural ultra-violet cut-off at the Planck energy $G^{1/2}$ where gravity becomes strong. On the other hand, in most of pregauge theories of electomagnetism, in which the Maxwell action is induced as an effective and approximate action at long distances by quantum effects of charged particles, the fine-structure constant is naturally related to the ultra-violet cut-off as

$$\alpha \approx 1/\ln(\lambda_{e.m}^2/M^2),$$

where M is a parameter of mass dimension. If this is the case, these two relations lead to the relation of

$$\alpha \approx 1/\ln(GM^2).$$

This is the so-called α-G relation first derived by us in 1977 in the unified pregauge and pregeometric theory of all fundamental forces [36]. Note, however, that in some pregauge and pregeometric theories these fundamental constants are determined as $G \approx M^2/\lambda^4$ and $\alpha \approx (M/\lambda)^4$ so that the α-G relation becomes

$$\alpha \approx GM^2 .$$

Hereafter, I will concentrate on the α-G relation of the former type, leaving the α-G relation of the latter type for later discussion.

For definiteness let us write the α-G relation as

$$\alpha = 1/A\ln(1/GM^2) ,$$

where A is a constant parameter depending on a particular unified pregauge and pregeometric model of all fundamental forces. In our unified pregauge and pregeometric model of all elementary particle forces including gravity [36], for example, the constant is simply given by $A = \Sigma Q^2/3\pi$, where ΣQ^2 is the sum of squared charges over all fundamental fermions. For N generations of quarks and leptons, $\Sigma Q^2 = 8N/3$, so that $A = 8N/9\pi$. Also, the mass parameter is approximately given by $M^2 \cong 5Nm_W^2/24\pi$ for N generations, so that the α-G relation approximately becomes $\alpha \cong 9\pi/8N\ln(24\pi/5NGm_W^2)$, where m_W is the charged weak boson mass. Furthermore, we also know that for six generations of quarks and leptons ($N = 6$) the α-G relation of

$$\alpha \cong 3\pi\Big/16\ln\left(4\pi/5Gm_W^2\right)$$

is very well satisfied by the experimental values of $\alpha \cong 1/137$, $G^{1/2} \cong 1.22 \times 10^{19}\,\text{GeV}$ and $m_W \cong 80\,\text{GeV}$. Therefore, from now on let us assume that there exist six generations of quarks and leptons, or three generations of quarks and leptons and their mirror particles, or that there exist three generations of quarks and leptons, their super-partners and more, so that $A = 16/3\pi$.

We now suppose that the fundamental length scale $1/\lambda$ be time-varying with respect to the mass scale related to the mass parameter M. Then, we expect that both the fine-structure and gravitational constants α and G are time-varying [37] and their time-derivatives $d\alpha/dt$ and dG/dt may satisfy the relation of

$$(d\alpha/dt)\alpha^2 = AdG/dt ,$$

which can be derived by differentiating the both hand sides of the α-G relation with respect to time. If instead $1/\lambda$ stays constant and if M varies, dG/dt must vanish but the time-derivatives $d\alpha/dt$ and dM/dt may satisfy the other relation of

$$(d\alpha/dt)\alpha^2 = 2A(dM/dt)/M .$$

For $A = 16/3\pi$, the above relations become

$$(dG/dt)/G = (3\pi/16)(d\alpha/dt)/\alpha^2$$

and

$$(dM/dt)M = (3\pi/32)(d\alpha/dt)/\alpha^2 .$$

Now the first relation together with the latest result of $<d\alpha/dt>/\alpha = (2.25 \pm 0.56) \times 10^{-15}/\text{yr}$ for redshift of $0.5 < z < 3.5$ by Webb et al. [38] immediately leads to our prediction of

$$(dG/dt)/G = (0.181 \pm 0.045) \times 10^{-12} / \text{yr}$$

We find that this prediction is not only consistent with the most precise limit of $(dG/dt)/G = (-0.6 \pm 2.0) \times 10^{-12} / \text{yr}$ by Thorsett [39] but also feasible for future experimental test.

If the α-G relation of the latter type holds instead of the one of the former type, it leads to either the relation

$$(d\alpha/dt)/\alpha = (dG/dt)/G$$

or

$$(d\alpha/dt)/\alpha = 2(dM/dt)/M ,$$

depending on whether the length scale $1/\lambda$ or the mass scale M varies while M or $1/\lambda$ stays constant. Then, the first relation together with the result of Webb et al. [38] immediately leads to another prediction of

$$(dG/dt)G = (2.25 \pm 0.56) \times 10^{-15} / \text{yr}.$$

We find that this predicted value for $(dG/dt)G$ seems too small to be feasible for experimental tests in the near future although it is consistent with the limit of Thorsett [39]. On the other hand either one of the second relations together with the result of Webb et al. [38] immediately leads to another prediction of

$$(dM/dt)M = (0.081 \pm 0.023) \times 10^{-12} / \text{yr},$$

or

$$(dM/dt)M = (1.13 \pm 0.28) \times 10^{-15} / \text{yr}.$$

However, I suspect that either one of these predicted values for $(dM/dt)M$ is too small to be feasible for experimental tests in the near future, although a prediction for the possible time-varying particle masses seems extremely interesting at least theoretically.

In concluding this Section, I would like to emphasize that the recent result of Webb et al. [38] suggesting a varying fine-structure constant may indicate not only a varying gravitational constant but also a varying cosmological constant [40], if our picture for varying constants of nature is right and future experiments to test our predictions for $(dG/dt)G$ in this Section may check not only the α-G relation but also the unified pregauge and pregeometric theory of all fundamental forces. A few questions would still remain: What is the origin of the varying length scale $1/\lambda$ or of the varying mass scale M? Is it related to the mass field [41], the "quintessence" [42] or the Kaluza-Klein extra space in extra dimensions? Are no "constants" of nature constant? After all, it may be that nothing is constant or permanent, as emphasized by the Greek and Indian philosophers some two and a half millennia ago!

V. Conclusions and further discussion

I have explained almost all the properties of the electron including the charge, spin, mass, mixing-angle and coupling constants in the unified composite model of all fundamental particles and forces. There remain some other important properties such as the electric and magnetic moments and the possible non-vanishing size of the electron. First of all, the latest experimental value for the electron mass is [43]

$$m_e = 0.0005485799092 \pm 0.0000000000004 \text{ u},$$

where u is the unified atomic mass unit $[= (931.494013 \pm 0.000037) \text{ MeV} = (1.66053873 \pm 0.00000013)10^{-27} \text{ kg}]$, while that for the electron charge magnitude is [10]

$$e = (1.602176462 \pm 0.000000063) \ 10^{-19} \text{ C}$$
$$= (4.80320420 \pm 0.00000019) \text{ esu}.$$

The fine structure constant is given by [10]

$$\alpha (\equiv e^2/4\pi\hbar c) = 1/(137.03599976 \pm 0.00000050).$$

The experimental upper bound on the charge difference between the electron and the positron is [10]

$$\left| q_{e+} + q_{e-} \right| / e \leq 4 \times 10^{-8}.$$

The experimental upper bound on the mass difference between the electron and the positron is [10]

$$\left| m_{e+} - m_{e-} \right| / m_e < 8 \times 10^{-9}|,$$

which gives strong constraint on possible violation of CPT invariance. On the other hand, the current experimental constraint on the electric dipole moment of the electron is [10, 44]

$$d_e \leq (0.07 \pm 0.07) \times 10^{-26} \text{ e cm},$$

which may allow a small violation of CP or T invariance in the electron sector. Furthermore, the experimental value for the electron magnetic moment is [10]

$$\mu_e = (1.001159652187 \pm 0.000000000004) e/2m_e,$$

which is consistent with the standard model. The experimental data on the difference between the electron and positron g-factor is [10]

$$\left| g_{e+} - g_{e-} \right| / g_{average} = (-0.5 \pm 2.1) \times 10^{-12}|,$$

which gives another strong constraint on possible violation of CPT invariance.

In 1996, the CDF Collaboration at Tevatron [45] released their data on the inclusive jet differential cross section for jet transverse energies E_T from 15 to 440 GeV with the significant excess over current predictions based on perturbative QCD calculations for $E_T > 200 GeV$, which may indicate the presence of quark substructure at the compositeness energy scale λ_c of the order of 1.6T eV. This could be taken as an exciting and intriguing historical discovery of the substructure of quarks (and leptons), which had been long predicted, or as the first evidence for the composite model of quarks (and leptons), which had been proposed since the middle 1970s [1]. It might dramatically change not only so-called "common sense" in physics or science but also that in philosophy, which often states that quarks (and leptons) are the smallest and most fundamental forms (or particles) of matter in mother nature. Note that such a relatively low energy scale for λ_C of the order of 1 TeV had been anticipated theoretically [46] or by precise comparison between currently available experimental data and calculations in the composite model of quarks (and leptons) [47]. In 1997, the H1 and ZEUS Collaborations at HERA [48] reported their data on the deep inelastic $e + p$ scattering with a significant excess of events over the expectation of the standard model of electroweak and strong interactions for high momentum-transfer squared $Q^2 > 15000\,(\text{GeV})^2$, which might indicate new physics beyond the standard model. Although neither one of these indications have been confirmed by the other experiments and the significance of the HERA anomaly has decreased with higher statistics, not only the substructure of quarks and leptons as well as Higgs scalars and gauge bosons, but also the possible existence of leptoquarks has been extensively reinvestigated [49]. As it stands now, I must emphasize that both the CDF and HERA anomalies are still with us, and that the explanation of the latter anomaly either by the leptoquark with the mass between 280 GeV and 440 GeV, or by the excited electron with the mass between 300 GeV and 370 GeV [50] is still very viable. The current lower bound on the mass of the excited electron is $m_e^* > 223\,\text{GeV}$ [10] while that on the compositeness energy scale of the electron is $\lambda_{LL}^+ (eeee) > 8.3$ TeV and $\lambda_{LL}^- (eeee) > 10.3$ TeV [10], which means that the size of the electron $(1/\lambda_e)$ is smaller than the order of 10^{-18} cm .

The possible substructure of fundamental fermions such as the electron was considered in some detail by McClure-Drell and Kroll [51] and by Low and myself [52] already in the middle of nineteen sixties, while that of quarks was pointed out by Wilson and others [53] in the early nineteen seventies. Also, the possible substructure of fundamental bosons such as the weak bosons was discussed in great detail by myself and others [54] in the mid-nineteen seventies. In conclusion, let me repeat what I said in my talks at the Paris Conference in 1982 [55] and at the Leipzig Conference in 1984 [56]. "It seems to me that it has taken and will take about a quarter century to go through one generation of physics: atomic physics in 1900-1925, nuclear physics in 1925-1950, hadron physics in 1950-1975, quark-lepton physics in 1975-2000, "subquark physics" in 2000-2025 and so on." "I would like to emphasize that the idea of composite models of quarks and leptons (and also gauge bosons as well as Higgs scalars), which was proposed by us, theorists, in the mid-seventies, has just become a subject of experimental relevance in the mid-eighties." A century has past since the discovery of the electron, the "first elementary particle," and, hopefully, the compositeness of "elementary particles" will soon be found.

Acknowledgements

The author would like to thank Professor Volodimir Simulik, the editor of this book, entitled *What is the electron? Modern structures, theories, hypotheses*, for inviting him to publish the present paper in the book and for encouraging him until it was written. He also wishes to thank Mr. Yoshizumi Terazawa, his brother, to whom he dedicates the present paper, for having introduced him to science and technology in his childhood.

References

[1] H. Terazawa, Y. Chikashige and K. Akama, *Phys. Rev. D* **15**, 480 (1977); H. Terazawa, *ibid.* **22**, 185 (1980). For a classic review see for example H. Terazawa in *Proc. XXII International Conf. on High Energy Physics*, Leipzig, 1984, edited by A. Meyer and E. Wieczorek (Akademie der Wissenschaften der DDR, Zeuthen, 1984), Vol. I, p. 63. For a latest review see H. Terazawa, in *Proc. Crimean Summer School-Conference on New Trends in High-Energy Physics*, Crimea (Ukraine), 2000, edited by P.N. Bogolyubov and L.L. Jenkovszky (Bogolyubov Institute for Theoretical Physics, Kiev, 2000), p. 226.

[2] J.C. Pati and A. Salam, *Phys. Rev. D* **10**, 275 (1974)

[3] H. Miyazawa, *Prog. Theor. Phys.* **36**, 1266 (1966); Yu.A. Gol'fand and E.P. Likhtman, *ZhETF Pis. Red.* **13**, 452 (1971); *JETP Lett.* **13**, 323 (1971); D.V. Volkov and V.P. Akulov *ibid.* **16**, 621 (1972); *JETP Lett.* **16**, 438 (1972); *Phys. Lett.* **46B**, 109 (1973); J. Wess and B. Zumino, *Nucl. Phys.* **B70**, 39 (1974).

[4] G. 't Hooft, in *Recent Developments in Gauge Theories*, edited by G't Hooft (Plenum, New York, 1980, p. 135; H. Terazawa, *Prog. Theor. Phys.* **64**, 1763 (1980).

[5] H. Terazawa, in Ref. [4].

[6] W. Buchmuller, R.D. Peccei and T. Yanagida, *Phys. Lett.* **B124** , 67, (1983); R. Barbieri, A. Masierro and G. Veneziano, *ibid.* **B124**, 179 (1983); .O. Greenberg, R.N. Mohapatra and M. Yasue, *ibid.* **B128**, 65 (1983).

[7] H. Terazawa, *J. Phys. Soc. Jpn.* **55**, 4249 (1986); H. Terazawa and M. Yasue, *Phys. Lett.* **B206**, 669 (1988); H. Terazawa in *Perspectives on Particle Physics*, edited by S. Matsuda *et al.* (World Scientific, Singapore, 1988), p. 193.

[8] H. Terazawa and M. Yasue, *Phys. Lett.* **B307**, 383 (1993); H. Terazawa, *Mod. Phys. Lett.* **A10**, 199 (1995).

[9] H. Terazawa, *Mod. Phys. Lett.* **A7**, 1879 (1992).

[10] K. Hagiwara *et al.* (Particle Data Group), *Phys. Rev. D* **66**, 010001 (2002)

[11] Y. Nambu, *Nucl. Phys.* **A629**, 3c (1998); **A638**, 35c (1998).

[12] A. Davidson, T. Schwartz and K.C. Wali, *J. Phys. G. Nucl. Part. Phys.* **24**, L55 (1998).

[13] S.L. Glashow, *Nucl. Phys.* **22**, 579 (1961); A. Salam, in *Elementary Particle Physics*, edited by N. Svartholm (Almqvist and Wiksell, Stockholm, 1968) p. 367; S. Weinberg, *Phys. Rev. Lett.* **19**, 1264 (1967).

[14] H. Georgi and S.L Glashow, *Phys. Rev. Lett.* **32**, 438 (1974).

[15] H. Georgi, H.R. Quinn and S. Weinberg, *Phys. Rev. Lett.* **33**, 451 (1974).

[16] N. Cabibbo, *Phys. Rev. Lett.* **10**, 531 (1963); S.L. Glashow, I. Iliopulos and L. Maiani, *Phys. Rev. D* **2**, 1285 (1970); M. Kobayashi and T. Maskawa, *Prog. Theor. Phys.* **49**, 652 (1973).

[17] H. Terazawa, *Prog. Theor. Phys.* **58**, 1276 (1977); V. Visnjic-Triantafillou, *Fermilab Report No. FERMILAB-Pub-80/34-THY*, 1980 (unpublished); H. Terazawa, in Ref. [4]; O.W. Greenberg and J. Sucher, *Phys. Lett.* **99B**, 339 (1981); H. Terazawa, *Mod. Phys. Lett.* **A7**, 3373 (1992). For the superficial non-unitarity of *V* as a possible evidence for the substructure of quarks, see H. Terazawa, *Mod. Phys. Lett.* **A11**, 2463 (1996).

[18] H. Terazawa, in Ref. [1].

[19] H. Terazawa, *Mod. Phys. Lett.* **A7**, 3373 (1992).

[20] Y. Fukuda *et al.*, *Phys. Rev. Lett.* **81**, 1562 (1998). For a review see, T. Kajita and Y. Totsuka, *Rev. Mod. Phys.* **73**, 85 (2001).

[21] B. Pontecorvo, *Zh. Eksp. i Teor. Fiz.* **33**, 549 (1957) [*Soviet Phys. JETP* **6**, 429 (1958)]; **34**, 247 (1958) [**7**, 172 (1958)]; **53**, 1717 (1967) [**62**, 984 (1968)]; Z. Maki, M. Nakagawa and S. Sakata, *Prog. Theor. Phys.* **28**, 870 (1962).

[22] S.H. Ahn *et al.* (K2K Collaboration), *Phys. Lett.* **B511**, 178 (2001); M.H. Ahn *et al.* (K2K Collaboration), *Phys. Rev. Lett.* **90**, 041801 (2003).

[23] Q.R. Ahmad *et al.* (SNO Collaboration), *Phys. Rev. Lett.* **87**, 071301 (2001); **89**, 011301 (2002); **89**, 011302 (2002).

[24] K. Nishijima, *Phys. Rev.* **108**, 907 (1957); J. Schwinger, *Ann. Phys.* **2**, 407 (1957); S. Bludman, *Nuovo Cimento* **9**, 433 (1958).

[25] H. Terazawa, in *Proc. International Conf. on Modern Developments in Elementary Particle Physics*, Cairo, Helwan and Assyut, 1999, edited by A. Sabry (Ain Shams University, Cairo, 1999), p. 128.

[26] C. Athanassopoulos *et al.* (LSND Collaboration), *Phys. Rev. Lett.* **75**, 2650 (1995); **81**, 1774 (1998); *Phys. Rev. C* **58**, 2489 (1998); A. Aguilar *et al.* (LSND Collaboration); *Phys. Rev. D* **64**, 112007 (2001). See, however, A. Romosan *et al.* (CCFR Detector Group), *Phys. Rev. Lett.* **78**, 2912 (1997) in which the high $\Delta m_{e\mu}^2$ region has been excluded. Furthermore the latest NuTeV data exclude the high $\Delta m_{e\mu}^2$ end favoured by the LSND experiment. See S. Avvakumov *et al. Phys. Rev. Lett.* **89**, 011804 (2002). Also see B. Armbruster *et al.* (KARMEN Collaboration), *hep-ex/0203021*, 14 Mar 2002, which does not confirm the LSND result.

[27] For recent reports see S. Fukuda *et al.* (Super-Kamiokande Collaboration), *Phys. Rev. Lett.* **86**, 5651 (2001); **86**, 5656 (2001); *Phys. Lett.* **B 535**, 179 (2002).

[28] M. Apollonio *et al.* (CHOOZ Collaboration), *Phys. Lett.* **B 466**, 415 (1999).

[29] K. Eguchi *et al.* (KamLAND Collaboration), *Phys. Rev. Lett.* **90**, 021802 (2003).

[30] H.V. Klapdor-Kleingrothaus *et al.*, *Mod. Phys. Lett.* **A16**, 2409 (2001).

[31] Z. Fodor, S.D. Katz and A. Ringwald, *Phys. Rev. Lett.* **88**, 171101 (2002).

[32] Elgaroy *et al.*, *Phys. Rev. Lett.* **89**, 061301 (2002).

[33] J.D. Bjorken, *Ann. Phys. (N.Y)* **24**, 174 (1963); H. Terazawa, Y Chikashige and K. Akama, in Ref. [1]. For a classic review see, for example, H. Terazawa, in *Proc. XIX International Conf. on High Energy Physics*, Tokyo, 1978, edited by S. Homma, M. Kawaguchi and H. Miyazawa (Physical Society of Japan, Tokyo, 1979) p. 617. For a recent review see H. Terazawa in Ref. [1].

[34] A.D. Sakharov, *Dokl. Akad. Nauk SSSR* **177**, 70 (1967) [*Sov. Phys. JETP* **12**, 1040 (1968)]; K. Akama, Y. Chikashige, T. Matsuki and H. Terazawa, *Prog. Theor. Phys.* **60**, 868 (1978); S.L. Adler, *Phys. Rev. Lett.* **44**, 1567 (1980); A. Zee, *Phys. Rev. D* **23**, 858 (1981); D. Amati and G. Veneziano, *Phys. Lett.* **105 B**, 358 (1981). For a review see, for example, H. Terazawa, in *Proc. First A.D. Sakharov Conf. on Physics*, Moscow, 1991, edited by L.V. Keldysh and V.Ya. Fainberg (Nova Science, New York, 1992) p.1013.

[35] L. Landau, in *Niels Bohr and the Development of Physics*, edited by W. Pauli (McGraw-Hill, New York, 1955), p. 52.

[36] H. Terazawa, Y. Chikashige, K. Akama and T. Matsuki, *Phys. Rev. D* **15**, 1181 (1977); *J. Phys. Soc. Jpn.* **43**, 5 (1977); H. Terazawa, *Phys. Rev. D* **16**, 2373 (1977); K. Akama, Y. Chikashige, T. Matsuki and H. Terazawa, in Ref. [34]; H. Terazawa, *Phys. Rev. D* **22**, 1037 (1980); **41**, 3541 (E) (1990); H. Terazawa and K. Akama, *Phys. Lett.* **69 B**, 276 (1980); **97 B**, 81(1980); H. Terazawa, *ibid.* **133 B**, 57 (1983). For a classic review see H. Terazawa in Ref. [33]. For a latest review see H. Terazawa in Ref. [1].

[37] P.M.A. Dirac, *Nature* **139**, 323 (1937); *Proc. R. Soc. London* **A165**, 199 (1938); **A333**, 403 (1973); **A338**, 439 (1974).

[38] J.K. Webb *et al.*, *Phys. Rev. Lett.* **82**, 884 (1999); **87**, 091301 (2001).

[39] S.E. Thorsett, *Phys. Rev. Lett.* **77**, 1432 (1996).

[40] A. Vilenkin, *Phys. Rev. Lett.* **81**, 5501 (1998); A.A. Starobinsky, *JETP Lett.* **68**, 757 (1998); H. Terazawa, in *Proc. 2-nd Biannual International Conf. on Non-Euclidean Geometry in Modern Physics*, Nyiregyhaza (Hungary), 1999, edited by I. Lovas *et al.*, *Acta Physica Hungarica, New Series: Heavy Ion Physics* **10**, 407 (1999).

[41] F. Hoyle and J.V. Narlikar, *Nature* **233**, 41 (1971); V. Canuto, P.J. Adams, S.-H. Hsieh and E. Tsiang, *Phys. Rev. D* **16**, 1643 (1977); H. Terazawa, *Phys. Lett.* **101 B**, 43 (1981).

[42] See, for example, B. Ratra and P.J.E. Peebles, *Phys. Rev. D* **37**, 3406, (1988); P.J.E. Peebles and B. Ratra, *Astrophys. J.* **325**, L17 (1988); C.T. Hill, D.N. Schramm and J.N. Fry, *Comments Nucl. Part. Phys.* **19**, 25 (1989); J. Frieman, C. Hill, A. Stebbins and I. Waga, *Phys. Rev. Lett.* **75**, 2077 (1995). T. Chiba, N. Sugiyama and T. Nakamura, *Mon. Not. R. Astron. Soc.* **289**, L5, (1997); **301**, 72 (1998); M.S. Turner and M. White, *Phys. Rev. D* **56**, 4439 (1997); R.R. Caldwell, R. Dave and P.J. Steinhardt, *Phys. Rev. Lett.* **80**, 1582 (1998); S.M. Carrol *ibid.* **81**, 3067 (1998); M. Bucher and D.N. Spergel, *Phys. Rev. D* **60**, 043505 (1999); *Sci. Am.* **280**, 42 (1999). For latest proposals see G. Dvali and M. Zaldarriaga, *Phys. Rev. Lett.* **88**, 091303 (2002); J.K. Erickson *et al.*, *ibid.* **88**, 121301 (2002). For the latest argument against models in which an observable variation of the fine-

structure constant is explained by motion of a cosmic scalar field see T. Banks, M. Dine and M. Douglas, *Phys. Rev. Lett.* **88**, 131301 (2002).

[43] T. Baier *et al.*, *Phys. Rev. Lett.* **88**, 011603 (2002).

[44] B.C. Regan *et al.*, *Phys. Rev. Lett.* **88**, 071805 (2002).

[45] CDF Collaboration, F. Abe *et al.*, *Phys. Rev. Lett.* **77**, 438 (1996); *ibid.* **77**, 5336 (1996); *Phys. Rev. D* **55**, R5263 (1997); T. Affolder *et al.*, *ibid.* **64**, 032001 (2001). For a latest report on search for new physics see CDF Collaboration, D. Acosta *et al.*, *Phys. Rev. Lett.* **89**, 041802 (2002). "The result are consistent with standard model expectations, with the possible exception of photon-lepton events with large missing transverse energy, for which the observed total is 16 events and the expected mean total is 7.6±0.7 events."

[46] See, for example, H. Terazawa, *Prog. Theor. Phys.* **79**, 734 (1988); *Phys. Rev. Lett.* **65**, 823 (1990); *Mod. Phys. Lett.* **A6**, 1825 (1991); K. Akama, H. Terazawa and M. Yasue, *Phys. Rev. Lett.* **68**, 1826 (1992).

[47] See, for example, K. Akama and H Terazawa, *Phys. Lett.* **B 321**, 145 (1994); *Mod. Phys. Lett.* **A 9**, 3423 (1994); H. Terazawa, *ibid.* **A 11**, 2463 (1996); K. Akama and H. Terazawa, *Phys. Rev. D* **55**, R2521 (1997); K. Akama, K. Katsuura and H. Terazawa, *ibid.* **56**, R2490 (1997).

[48] H1 Collaboration, C. Adloff *et al.*, *Z. Phys.* **C74**, 191 (1997); ZEUS Collaboration, J. Breitweg *et al.*, *ibid.* **C74**, 207 (1997).

[49] See, for example, H1 Collaboration, C. Adloff *et al.*, *Phys. Lett.* **B479**, 358 (2000) and V.M. Abasov *et al.*, D0 Collaboration, *Phys. Rev. Lett.* **87**, 061802 (2001).

[50] K. Akama, K. Katsuura and H. Terazawa in Ref. [47].

[51] J.A. McClure and S.D. Drell, *Nuovo Cimento* **37**, 1638 (1965); N.M. Kroll, *ibid.* **45**, 65 (1966).

[52] F.E. Low, *Phys. Rev. Lett.* **14**, 238 (1965); H. Terazawa, *Prog. Theor. Phys.* **37**, 204 (1967); H. Terazawa, M. Yasue, K. Akama and M. Hayashi, *Phys. Lett.* **112 B**, 387 (1982).

[53] K.G. Wilson in *Proc. 1971 International Symposium on Electron and Photon Interactions* at High Energies, Cornell Univ., Ithaca, 1971, edited by N.B. Mistry (Cornell Univ., Ithaca, N.Y., 1971) p. 115; K. Matumoto, *Prog. Theor. Phys.* **47**, 1795 (1972); M. Chanowitz and S.D. Drell, *Phys. Rev. Lett.* **30**, 807 (1973); *Phys. Rev. D* **9**, 2078 (1974); K. Akama, *Prog. Theor. Phys.* **51**, 1879 (1974); K. Akama and H. Terazawa, *Mod. Phys. Lett.* **A 9**, 3423 (1994).

[54] H. Terazawa, *Phys. Rev. D* **7**, 3663 (1973); *D* **8**, 1817 (1973); J.D. Bjorken, *ibid. D* **19**, 335 (1979); P.Q. Hung and J.J. Sakurai, *Nucl. Phys.* **B 143**, 81 (1978); **B 148**, 539 (1979).

[55] H. Terazawa, in *Proc. XXI International Conf. on High Energy Physics*, Paris, 1982, edited by P. Petiau and M. Porneuf, *J. de Phys.* **C3**-191 (1982).

[56] H. Terazawa, in Ref. [1].

Prospects for the Point Electron

Thomas E. Phipps, Jr.
908 South Busey Avenue
Urbana, Illinois 61801

Introduction: the relativity of physical size

Many ingenious models for the electron, perhaps the most stable and funda-mental of known particles, have been devised. Some of these are discussed elsewhere in this book. Here we shall examine the advantages and prospects of the simplest model of all, the mathematical point. It might seem that this could be dismissed at the outset on general philosophical grounds; *e.g.*, that a point is "infinitely small," therefore (like the "infinitely large") operationally indefin-able and hence non-physical. This overlooks the fact that models are not to be confused with that which is modeled, but are to be judged by results rather than by inferred resemblances to truth. So, I shall not address the imponderable on-tology of what the electron "is," but only what a point-particle model of it might accomplish. As a point of my own philosophy, I claim that this sort of metaphorical approximation to reality is what the science of physics at best provides. When it pretends to do more, it trespasses on the territories of phi-losophy, religion, and faith.

On the scale of Newtonian physics the point particle has done yeoman service as an approximation to everything from planets to bullets, and has given us a Newtonian principle of relativity of physical size. Dirac believed that the granularity of atoms brought an end to this Newtonian relativity. He wrote:[1] "So long as *big* and *small* are merely relative concepts, it is no help to explain the big in terms of the small. It is therefore necessary to modify classi-cal ideas in such a way as to give an absolute meaning to size." But if we ex-amine not the words Dirac used but the parameters, we discover that his own most wildly successful and seminal *theory of the electron* describes not an ex-tended particle but a mathematical point! Therefore he himself was the active agent in saving Newtonian relativity on all size scales. It is upon Dirac's suc-cess in describing the point electron that we shall build here in seeking con-sciously to implement a universal Principle of Relativity of Physical Size. It will be evident that such a principle by no means implies the physical proposi-tion that "size does not matter." It merely implies the hopeful view that the point-particle approximation can be descriptively useful on all size scales... and, more cogently, that *the form or the parameterization of physical descrip-tive equations is invariant; i.e., does not abruptly change at some threshold such as the "atomic" or "nuclear."* Thus, like another better-known relativity principle, it implies mathematical form preservation or invariance. Stated in that modest way, it seems to make a good deal of sense, does it not? For if there were an abrupt change in parameterization, would this not have to reflect an (unobserved) abrupt change in the physics? How else than by *form preservation* are we to make useful inferences from known physics about the topography of unknown physical-descriptive territory? With such questions for clues, we should be able to do a bit of elementary detective work that would not strain Sherlock. Indeed, the implications of size relativity are the only real clue we have to guide us in bettering our pretend-knowledge (as embodied in the ludicrously over-hyped Standard Theory) of particles on nuclear and subnuclear size scales.

A rigorized formal correspondence

If size relativity is to be a useful guide in physical exploration, it must apply to the descriptive transition between the Newtonian and atomic descriptive realms. And it must embody rigorous form preservation. That is the first test that the principle must pass before its "universality" can be substantiated. Here an immediate breakdown of the principle occurs, if Dirac[1] is to be believed. Exercising equal parts of optimism and scepticism, let us set aside Dirac's judgment and look closely at the best that can be done in formally aligning classical and quantum mechanics … that is, in setting up a "formal Correspondence" between the two. We need to inquire: "Form preservation" under what transformation or group? At once we see that our best chance is to improve the existing formal Correspondence between the Hamilton-Jacobi (H-J) mechanics (of point particles) and the Schrödinger equation; for these two already bear to each other a remarkable formal resemblance. Regrettably, it may be that the reader's education has been skimped in regard to the H-J formalism and classical canonical mechanics[2] in favour of something more trendy, such as string theory[3]. We shall not attempt a proper tutorial, but merely provide a reminder that the classical equations of motion of n point particles take the H-J form,

$$H = -\frac{\partial S}{\partial t}, \quad p_j = \frac{\partial S}{\partial q_j}, \quad -P_j = \frac{\partial S}{\partial Q_j}, \quad j = 1, 2, \cdots, 3n, \qquad (1a,b,c)$$

where $H = H\left(q_j, p_j, t\right)$ is the Hamiltonian or energy function, and $S = S\left(q_j, Q_j, t\right)$ is a scalar known as *Hamilton's principal function*. These equations completely describe the point-particle mechanics of the classical domain. Observe that there are two complementary sets of descriptive parameters, apart from time; namely, the so-called "old canonical variables" $\left(q_j, p_j\right)$ and the "new canonical variables" $\left(Q_j, P_j\right)$. The transformations between these two are termed "canonical." Just as special relativity preserves form under coordinate transformations, we might expect the principle of size relativity to imply "form" preservation under canonical transformations. That is manifestly impossible if we accept the universal opinion that Schrödinger's equations,

$$H\Psi = -\frac{\hbar}{i}\frac{\partial}{\partial t}\Psi \;,\; p_j\Psi = \frac{\hbar}{i}\frac{\partial}{\partial q_j}\Psi \;,\; \Psi = \Psi\left(q_j,t\right), \tag{2a,b}$$

tell the complete story of mechanics in the quantum domain. For, observe that there are three differences between the H-J and Schrödinger equations: (a) The latter contain a wave function or operand Ψ; hence the "old canonical variables" have morphed into "dynamical variables" $\left(q_j, p_j\right)$ that are operators. (b) The S-function has disappeared and been formally replaced by $\hbar/i$. (c) The "new canonical variables" $\left(Q_j, P_j\right)$ have disappeared and been replaced by nothing. It is this latter abrupt disappearance that blocks all possibility of invariance under the formal Correspondence transformation—and that must be corrected if a size relativity principle, implying a rigorized Correspondence, is to be implemented. How can we formulate operator equations that combine all features of both (1) and (2), when applied in their appropriate physical domains? The following set of equations does exactly that:

$$H\Psi_f = -\frac{\partial S}{\partial t}\Psi_f \;,\; p_j\Psi_f = \frac{\partial S}{\partial q_j}\Psi_f \;,\; -P_j\Psi_f = \frac{\partial S}{\partial Q_j}\Psi_f \;. \tag{3a,b,c}$$

In order to recover quantum mechanics from (3) it is necessary to postulate $S = \hbar/i$, which is equivalent to the Heisenberg postulate in view of

$$\left[p_k q_j - q_j p_k\right]\Psi_f = \left[\left(\frac{\partial}{\partial q_k}\right)Sq_j - q_j\left(\frac{\partial}{\partial q_k}\right)S\right]\Psi_f = S\delta_{jk}\Psi_f$$
$$\rightarrow \left(\frac{\hbar}{i}\right)\delta_{jk}\Psi_f \tag{4}$$

where δ_{jk} is 1 if $j = k$, 0 otherwise. Here $H = H\left(q_j, p_j, t\right)$, $\Psi_f = \Psi_f\left(q_j, Q_j, P_j, t\right)$, and in general $S = S\left(q_j, Q_j, t\right)$, if we set aside the specialization $S = \hbar/i$ that describes atomic physics. More generally, we shall find it advantageous to assume

$$S = \frac{\hbar}{i}s\left(q_j, Q_j, t\right), \tag{5}$$

where s is some real scalar function to be determined... in case we might want to generalize beyond the atomic case $\left(s \rightarrow 1\right)$ to describe, e.g., point particles in nuclear states.

The element of "innovation" here, Eq. (3c), is not really new. It is the restoration of an operator analog of (1c)—as is obviously essential in order to avoid abrupt changes of parameterization of the mechanical formalism—to reflect an absence of abrupt discontinuities in the physics. Eq. (3c) could never have been dropped from a *rigorous* formal Correspondence. Eq. (3) constitutes an operator analog of (1) and thus embodies both the size relativity principle and a rigorized formal Correspondence. Consequently, in the transition from "c-numbers" (the commuting real numbers of ordinary arithmetic) to "q-numbers" (operators), there is no abrupt change in form or parameterization. The classical canonical H-J theory is recovered from Eq. (3) by treating the formal operand Ψ_f as a constant and cancelling it from all three of (3a,b,c). Hence the "Correspondence" becomes a two-way one... it works just as well going from the quantum to the classical side as going the other way. This is not true of Eq. (2), which allows a transition from classical to quantum, but not the other way [for no analogs of the new canonical variables $\left(Q_j, P_j \right)$ are present in (2), so quantum straw is lacking to make classical bricks—sweeping statements to the contrary by some of the modern era's most famous physicists to the contrary notwithstanding]. H-J theory is *not* a formal limit of accepted quantum theory.

If we postulate Eq. (3) as holding for all mechanics—classical, quantum, and beyond—and thus avoid any formal difference among these quite different physical descriptive realms, how is our theorizing to reflect the vast differences we know to exist in nature? The answer exploits the fact that Eq. (3), a more complicated mathematical form than any set of mechanical equations previously considered by physicists, offers more solution options. Eq. (2), the Schrödinger equation, is really one equation in one unknown function Ψ. But in its most general form Eq. (3) is two equations in two unknowns, Ψ_f and S. Only in the special case of atomic solutions is it permissible to specialize to $S = \hbar / i = constant$. In describing nuclear states it may prove advantageous to treat the "commutator" value S [see Eq. (4)] as a function of space coordinate values—in particular as a function of distance from a nuclear "force center." We see thus that there are three distinct classes of solution of (3):

Class I. $\Psi_f = constant$. The solutions for Hamilton's principal function S describe the Newtonian states of motion (continuous trajectories).

Class II. $S = \hbar / i$, $\Psi_f = \Psi$. These are the ordinary quantum states descriptive of atoms.

Class III.   Both S and Ψ_f non-constant. These are states possibly descriptive of point particles within nuclei or "elementary particles." (This is speculative.)

There is an abrupt discontinuity among these three solution-class options. But it is not a physical discontinuity—it is a *descriptive choice* discontinuity. Only in that altered sense can we agree with Dirac that quantum mechanical discontinuity sets a size scale to the world. The formalism itself, the equations of motion, set no such size scale. They are size invariant. Only our decision, our choice to pick one class of solutions or another, reflects a passage from one descriptive realm to another. And this is a good thing, since every new bit of empirical knowledge we acquire further blurs the line between quantum and classical worlds. There are observable particle-wave phenomena in the centimetre range, and non-localities of quantum action on the inter-stellar scale. We simply cannot rely on "size" to distinguish these worlds. One must know enough physics to use Eq. (3) wisely, to make the right solution choice to match the particular physical problem at hand—no formalism being foolproof. Such a necessity to make intelligent choices is nothing new. In treating Maxwell's equations, for instance, we need to know enough physics to choose between advanced and retarded solutions. And such descriptive choices are always discontinuous, although the physics is not.

Class-II solutions: atomic-level description

We passed a bit too quickly over the atomic solutions. Let us examine in more detail how ordinary quantum mechanics is extracted from Eq. (3), the form postulated for all mechanics. On setting $S = \hbar / i$ in Eq. (3), we see that it can be written as

$$H\Psi_f = -\frac{\hbar}{i}\frac{\partial}{\partial t}\Psi_f \ , \ p_j\Psi_f = \frac{\hbar}{i}\frac{\partial}{\partial q_j}\Psi_f \ , \ -P_j\Psi_f = \frac{\hbar}{i}\frac{\partial}{\partial Q_j}\Psi_f \quad \text{(6a,b,c)}$$

The first two of these equations are of the familiar form (2). The third is new. We recall that in H-J mechanics the new canonical variables $\left(Q_j, P_j\right)$ are constants of the motion—unlike the old canonical variables $\left(q_j, p_j\right)$, which morph here into "dynamical variable" operators. Naturally, by the arguments we have already given, there can be no discontinuity in the interpretation of $\left(Q_j, P_j\right)$ as constants; so constants they remain in the operator calculus—constants being good operators. By inspection Eq. (6c) has the solution

$$\Psi_f = e^{-\frac{i}{\hbar}\sum_j Q_j P_j}\Psi\left(q_j, t\right) \ . \quad (7)$$

[Partially differentiate (7) with respect to Q_j, to verify that it satisfies (6c)]. Thus the wave function Ψ_f satisfying Eq. (3) differs from the standard Schrödinger wave function Ψ only by the *constant phase factor*

$$e^\alpha = e^{-\frac{\hbar}{i}\sum_j Q_j P_j} \quad (8)$$

attached to the Schrödinger function Ψ. After cancelling this phase factor from both sides of Eqs. (6a,b), we get exactly Eq. (2), the Schrödinger equation. So it would appear that the Class-II solutions of Eq. (3) precisely duplicate ordinary quantum mechanics (OQM). (These solutions, then, constitute a "covering theory" of OQM.) But that is true only in a formal or mathematical sense. On the interpretational side the (uncancelled) phase factor e^α makes a great difference, through its ability to affect quantum phases by undergoing abrupt changes. Let us examine this more closely.

Class-II solutions: quantum measurement theory

Einstein objected that quantum theory is "incomplete." He probably meant that it lacked trajectories. Here we have instead asserted the accepted quantum formalism to be *parametrically* incomplete. This is an altogether different affair. Since we have recovered OQM as a viable class of solutions, it is apparent that we are just as far as ever from trajectories. But we may be in a position to correct another more serious loss occasioned by OQM—the loss of *objectivity*. Objectivity does not necessarily require trajectories, but it does need event points. That is, real, localized observable happenings must be described by any physically valid theory... and OQM wholly lacks parameters to describe such objective point events. If we think of a quantum phase space in which a point is specified by $\left(q_j, p_j\right)$, we have to recognize that a point so specified is *not* an observable event. Rather, $\left(q_j, p_j\right)$ is a running variable describing—if you like—a virtual event or *presence* or one of a sequence of virtual events forming a Feynman-like pseudo-trajectory. But it is not an *observable* localized happening in real space or "real" phase space... for the simple reason that the $\left(q_j, p_j\right)$ are *operators*. To be observable in real phase space, whose axes are labelled with ordinary real c-numbers, it is necessary that c-number parameters be included in the descriptive formalism. Otherwise the formalism is powerless to describe observable occurrences, or even to recognize that anything happens in the world. That is in fact the situation of OQM, based on Eq. (2). Logically, according to (2), nothing can happen in the quantum world, because the accepted formalism lacks c-number parameters to describe point events representing observable occurrences.

What follows from such a lack? History has witnessed a great proliferation of ever more ingenious "interpretational" makeshifts, Band-Aids, and substitutes for a valid formalism. We have had three-quarters of a century of it now, and counting. This remarkable social phenomenon, only nominally related to physics, is known as "quantum measurement theory." It has become a way of life, a source of steady income, an endless intellectual challenge, for a whole sub-culture among physicists. Its adherents dedicate their lives to avoiding recognition of the obvious: that OQM is under-parameterized. Their basic dogma is that *mathematically* OQM is an immaculate conception that must not be altered in any way. By contrast, *the world* (being defenceless) is their plaything... the rules of their game allow the world to be distorted into any shape that will fit their rigidly unyielding mathematics. I wish I could say that the Many Worlds Interpretation of OQM represents the apogee of their flights of fancy, but in fact there is no limit... they literally stop at nothing. By now the amount of professional *interest* vested in measurement-theory nonsense rivals that vested in string-theory nonsense[3]. Indeed, the two groups of theorists could exchange professional concerns today without any externally detectable change—in either quality of product or effect on the rest of physics.

In the beginning these interpretational makeshifts were simpler and less sophisticated than they are today; therefore they were more perspicuous. There used to be things called "quantum jumps," and something called "severance of the von Neumann chain" of phase connections between observer and observed. Both of these approaches recognized that something had to be done about quantum phase connections—but both wandered in the wilderness because the theory had no parameters to do it with. Early on, there was a "Projection Postulate," contrived to do postulationally what needed to be done parametrically. This has had a phoenix-like rebirth with the latest jargon of obscuration, "quantum (phase) entanglement." The reason for this rebirth is that in order to make any connection at all with observation the OQM equation of motion has to be discarded in order to let things happen locally in the world. Having *postulated* an equation of motion, Eq. (2), the "bold interpreteers" judged it natural to say, "Oops, that was wrong, we now add a Projection postulate that contradicts our equation-of-motion postulate by replacing a pure state with a mixture." At one time there was even a vogue for carrying this one step farther with a "Selection Postulate"—*via* another, "Oops, that was wrong, we further postulate a single-state Selection from the projected state mixture, thus contradicting our Projection that contradicted our postulated equations of motion." But for some reason this busy postulational first-aid work, perhaps because it produced a structure resembling The House That Jack Built, fell into desuetude until observational necessity eventually forced something (anything) to be done about cutting phase connections. Aided by new jargon, these present-day champions of logic, mathematicians *manqué*, are still vying among themselves to build inconsistent axiomatic systems. The more postulates the merrier... for postulates whose only logical obligation is to contradict one another are always available in any number at no charge.

It should be obvious to every child that the only way out of this thicket of obscuration is to get the equations of motion right in the first place. That is where and only where the postulational "corrections" should be applied. This can happen, as a social phenomenon, only if enough children find enough things laughable about the Emperor's Parade. The necessary open-eyed innocents are going to have to come from the gene pool of uncommitted physicists. But with ever-more specialization and ever-more "professionalism" of mutual back-scratching among specialties, we have seen a steady trend toward less, rather than more, probability of an outcome favourable to physics as science. There is just not the requisite laughter in the air. As physics grows less respectable, the need of physicists for respect grows more urgent. In these conditions physics becomes a very serious business, indeed, or rather profession. Look at other professions. In my youth, when medicine was a calling, doctors made house calls. They went where needed. Then medicine became a profession. The word *profession* says it all.

Well... has Eq. (3) finally got the equations of motion right? If so, it is a new ballgame, interpretionwise. For there are now extra parameters, constants of the motion, explicitly present in the theory. Moreover, those c-number constants appear in a phase factor (8) on the wave function (7). That is just where we should want them to be, if phase-connection severance or the description of "loss of phase knowledge" is our objective. And that, indeed, is precisely our objective. It is what all the postulational fuss was about—"Projection," "Selection," and the rest. But that must now be forgotten, if the new paradigm is to receive a fair trial. The questions will be the same, but the solutions will be new and the methods of arriving at them will be somewhat unfamiliar.

Consider quantum particles "basking" in an atomic pure state. They obey Eq. (6). At some time in the past, we may suppose, this pure state was "prepared." At that time or earlier the parameters $\left(Q_j, P_j\right)$ received some fixed values, which they maintain throughout the duration of the pure state. They must stay constant in order that the phase factor e^{α}, given by Eq. (8), may stay constant. During this basking period all processes may be considered virtual and all "time flow" in abeyance (without responsibility to causality), all phase connections remaining intact. Nothing observable can happen anywhere in the system. Then something happens. This occurs purely at nature's initiative. Human thought or "mind" has nothing to do with it. The job of physics is to describe the happening, after the fact. The happening is localized—of the nature of a point event—to fit with relativity. But we may push beyond that in supposing the localization to take place in phase space, so that both Q_j and P_j acquire numerical values. The result is an abrupt, unknowable change $\Delta\alpha$ in the phase angle of the wave function. (Heisenberg's "Uncertainty" is not violated, since it concerns the old canonical variables, not the new ones.)

The unknowability of $\Delta\alpha$ implies severance of phase connections and loss of phase "knowledge," although this is a poor way of speaking, inasmuch as one cannot lose what one never had. So, here we have the "quantum jump," a severance of the von Neumann chain, effected by a jump in numerical values of the new canonical variables. Something has happened locally in nature, described by a sudden change in c-number values of event-descriptive parameters. To accomplish this, the wave function phase discontinuously changes in an unpredictable way—thus severing the past from the future and ratcheting time flow at the most basic quantum descriptive level. As long as the pure-state phase stays constant, we cannot say that time flows at all. But when the phase jumps in a way we cannot know, it is allowable to say that phase "knowledge" is lost in an irreversible way, and that time flows irreversibly "forward," in conformity with a postulated observance of causality. As an extra dividend for rigorizing formal Correspondence and thereby "completing" the equations of motion of quantum mechanics, we gain an accounting for the "arrow of time" at the quantum level. Like the Scarlet Pimpernel, that arrow has been sought high and low, even in the farthest reaches of the cosmos... and all the time it has been hiding in our sub-basement right at home. We also dispose of all those versions of quantum measurement theory that rest on deep Wignerian speculations about "mind" intervening as a causal agent in nature. Calling on mind to sever quantum phase connections is just as silly as calling on mutually contradictory postulates to do the job. Mind-fans will certainly prefer their "insights" to the more prosaic notion that the job of the mind, as applied to physics, is to describe nature, not to actuate it. To accomplish the description of nature requires descriptive parameters. Where parameters are lacking, mind and postulates are equally poor surrogates.

What we have said so far about phase jumps reflecting locally completed processes applies strictly only to the simplest one-body and one-component (one-channel) problems. In many experimental situations, or generally in many-body problems, a quantum system may be described by numerous component wave functions, among which may occur only partial reductions of the total system wave function. I have suggested[4] that these partial reductions be termed "virtual events," and that they might be used to describe the type of absorber action whereby some observable effect[5] is "frustrated" by the mere presence of a potential absorber, without any actual localized absorption occurring. (The accepted jargon calls these "interaction-free" or "non-demolition" measurements. What it amounts to is the introduction of phase incoherence[4] into some but not all channels of a multi-channel pure-state process.) By the view I have suggested, *the absorber always absorbs...* but not always by a locally-completed "real" event; possibly by a virtual event that imposes phase incoherence on a single component or channel of a many-channel process. That virtual occurrence is not observable directly because the phase jump does not affect the quantum system as a whole (quanta by definition act only as a whole!); but it is indirectly inferred[5] through the observable aspects of "frustration." This large and somewhat subtle subject is at present speculative and in need of development. I have been able to give only the most crude and fallible introduction to it[4].

In summary concerning measurement theory: OQM endows factual history with ensemble attributes that have no basis in experience. To correct this, the theory needs to acquire *c*-number parameters descriptive of unique, factual point events. The best way to do this is to rigorize the formal Correspondence with H-J theory, thereby restoring analogs of the new canonical variables, or constants of the motion. The same is mandated by a Principle of Relativity of Physical Size, applied in the context of a point particle model.

Class-II solutions: the Dirac electron

By linearizing the classical one-body relativistic energy expression,

$$E^2 = m^2c^4 + p_1^2 + p_2^2 + p_3^2 , \tag{9}$$

through use of 4×4 anti-commuting unit matrices $(\alpha_1, \alpha_2, \alpha_3, \alpha_m)$, Dirac[1] obtained an operator description of the free electron at the atomic level of description. On introducing electromagnetism *via* the potentials, $p_j \to p_j + (e/c)A_j$, $j = 1, 2, 3$ (where e is the unsigned charge of the electron), he obtained a Hamiltonian (energy function) of the form

$$H = -c\boldsymbol{\alpha} \cdot \left(\mathbf{p} + \frac{e}{c}\mathbf{A} \right) - mc^2\alpha_m - eV , \tag{10}$$

where $V = A_0$ is the scalar potential, *etc.* This Hamiltonian, which is a perfectly good classical one for any point particle, proved fabulously successful when applied on the quantum side to description of the electron-positron, by means of the operator identifications $p_j \to (\hbar/i)(\partial/\partial q_j)$, in accordance with Eq. (2b). We need say no more about this, since it forms a cornerstone of modern physical theory, and is doubtless taught everywhere.

Just one point need detain us. This is that the same formalism (of H-J pedigree) that on the classical side describes any point particle suddenly turns out on the quantum side to describe only one species of particle—the electron.

What could this mean? It seems a very startling constriction of physical descriptive purview. Does it mean that our size relativity principle fails? That a sudden discontinuity at the quantum boundary is real and reflects an absoluteness of physical scale, as Dirac[1] thought? So it would seem... but I have ventured to suggest an alternative[6] appraisal that is both simple and drastic. This is embodied in the *beta-structure hypothesis*; namely, that *all physical particles and the vacuum are composed of electrons*. The steps of reasoning behind this are simple: (1) If a principle of relativity of physical size is valid, then a single mechanical equation of motion [*viz*., Eq. (3)] must govern particle mechanics on all size scales. (2) If Eq. (3) governs, then its atomic-level specialization $S = \hbar / i$ manifestly describes the Dirac electron. (3) If no physical discontinuity marks the transition between classical and atomic levels, as to either equations of motion or form (10) of the Hamiltonian, then there can be no abrupt change in physical nature of the particles described. (4) If there is no distinct boundary between the realms inhabited by "all particles" and by "electrons," then *all particles* must in fact *be* electrons or their composites. (Here I do not distinguish between electrons and positrons.)

This idea meets at once certain troubles, such as that protons seem quite different from positrons. But it is implicit in the beta-structure hypothesis that positrons (vacant electron states) are somehow captured within protons and held permanent prisoners there; and that the proton is in fact a very many-body relativistic system composed ultimately of electrons. Similarly the vacuum, as Dirac originally thought ("hole theory") before he was brain-washed, is composed of electrons in negative-energy states. The relativistic very many-body problem is so difficult, so little explored, and so cleverly dodged (*e.g.*, through field-theoretical devices such as second quantization), that I do not see how any prudent physicist, not adept at dodging, can dogmatically reject the beta-structure hypothesis. Not included among prudent physicists is that vast majority who unhesitatingly bet the farm on field theory. It is my belief, to mix metaphors, that pure point particle mechanics has still plenty of mileage left in it. Admittedly, the leap from size relativity and form preservation to the above wild guess about world structure is a bold one... but the fact that field theory totally rejects it rather prejudices me in its favour. Is it not high time that a few physical theorists began to think outside the field-theory box?

Class-III solutions: formalism

From Eqs. (3a) and (5) we get

$$H\Psi_f = -\frac{\hbar}{i}\frac{\partial}{\partial t}s\Psi_f \ . \tag{11}$$

The Class-III solutions are those that treat S or s and Ψ_f as non-constant. This shows at once that there is a problem: Since s is to be treated as real (Hermitian), and $-(\hbar / i)\partial / \partial t$ is known to be Hermitian, we have H represented in (11) as a product of two Hermitian operators. It is a well-known theorem that the product of two Hermitian operators is non-Hermitian. The reader will have to take my word for it that it is simply not physics to represent a physical energy as a non-Hermitian operator. That would say energy is unreal, which is physically incorrect, as far as is known. Fortunately, there is a ready fix for this that works like magic. We simply introduce a new Hermitian Hamiltonian $\mathfrak{H}$ in place of the classical-analog Hamiltonian H, a Hermitian momentum p_j in place of p_j, and a transformed wave function Ψ, by means of the definitions

$$\mathfrak{H} = Hs^{-1}, \quad p_j = p_j s^{-1}, \quad \Psi = s\Psi_f . \qquad (12a,b,c)$$

Here, for the one-body problem, $j = 1,2,3$. This transforms Eq. (3) into

$$\mathfrak{H}\Psi = -\frac{\hbar}{i}\frac{\partial}{\partial t}\Psi, \quad p_j\Psi = \frac{\hbar}{i}\frac{\partial}{\partial q_j}\Psi, \quad -P_j s^{-1}\Psi = \frac{\hbar}{i}\frac{\partial}{\partial Q_j}\Psi . \qquad (13a,b,c)$$

From this we see that the basic formalism of OQM is recovered, even for the Class-III solutions, but with a transformed Hamiltonian and an extra relation, Eq. (13c). In terms of the Hermitian operator $\mathfrak{H}$ the equation of motion of a Heisenberg variable X,

$$\frac{dX}{dt} = \frac{\partial X}{\partial t} + \frac{1}{i\hbar}\left(X\mathfrak{H} - \mathfrak{H}X\right) , \qquad (14)$$

is also recovered. Thus all the standard OQM techniques employing Hermitian operators are applicable [the operators on both sides of (13c) being Hermitian, as well as those of (13a,b)]. The formal operand Ψ_f is not useful for calculating observable probability distributions; but its transformed analog Ψ is [and is understood as the operand in (14)]. Similarly, any classical-analog Hamiltonian H, such as Eq. (10), is not the observable physical energy, but $\mathfrak{H}$ is, since it is the generator of infinitesimal time displacements of the system [Eq. (13a)]. In short, the non-Hermitian *classical-analog* quantities entering Eq. (3) have served their purpose of form preservation over the whole physical range, and *in the particular case of Class-III solutions* are to be discarded in favour of their transformed Hermitian counterparts. This has proven a disappointment to mathematicians, who feel cheated of novelty by this return to familiar forms [Eqs. (13a,b), (14)]. But physicists will recognize that any alteration of the Hamiltonian [Eq. (12a)] entails "new physics"—which should console them for the lack of "new math."

Class-III solutions: the electron on the nuclear scale

Given Eq. (3) as descriptive of particle mechanics on all size scales, we have seen that its Class-I solutions describe the motions of any classical point particles (possessed of trajectories). The Class-II solutions, descriptive of the atomic realm (without trajectories but with objective point events), given Eq. (10) as the (relativistic one-body) Hamiltonian, describe only electron-positrons. We may suppose that the same classical-analog Hamiltonian, with the help of Class-III solutions, might describe the same particle (electron) on a still smaller size scale. This is a speculation, but it proves fruitful. We recall that the Class-III solutions obey a commutation rule [Eq. (4), left of arrow] that generalizes the Heisenberg postulate. Thus we are contemplating a new physics of the nuclear realm, whereby the Heisenberg postulate may be locally *disobeyed*. That is, the commutator of the position and momentum dynamical variables may become non-constant in the vicinity of a nuclear force center. This gives an entirely new meaning to the concept of "nuclear force," and implies that it is not like other "forces" known from larger-scale (including atomic-scale) experience.

To probe the general nature of point-electron dynamics in this nuclear or sub-nuclear domain, let us consider a relativistic central-force one-body problem. In this case the classical-analog Hamiltonian is given by Eq. (10) with $V = Ze/r$, where we allow for Z positive charges on the attractive force center, assumed to be a fixed point. The mathematics has been given in some detail in earlier references[6-8], and will only be summarized here. The Hermitian Hamiltonian is [from Eqs. (10), (12), (13)]

$$\mathfrak{H} = Hs^{-1} = -c\boldsymbol{\alpha} \cdot \left(\frac{\hbar}{i} \nabla + \frac{e}{c} \mathbf{A} s^{-1} \right) - mc^2 \alpha_m s^{-1} - \frac{Ze^2}{r} s^{-1} . \tag{15}$$

We consider the conservative case in which the Hamiltonian and s are time-independent, so that the substitution $\Psi = e^{-(i/\hbar)E't} \psi$ reduces (13a) to the eigenvalue equation

$$\mathfrak{H}\psi = E'\psi . \tag{16}$$

The assumption of spherical symmetry, $s = s(r)$, together with an identity given by Dirac[1], allows this (with $\mathbf{A} = 0$) to be reduced to

$$\mathfrak{H} = i\hbar c\varepsilon \left(\frac{\partial}{\partial r} + \frac{1}{r} \right) - \frac{i\hbar cj}{r} \varepsilon \rho_3 - mc^2 s^{-1} \rho_3 - \frac{Ze^2}{r} s^{-1} , \tag{17}$$

where j is an operator that commutes with any function of r (hence with s) and Dirac gives the representations

$$\varepsilon = \begin{pmatrix} 0 & -i \\ i & 0 \end{pmatrix} \text{ and } \rho_3 = \begin{pmatrix} 1 & 0 \\ 0 & -1 \end{pmatrix} \tag{18}$$

The eigenvalue equation (16), with ψ a two-component wave function, then yields the two simultaneous equations

$$\psi_2' + \frac{j+1}{r}\psi_2 + \left[\frac{s^{-1}}{\hbar c} \left(-mc^2 - \frac{Ze^2}{r} \right) - \frac{E'}{\hbar c} \right] \psi_1 = 0 , \tag{19a}$$

$$-\psi_1' + \frac{j-1}{r}\psi_1 + \left[\frac{s^{-1}}{\hbar c} \left(mc^2 - \frac{Ze^2}{r} \right) - \frac{E'}{\hbar c} \right] \psi_2 = 0 , \tag{19b}$$

for the two ψ-components. We need another equation to determine s, and this is furnished by (13c) with $j = 1,2,3$. Introducing formal spherical polar coordinates by means of $R = \sqrt{Q_1^2 + Q_2^2 + Q_3^2}$, $Q_1 = R\sin(\theta)\cos(\phi)$, $Q_2 = R\sin(\theta)\sin(\phi)$, $Q_3 = R\cos(\theta)$, we find

$$\frac{\partial}{\partial Q_j} = \left(\frac{\partial R}{\partial Q_j}\right)\frac{\partial}{\partial R} = \xi_j\left(\theta,\phi\right)\frac{\partial}{\partial R} \ , \ j = 1,2,3, \tag{20}$$

the ξ_j being direction cosines. Let θ,ϕ specify an arbitrary fixed direction from the coordinate origin. Then the ξ_j are constants, and in view of spherical symmetry we can consider the P_j in Eq. (13c) to obey $P_j = P\xi_j$, where P is some constant. By this means the three relations (13c) are reduced to the single (two-component) equation

$$Ps^{-1}\psi = -\frac{\hbar}{i}\frac{\partial}{\partial R}\psi \ . \tag{21}$$

Recall that R is related to the Q_j and thus is a constant of the motion. Both s and ψ may be considered to depend at least implicitly on this constant. However, we choose to eliminate the explicit appearance of R by seeking a solution that depends on $\mathbf{r}$ and $\mathbf{R} = (Q_1,Q_2,Q_3)$ only through the combination $|\mathbf{r} - \mathbf{R}|$. If such a solution exists, we can replace $\partial/\partial R$ by $-\partial/\partial r$, so that (21) becomes

$$Ps^{-1}\psi = \frac{\hbar}{i}\frac{\partial}{\partial r}\psi \ . \tag{22}$$

For consistency with our assumption of spherical symmetry, it is evidently necessary to impose a particular initial condition; namely,

$$\mathbf{R} = (Q_1,Q_2,Q_3) = (0,0,0) \ . \tag{23}$$

This initial condition will unfortunately limit the usefulness of our solution, since it implies the presence at the coordinate origin of an infinitely massive force center. (Strictly speaking, we assume the electron to be "found" at an event point $\mathbf{R}$ coincident with the origin.) We accept this limitation and proceed, because we are here more interested in proving the existence of some solution than in finding the most general one. Since now $\psi = \psi(r)$, we can replace all partial derivatives by total ones, so that (22) becomes

$$s^{-1}\psi = \frac{\hbar}{iP}\frac{d}{dr}\psi \ . \tag{24}$$

The assumptions that s is some scalar (spin-independent) real function $s = s(r)$, possessing an inverse, and that P is a constant suffice, with (24), to establish the equality of logarithmic derivatives of the two ψ-components. Thus

$$\psi_1^{-1}\left(\frac{d\psi_1}{dr}\right) = \psi_2^{-1}\left(\frac{d\psi_2}{dr}\right) \rightarrow \ln\left(\frac{\psi_1}{\psi_2}\right) = const. \rightarrow \psi_2 = C\psi_1 \ , \tag{25}$$

where C is some constant. Using (24), (25) to eliminate s and ψ_2 from (19), we obtain from the two parts of the latter equation

$$\psi_1'\left[C + \frac{imc}{P} + \frac{iZe^2}{cPr}\right] + \psi_1\left[-\frac{E'}{\hbar c} + \frac{(j+1)C}{r}\right] = 0 \ , \tag{26a}$$

$$\psi_1' \left[-1 - \frac{imcC}{P} + \frac{iZe^2C}{cPr} \right] + \psi_1 \left[-\frac{E'C}{\hbar c} + \frac{j-1}{r} \right] = 0 \ . \tag{26b}$$

As can be seen by multiplying (26a) by C, these two equations for ψ_1 are compatible if and only if

$$C^2 = \frac{j-1}{j+1} \quad \text{and} \quad \frac{imcC}{P} = \frac{-j}{j+1} \ . \tag{27a,b}$$

In order to recover the Heisenberg postulate ($s = 1$) at long distances from the force center, we shall require unity as the asymptotic value of s,

$$\lim_{r \to \infty} s(r) = 1 \ . \tag{28}$$

From (24) we see that in this limit both components of ψ behave like $exp(iPr/\hbar)$. Since the wave function components must be bounded at infinity, P must have a positive imaginary part. In Dirac's electron theory the operator j takes any positive or negative integral eigenvalues. Since j commutes with any function $s(r)$, we may expect j to have somewhat similar commutation properties and eigenvalues in the present formalism. However, it is easily seen from (27) that the eigenvalues ± 1 must be excluded here. It follows from this exclusion that C is real; thus that P is pure imaginary and positive. Hence P can be written as $P = iK^2$, where K is real and non-zero. Then Eq. (27) yields

$$\frac{mc}{K^2} = \frac{|j|}{\sqrt{j^2 - 1}} \qquad C = -\frac{|j|}{j} \sqrt{\frac{j-1}{j+1}} \tag{29a,b}$$

Eq. (29a) makes it obvious that $j = \pm 1$ must be forbidden. All other non-zero integral eigenvalues of j are allowed. Substituting $P = iK^2$ and (26a) into (24), written for the first component ψ_1, we obtain

$$s^{-1} = -\frac{\hbar}{K^2} \left(\frac{\psi_1'}{\psi_1} \right) = \frac{\hbar}{K^2} \left[-\frac{E'}{\hbar c} + \frac{(j+1)C}{r} \right] \Big/ \left[C + \frac{mc}{K^2} + \frac{Ze^2}{cK^2 r} \right]. \tag{30}$$

Applying to this the asymptotic condition (28), we find

$$\lim_{r \to \infty} s(r) = 1 = \left[C + \frac{mc}{K^2} \right] \Big/ \left[-\frac{E'}{K^2 c} \right] \ . \tag{31}$$

Solved for the energy eigenvalues E', this yields with the help of (29)

$$E' = E'_j = -cK^2C - mc^2 = -\frac{mc^2}{j}, \quad j = \pm 2, \pm 3, \cdots \ , \tag{32}$$

as the eigenvalue spectrum. The corresponding eigenfunctions can be found by integrating Eq. (30). However, it is easier to guess a solution of the form

$$\psi_1 = \psi_{1j} = A_j e^{-\alpha r} (r + \beta)^\gamma \ . \tag{33}$$

From this we obtain

$$\frac{\psi_1'}{\psi_1} = -\alpha + \frac{\gamma}{r + \beta} = \frac{-\alpha + \dfrac{\gamma - \alpha\beta}{r}}{1 + \beta/r} \ . \tag{34}$$

On putting (29) and (32) into (30) we evaluate s finally as

$$s = s_j(r) = \left[1 + \frac{jZe^2}{mc^2 r}\right] \Big/ \left[1 - \frac{\hbar}{mc} \frac{|j|\sqrt{j^2 - 1}}{r}\right] . \tag{35}$$

Comparison of the ratio ψ_1' / ψ_1 from (30) with (34) then yields the following evaluation of the constants in (33):

$$\alpha = \frac{mc}{\hbar} \frac{\sqrt{j^2 - 1}}{|j|}, \quad \beta = \frac{jZe^2}{mc^2}, \quad \gamma = j^2 - 1 + \frac{Ze^2}{\hbar c} \frac{j}{|j|} \sqrt{j^2 - 1} . \tag{36}$$

Eq. (33) then evaluates the first eigenfunction component ψ_1. The second, ψ_2, follows from (25) and (29b). The multiplier A_j in Eq. (33) is arbitrary and can be used for wave function normalization.

This completes our formal demonstration of the existence of localized bound-state Class-III solutions on the sub-atomic scale. These are electron states, since m is the electron mass. It is seen that α is of the order of the electron Compton wavelength, and by (33) that this controls the "size" of the electron wave function. But the fact that the canonical "momentum" parameter P is pure imaginary [$P = iK^2$] seems to imply that the point electron cannot be "found" or "detected" on any size scale larger than a mathematical point—which according to Eq. (23) is collocated with the force center at the origin. In other words, any event of "finding" is described by the new canonical variables $Q_1 = Q_2 = Q_3 = 0$. We know that nucleons have non-zero sizes. Therefore this solution is of no direct use for describing them. It treats merely an idealized limiting case of the infinite-mass force center localized at a point. Some of the simplifications we have pointed out along the way would have to be corrected in order to describe a finite-mass nucleon. That would ultimately involve solving a relativistic very many-body problem, and is beyond this writer's capabilities. Still, the results so far seem encouraging.

Summation

We have seen that point particle mechanics is not dead, and that a nuclear dynamics founded on the Class-III solutions, which locally violate the Heisenberg postulate, lies easily within the realm of formal descriptive possibility. Such an enhancement of dynamics seems limited on the sub-atomic scale to a description of the electron-positron—a fact that suggests a "beta structure hypothesis," viz., that only electrons exist on the finer scales in nature. Our derivation of eigenvalues, Eq. (32), and of eigenfunctions, Eq. (33), establishes that stable bound states exist, beyond any known on the basis of classical (Class-I) or atomic (Class-II) solutions of our postulated equations of motion for all mechanics, Eq. (3). The eigenvalues in question lie within what was termed by Pauli the "Zwischengebiet"—the region of real mass-energy, but imaginary momentum, lying between particle total energies $\pm mc^2$. (That fits also with the imaginary value $P = iK^2$ of the canonical momentum parameter.) This furnishes a ready mechanical explanation for nuclear beta processes and encourages further speculations, for instance that all heavy particles may be aggregates of many imaginary-momentum electrons in real-mass states, *i.e.*, electronic states within the *Zwischengebiet*... and that neutrinos (if found to be of zero rest mass) may be energy quanta associated with electronic transitions between states of real and imaginary momentum, as distinguished from photons, which are zero-mass quanta associated with electronic transitions between states of real momentum.

The Class-I (Hamilton-Jacobi) solutions are exact solutions of Eq. (3), hence are as valid *approximate descriptors* of nature as the atomic (Class-II) solutions. Each solution class is available to describe its own appropriate aspect (and scale) of experience. None is subordinate to any other. Therefore we do not have to use the de Broglie wavelength of a planet to get a more "accurate" description of its motion; nor do we have to picture a "wave function of the universe." Our basic theme has been a rigorization of formal Correspondence, motivated by a Principle of Relativity of Physical Size. An immediate consequence has been the parametric restoration of formal analogs of the new canonical variables (constants of the motion). The parameter count *must not change* under formal Correspondence, there being no corresponding discontinuity in nature. The restoration of *c*-number parameters in the Class-II equations of motion clears up all the OQM mysteries that have provided full employment for quantum measurement theorists (by providing a *parametric mechanism* for phase-connection severance that replaces "Projection"). Prospects for a resurgence of the dynamics of the point electron have never been brighter. Still, I have found during forty years that such ideas are of little interest to professional physicists... who remain supremely assured that quantum field theory, not particle dynamics, is the mathematical language by which nature communicates her inmost secrets. Thus they conform to the definition of an expert, as one who makes no small mistakes.

References

1.   P. A. M. Dirac, *The Principles of Quantum Mechanics* (Oxford, Clarendon, 1947)

2.   H. Goldstein, *Classical Mechanics* (Addison-Wesley, Cambridge, 1950)

3.   P. Woit, *Am. Scientist* **90**, 110-112 (2002)

4.   T. E. Phipps, Jr., *Phys. Essays* **11**, 155-163 (1998)

5.   P. Kwiat, H. Weinfurter, T. Herzog, and A. Zeilinger, *Phys. Rev. Lett.* **74**, 4763 (1995)

6.   T. E. Phipps, Jr., *Heretical Verities: Mathematical Themes in Physical Description* (Classic Nonfiction Library, Urbana, IL, 1986)

7.   T. E. Phipps, Jr., *Phys. Rev.* **118**, 1653 (1960)

8.   T. E. Phipps, Jr., *Found. Phys.* **3**, 435 (1973); **5**, 45 (1975); **6**, 71 (1976); **6**, 263 (1976).

The Spinning Electron

Martin Rivas
Theoretical Physics Department
University of the Basque Country
Apdo. 644-48080 Bilbao, Spain
e-mail:wtpripem@lg.ehu.es

A classical model for a spinning electron is described. It has been obtained within a kinematical formalism proposed by the author to describe spinning particles. The model satisfies Dirac's equation when quantized. It shows that the charge of the electron is concentrated at a single point but is never at rest. The charge moves in circles at the speed of light around the centre of mass. The centre of mass does not coincide with the position of the charge for any classical elementary spinning particle. It is this separation and the motion of the charge that gives rise to the dipole structure of the electron. The spin of the electron contains two contributions. One comes from the motion of the charge, which produces a magnetic moment. It is quantized with integer values. The other is related to the angular velocity and is quantized with half integer values. It is exactly half the first one and points in the opposite direction. When the magnetic moment is written in terms of the total observable spin. one obtains the $g = 2$ gyromagnetic ratio. A short range interaction between two classical spinning electrons is analysed. It predicts the formation of spin 1 bound states provided some conditions on their relative velocity and spin orientation are fulfilled, thus suggesting a plausible mechanism for the formation of a Bose-Einstein condensate.

1. Introduction

The spin of the electron has for many years been considered a relativistic and quantum mechanical property, mainly due to the success of Dirac's equation describing a spinning relativistic particle in a quantum context. Nevertheless, in textbooks and research works one often reads that the spin is neither a relativistic nor a quantum mechanical property of the electron, and that a classical interpretation is also possible. The work by Levy-Leblond [1] and subsequent papers by Fushchich *et al.* [2], which show that it is possible to describe spin ½ particles in a pure Galilean framework, with the same $g = 2$ gyromagnetic ratio, spin-orbit coupling and Darwin terms as in Dirac's equation, lead to the idea that spin is not strictly a relativistic property of the electron.

The spin is the angular momentum of the electron, and the classical and quantum mechanical description of spin is the main subject of the kinematical formalism of elementary spinning particles published by the author [3]. This work presents the main results of this formalism and, in particular, an analysis of a model of a classical spinning particle whose states are described by Dirac's spinors when quantized. Other contributions are also discussed.

2.   Classical elementary particles

To understand what a classical elementary particle is from the mathematical point of view, we consider first the example of a point particle. It is the simplest geometrical object with which we can build any other geometrical body of any size and shape. The point particle is the classical elementary particle of Newtonian mechanics and has no spin. Yet we know today that spin is one of the intrinsic properties of all known elementary particles. The description of spin is related to the representation of the generators of the rotation group, and we know it is an intrinsic property since it is related to one of the Casimir operators of the Galilei and Poincaré groups.

From the Lagrangian point of view, the initial (and final) state of the point particle is a point on the continuous space-time manifold. In fact what we fix as boundary conditions for the variational problem are the position r_1 at time t_1 and the position r_2 at the final time t_2. We call *kinematical variables* of any mechanical system the variables which define the initial (and final) configuration of the system in this Lagrangian description, and *kinematical space* the manifold covered by these variables. The point particle is a system of three degrees of freedom with a four-dimensional kinematical space.

In group theory, a homogeneous space of any Lie group is the quotient structure between the group and any of its continuous subgroups. The important property of the kinematical space of a point particle, from the mathematical viewpoint, is that it is a homogeneous space of the Galilei and Poincaré groups.

In the example of the point particle, the kinematical space manifold is the quotient structure between the Poincaré group and the Lorentz group in the relativistic case, and also the quotient between the Galilei group and the homogeneous Galilei group in the non-relativistic one.

We use this idea to arrive at the following definition.

> *Definition: A classical elementary particle is a mechanical system whose kinematical space is a homogeneous space of the kinematical group.*

The spinless point particle fulfils this definition, but it is not the most general elementary particle that can be described, because we have larger homogeneous spaces with a more complex structure. The largest structured particle is the one for which the kinematical space is either the Galilei or Poincaré group or any of its maximal homogeneous spaces.

With this definition we have a new formalism, based upon group theory, to describe elementary particles from a classical point of view. It will be quantized by means of Feynman's path integral method, where the kinematical variables are precisely the common end points of all integration paths. The wave function of any mechanical system will be a complex function defined on the kinematical space. In this way, the structure of an elementary particle is basically related to the kinematical group of space-time transformations that implements the Special Relativity Principle. It is within the kinematical group of symmetries that we must look for the independent and essential classical variables to describe an elementary object.

When we consider a larger homogeneous space than the space-time manifold, for both Galilei and Poincaré groups, we have variables additional to time and space to describe the states of a classical elementary particle. These additional variables will produce a classical description of spin.

3.   Main features of the formalism

When we write the Lagrangian of any mechanical system in terms of the introduced kinematical variables, and the dynamics is expressed in terms of some arbitrary evolution parameter τ (not necessarily the time parameter), we get the following properties:

- The Lagrangian is independent of the evolution parameter τ. The time evolution of the system is obtained by choosing $t(\tau) = \tau$.
- The Lagrangian is only a function of the kinematical variables x_i and their first τ derivatives $\dot{x}_i$.
- The Lagrangian is a homogeneous function of first degree in terms of the derivatives of the kinematical variables $\dot{x}_i$ and therefore Euler's theorem implies that it can be written as $L(x, \dot{x}) = F_i(x, \dot{x})\dot{x}_i$, where $F_i = \partial L / \partial \dot{x}_i$.
- If some kinematical variables are time derivatives of any other kinematical variables, then the Lagrangian is necessarily a generalised Lagrangian depending on higher order derivatives when expressed in terms of the essential or independent degrees of freedom. Therefore, the dynamical equations corresponding to these variables are no longer of second order, but, in general, of fourth or higher order. This will be the case for the charge position of a spinning particle.
- The transformation of the Lagrangian under a Lie group that leaves the dynamical equations invariant is $L(gx, g\dot{x}) = L(x, \dot{x}) + d\alpha(g; x)/d\tau$, where $\alpha(g; x)$ is a gauge function for the group G and the kinematical space X. It only depends on the parameters of the group element and on the kinematical variables. It is related to the exponents of the group [4].
- When the kinematical space X is a homogeneous space of G, then $\alpha(g; x) = \xi(g, g_x)$, where $\xi(g_1, g_2)$ is an exponent of G.
- When quantizing the system, Feynman's kernel is the probability amplitude for the mechanical process between the initial and final state. It will be a function, or more precisely a distribution, over the $X \times X$ manifold. Feynman's quantization establishes the link between the description of the classical states in terms of the kinematical variables and its corresponding quantum mechanical description in terms of the wave function.
- The wave function of an elementary particle is thus a complex square integrable function defined on the kinematical space.
- The Hilbert space structure of this set of functions is achieved by a suitable choice of a group invariant measure defined over the kinematical space.

- The Hilbert space of a classical system, whose kinematical space is a homogeneous space of the kinematical group, carries a projective, unitary, irreducible representation of the group. In this way, the classical definition of an elementary particle has a correspondence with Wigner's definition of an elementary particle in the quantum case.

4.   The classical electron model

The latest LEP experiments at CERN suggest that the electron charge is confined within a region of radius $R_e < 10^{-19}$ m. Nevertheless, the quantum mechanical effects of the electron appear at distances of the order of its Compton's wavelength $\lambda_C = \hbar / mc \cong 10^{-13}$ m, which are six orders of magnitude larger.

One possibility to reconcile these features is the assumption, from the classical viewpoint, that the charge of the electron is a point, but at the same time this point is never at rest and it is affected by an oscillating motion in a confined region of size λ_C. This motion is known in the literature as *Zitterbewegung*. This is the basic structure of spinning particle models that will be obtained within the proposed kinematical formalism, and also suggested by Dirac's analysis of the internal motion of the electron [5]. It is shown that the charge of the particle is at a single point r, but this point is not the centre of mass of the particle. Furthermore, the charge of the particle is moving at the speed of light, as shown by Dirac's analysis of the electron velocity operator. Here, the velocity corresponds to the velocity of the point r, which represents the position of the charge. In general, the point charge satisfies a fourth-order differential equation, which is the most general differential equation satisfied by any three-dimensional curve.

We shall see that the charge moves around the centre of mass in a kind of harmonic or central motion. It is this motion of the charge that gives rise to the spin and dipole structure of the particle. In particular, the classical relativistic model that when quantized satisfies Dirac's equation shows, for the centre of mass observer, a charge moving at the speed of light in circles of radius $R_0 = \hbar / 2mc$ and contained in a plane orthogonal to the spin direction [6,7]. This classical model of electron is what we will obtain when analysing the relativistic spinning particles.

To describe the dynamics of a classical charged spinning particle, we must therefore follow just the charge trajectory or, alternatively, the centre of mass motion and the motion of the charge around the centre of mass. In general the centre of mass satisfies second-order, Newton-like dynamical equations, in terms of the total external force. But this force has to be evaluated not at the centre of mass position, but rather at the position of the charge. We will demonstrate all these features by considering different examples.

5. Non-relativistic elementary particles

Let us first consider the non-relativistic formalism because the mathematics involved is simpler. In the relativistic case the method is exactly the same, [3,6,7] and we limit ourselves here to giving only the main results. We start with the description of the Galilei group to show how we obtain the variables that determine a useful group parameterization. These variables associated with the group will later be transformed into the kinematical variables of the elementary particles. We end this section with an analysis of some different kinds of classical elementary particles.

5.1 Galilei group

The Galilei group is a group of space-time transformations characterised by ten parameters $g \equiv (b, \vec{a}, \vec{v}, \vec{\alpha})$. The action of a group element g on a space-time point $x \equiv (t, \vec{r})$, represented by $x' = gx$, is considered in the following form

$$x' = \exp(bH)\exp(\vec{a} \cdot \vec{P})\exp(\vec{v} \cdot \vec{K})\exp(\vec{\alpha} \cdot \vec{J})x$$

It is a rotation of the point, followed by a pure Galilei transformation, and finally a space and time translation. Explicitly, the above transformation becomes

$$t' = t + b, \tag{1}$$

$$\vec{r}' = R(\alpha)\vec{r} + \vec{v}t + \vec{a}. \tag{2}$$

The group action (1)-(2) represents the relationship between the coordinates $(t, \vec{r})$ of a space-time event, as measured by the inertial observer O, and the corresponding coordinates $(t', \vec{r}')$ of the same space-time event as measured by another inertial observer O'. Parameter b is a time parameter, $\vec{a}$ has dimensions of space, $\vec{v}$ of velocity and $\vec{\alpha}$ is dimensionless, and these dimensions will be shared by the corresponding variables of the different homogeneous spaces of the group.

The variables b and $\vec{a}$ are the time and position of the origin of frame O at time $t = 0$ as measured by observer O'. The variables $\vec{v}$ and $\vec{\alpha}$ are respectively the velocity and orientation of frame O as measured by O'.

The composition law of the group $g'' = g'g$ is:

$$b'' = b' + b, \tag{3}$$

$$\vec{a}'' = R(\vec{\alpha}')\vec{a} + \vec{v}'b + \vec{a}', \tag{4}$$

$$\vec{v}'' = R(\vec{\alpha}')\vec{v} + \vec{v}', \tag{5}$$

$$R(\vec{\alpha}'') = R(\vec{\alpha}')R(\vec{\alpha}). \tag{6}$$

The generators of the group in the realization (1, 2) are the differential operators

$$H = \partial/\partial t, \quad P_i = \partial/\partial x^i, \quad K_i = t\partial/\partial x^i, \quad J_k = \varepsilon_{kli}x^l \partial/\partial x^i \tag{7}$$

and the commutation relations of the Galilei Lie algebra are

$$[\vec{J}, \vec{J}] = -\vec{J}, \quad [\vec{J}, \vec{P}] = -\vec{P}, \quad [\vec{J}, \vec{K}] = -\vec{K}, \quad [\vec{J}, H] = 0, \tag{8}$$

$$[H, \vec{P}] = 0, \quad [H, \vec{K}] = \vec{P}, \quad [\vec{P}, \vec{P}] = 0, \quad [\vec{K}, \vec{P}] = 0. \tag{9}$$

The Galilei group has the non-trivial exponents [4]

$$\xi(g,g') = m\left(\frac{1}{2}\vec{v}^2 b' + \vec{v}\cdot R(\vec{\alpha})\vec{a}'\right). \tag{10}$$

They are characterised by the non-vanishing parameter m. The gauge functions for the Lagrangians defined on the different homogeneous spaces of the Galilei group are of the form

$$\alpha(g;x) = m\left(\frac{1}{2}\vec{v}^2 t + \vec{v}\cdot R(\vec{\alpha})\vec{r}\right)$$

They all vanish if the boost parameter $\vec{v}$ vanishes. This implies that a Galilei Lagrangian for an elementary particle is invariant under rotations and translations, but not under Galilei boosts. In the quantum case this means that the Hilbert space for this system carries a unitary representation of a central extension of the Galilei group. In the classical case, the generating functions of the canonical Galilei transformations, with the Poisson bracket as the Lie operation, satisfy the commutation relations of the Lie algebra of the central extension of the Galilei group [4].

The central extension of the Galilei group [8] is an 11-parameter group with an additional generator I which commutes with the other ten,

$$[I,H] = [I,\vec{P}] = [I,\vec{K}] = [I,\vec{J}] = 0, \tag{11}$$

while the remaining commutation relations are the same as above (8, 9), the only exception being the last, which now appears as

$$[K_i, P_j] = -m\delta_{ij}I. \tag{12}$$

If the following polynomial operators are defined on the group algebra

$$\vec{W} = I\vec{J} - \frac{1}{m}\vec{K}\times\vec{P}, \qquad U = IH - \frac{1}{2m}\vec{P}^2, \tag{13}$$

we see that U commutes with all generators of the extended Galilei group and that $\vec{W}$ satisfies the commutation relations

$$[\vec{W},\vec{W}] = -I\vec{W}, \quad [\vec{J},\vec{W}] = -\vec{W}, \quad [\vec{W},\vec{P}] = [\vec{W},\vec{K}] = [\vec{W},H] = 0.$$

We find that $\vec{W}^2$ also commutes with all generators. It turns out that the extended Galilei group has three functionally independent Casimir operators. In those representations in which the operator I becomes the unit operator, for instance, in the irreducible representations they are, respectively, interpreted as the mass, $M = mI$, the internal energy $H_0 = H - P^2/2m$, and the absolute value of the spin

$$\vec{S} = \vec{J} - \frac{1}{m}\vec{K}\times\vec{P}, \quad \Rightarrow \quad S^2 = \left(\vec{J} - \frac{1}{m}\vec{K}\times\vec{P}\right)^2. \tag{14}$$

In what follows we take the above definition (14) as the definition of the spin of a nonrelativistic particle. In those representations in which I is the unit operator, the spin operator $\vec{S}$ satisfies the commutation relations:

$$[\vec{S},\vec{S}] = -\vec{S}, \quad [\vec{J},\vec{S}] = -\vec{S}, \quad [\vec{S},\vec{P}] = [\vec{S},\vec{K}] = [\vec{S},H] = 0,$$

i.e., it is an angular momentum operator, transforms like a vector under rotations and is invariant under space and time translations and under Galilei boosts, respectively.

Furthermore, it reduces to the total angular momentum operator $\vec{J}$ in those frames in which $\vec{P} = \vec{K} = 0$.

5.2 The spinless point particle

The kinematical variables of the point particle are $\{t, \vec{r}\}$, time and position, respectively. The nonrelativistic Lagrangian written in terms of the τ derivatives of the kinematical variables is the first order homogeneous function

$$L_{NR} = \frac{m}{2}\frac{\dot{\vec{r}}^2}{\dot{t}} = T\dot{t} + \vec{R}\cdot\dot{\vec{r}},$$

where we define $T = \partial L / \partial \dot{t}$ and $R_i = \partial L / \partial \dot{r}^i$. The constants of motion obtained through the application of Noether's theorem to the different subgroups of the Galilei group are

$$\text{energy} \quad H = -T = \frac{m}{2}\left(\frac{d\vec{r}}{dt}\right)^2,$$

$$\text{linear momentum} \quad \vec{P} = \vec{R} = m\frac{d\vec{r}}{dt},$$

$$\text{kinematical momentum} \quad \vec{K} = m\vec{r} - \vec{P}t,$$

$$\text{angular momentum} \quad \vec{J} = \vec{r} \times \vec{P}.$$

The spin for this particle is $\vec{S} = \vec{J} - \vec{K} \times \vec{P}/m = 0$.

5.3 A spinning elementary particle

According to the definition, the most general nonrelativistic elementary particle [9] is the mechanical system whose kinematical space X is the whole Galilei group $\mathcal{G}$. The kinematical variables are, therefore, the ten real variables $x(\tau) \equiv \{t(\tau), \vec{r}(\tau), \vec{u}(\tau), \vec{\rho}(\tau)\}$, with domains $t \in \mathbb{R}$, $\vec{r} \in \mathbb{R}^3$, $\vec{u} \in \mathbb{R}^3$ and $\vec{\rho} \in SO(3)$. The latter, with $\rho = \tan \alpha / 2$, is a particular parameterization of the rotation group. In this parameterization the composition law of rotations is algebraically simple, as shown below. All these kinematical variables have the same geometrical dimensions as the corresponding group parameters. The relationship between the values $x'(\tau)$ and $x(\tau)$ take, at any instant τ, for two arbitrary inertial observers

$$t'(\tau) = t(\tau) + b, \tag{15}$$

$$\vec{r}'(\tau) = R(\vec{\mu})\vec{r}(\tau) + \vec{v}t(\tau) + \vec{a}, \tag{16}$$

$$\vec{u}'(\tau) = R(\vec{\mu})\vec{u}(\tau) + \vec{v}, \tag{17}$$

$$\vec{\rho}'(\tau) = \frac{\vec{\mu} + \vec{\rho}(\tau) + \vec{\mu} \times \vec{\rho}(\tau)}{1 - \vec{\mu} \cdot \vec{\rho}(\tau)}. \tag{18}$$

The way the kinematical variables transform allows us to interpret them, respectively, as the time (15), position (16), velocity (17) and orientation (18) of the particle.

There exist three differential constraints among the kinematical variables: $\vec{u}(\tau) = \dot{\vec{r}}(\tau)/\dot{t}(\tau)$. These constraints, and the homogeneity condition on the Lagrangian L in terms of the derivatives of the kinematical variables, reduce from ten to six the essential degrees of freedom of the system. These degrees of freedom are the position $\vec{r}(t)$ and the orientation $\vec{\rho}(t)$. Since the Lagrangian depends on the derivative of $\vec{u}$ it thus depends on the second derivative of $\vec{r}(t)$. For the orientation variables the Lagrangian only depends on the first derivative of $\vec{\rho}(t)$. It can be written as

$$L = T\dot{t} + \vec{R} \cdot \dot{\vec{r}} + \vec{U} \cdot \dot{\vec{u}} + \vec{V} \cdot \dot{\vec{\rho}}, \tag{19}$$

where the functions written in capital letters are defined as before as $T = \partial L / \partial \dot{t}$, $R_i = \partial L / \partial \dot{r}^i$, $U_i = \partial L / \partial \dot{u}^i$, $V_i = \partial L / \partial \dot{\rho}^i$. In general they will be functions of the ten kinematical variables $(t, \vec{r}, \vec{u}, \vec{\rho})$ and homogeneous functions of zero degree of the derivatives $(\dot{t}, \dot{\vec{r}}, \dot{\vec{u}}, \dot{\vec{\rho}})$.

If we introduce the angular velocity $\vec{\omega}$ as a linear function of $\dot{\vec{\rho}}$, then the last term of the expansion of the Lagrangian (19), $\vec{V} \cdot \dot{\vec{\rho}}$, can also be written as $\vec{W} \cdot \vec{\omega}$, where $W_i = \partial L / \partial \omega^i$.

The different Noether constants of motion are related to the invariance of the dynamical equations under the Galilei group, and are obtained by the usual Lagrangian methods. They are the following observables:

$$\text{energy} \quad H = -T - \vec{u} \cdot \frac{d\vec{U}}{dt}, \tag{20}$$

$$\text{linear momentum} \quad \vec{P} = \vec{R} - \frac{d\vec{U}}{dt}, \tag{21}$$

$$\text{kinematical momentum} \quad \vec{K} = m\vec{r} - \vec{P}t - \vec{U}, \tag{22}$$

$$\text{angular momentum} \quad \vec{J} = \vec{r} \times \vec{P} + \vec{u} \times \vec{U} + \vec{W}. \tag{23}$$

From $\dot{\vec{K}} = 0$, comparing with (21), we find $\vec{R} = m\vec{u}$, and the linear momentum has the form $\vec{P} = m\vec{u} - d\vec{U}/dt$. We see that the total linear momentum does not coincide with the direction of the velocity $\vec{u}$. The functions $\vec{U}$ and $\vec{W}$ are what distinguishes this system from the point particle case. The spin structure is thus directly related to the dependence of the Lagrangian on the acceleration and angular velocity.

We see that $\vec{K}$ in (22) differs from the point particle case $\vec{K} = m\vec{r} - \vec{P}t$, in the term $-\vec{U}$. If we define the vector $\vec{k} = \vec{U}/m$, with dimensions of length, then $\dot{\vec{K}} = 0$ leads to the equation:

$$\vec{P} = m\frac{d(\vec{r} - \vec{k})}{dt}.$$

The vector $\vec{q} = \vec{r} - \vec{k}$, defines the position of the centre of mass of the particle. It is a different point from $\vec{r}$, whenever $\vec{k}$ (and thus $\vec{U}$) is different from zero.

In terms of $\vec{q}$ the kinematical momentum takes the form

$$\vec{K} = m\vec{q} - \vec{P}t,$$

which looks like the result in the case of the point particle, where the centre of mass and centre of charge are the same point.

The total angular momentum (23) has three terms. The first term $\vec{r} \times \vec{P}$ resembles an orbital angular momentum, and the other two $\vec{Z} = \vec{u} \times \vec{U} + \vec{W}$ can be taken to represent the spin of the system. In fact, the latter observable is an angular momentum. It is related to the new kinematical variables and satisfies the dynamical equation $d\vec{Z}/dt = \vec{P} \times \vec{u}$. Because $\vec{P}$ and $\vec{u}$ are not collinear vectors, $\vec{Z}$ is not a conserved angular momentum. This is the dynamical equation satisfied by Dirac's spin operator in the quantum case. The observable $\vec{Z}$ is the classical spin observable equivalent to Dirac's spin operator.

One important feature of the total angular momentum is that the point $\vec{r}$ is not the centre of mass of the system, and therefore the $\vec{r} \times \vec{P}$ part can no longer be interpreted as the orbital angular momentum of the particle. The angular momentum $\vec{Z}$ is the angular momentum of the particle with respect to the point $\vec{r}$, but not with respect to the centre of mass.

The spin of the system is defined as the difference between the total angular momentum $\vec{J}$ and the orbital angular momentum of the centre of mass motion $\vec{L} = \vec{q} \times \vec{P}$. It can assume the following different expressions:

$$\vec{S} = \vec{J} - \vec{q} \times \vec{P} = \vec{J} - \frac{1}{m}\vec{K} \times \vec{P} = \vec{Z} + \vec{k} \times \vec{P} = -m\vec{k} \times \frac{d\vec{k}}{dt} + \vec{W}. \qquad (24)$$

The second form of the spin $\vec{S}$ in (24) is exactly expression (14) which leads to one of the Casimir operators of the extended Galilei group. It is expressed in terms of the constants of the motion $\vec{J}$, $\vec{K}$ and $\vec{P}$, and it is therefore another constant of motion. Because the particle is free and there are no external torques acting on it, it is clear that the spin of the system is represented by this constant angular momentum and not by the other angular momentum observable $\vec{Z}$, which is related to Dirac's spin operator.

The third expression in (24) is the sum of two terms, one $\vec{Z}$, coming from the new kinematical variables, and another $\vec{k} \times \vec{P}$, which is the angular momentum, of the linear momentum located at point $\vec{r}$, with respect to the centre of mass. Alternatively we can describe the spin according to the last expression in (24) in which the term $-\vec{k} \times md\vec{k}/dt$ suggests a contribution of (anti) orbital type coming from the motion around the centre of mass. It is related to the Zitterbewegung, or more precisely to the function $\vec{U} = m\vec{k}$, which comes from the dependence of the Lagrangian on acceleration. The term $\vec{W}$ comes from the dependence on the other three degrees of freedom ρ_i, and thus on the angular velocity. This Zitterbewegung is the motion of the centre of charge around the centre of mass, as we shall see in an example in section 5.6. That the point $\vec{r}$ represents the position of the centre of charge has also been suggested in previous works for the relativistic electron [10].

To analyse the different contributions to the spin of the most general elementary particle we shall consider now two simpler examples. In the first one, the spin is related to the existence of orientation variables, and in the second, to the dependence of the Lagrangian on the acceleration.

5.4   Spinning particle with orientation

The kinematical space is $\mathcal{G}/G_K$, where G_K is the three-dimensional subgroup which consists of the commutative Galilei boosts, or pure Galilei transformations at a constant velocity. The kinematical variables are now $\{t, \vec{r}, \vec{\alpha}\}$, time, position and orientation, respectively. The possible Lagrangians are not unique in this case. They must be functions only of the velocity $\vec{u} = d\vec{r}/dt$ and of the angular velocity $\vec{\omega}$. They have the general form

$$L = T\dot{t} + \vec{R} \cdot \dot{\vec{r}} + \vec{W} \cdot \vec{\omega},$$

where $T = \partial L/\partial \dot{t}$, $\vec{R} = \partial L/\partial \dot{\vec{r}}$, $\vec{W} = \partial L/\partial \vec{\omega}$.

The basic conserved observables are:

$$\text{energy} \quad H = -T,$$

$$\text{linear momentum} \quad \vec{P} = m\vec{u},$$

$$\text{kinematical momentum} \quad \vec{K} = m\vec{r} - \vec{P}t,$$

$$\text{angular momentum} \quad \vec{J} = \vec{r} \times \vec{P} + \vec{W}.$$

For such a particle $\vec{r} = \vec{q}$, the centre of mass and centre of charge coincide and the spin $\vec{S} = \vec{W} \neq 0$. A particular Lagrangian which describes this system is the Lagrangian of a spherically symmetric body:

$$L = \frac{1}{2}m\left(\frac{dr}{dt}\right)^2 + \frac{I}{2}\omega^2,$$

where the spin is $\vec{S} = \vec{W} = I\vec{\omega}$.

5.5   Spinning particle with *Zitterbewegung*

The kinematical space is the manifold $\mathcal{G}/SO(3)$, where $SO(3)$, is the three-dimensional subgroup of rotations. The kinematical variables are $x(\tau) = \{t, \vec{r}, \vec{u}\}$, time, position and velocity, respectively. The possible Lagrangians are not unique as in the previous case, and must be functions of the velocity $\vec{u} = d\vec{r}/dt$ and the acceleration $\vec{a} = d\vec{u}/dt$.

The Lagrangians have the general form when expressed in terms of the kinematical variables and their τ-derivatives

$$L = T\dot{t} + \vec{R} \cdot \dot{\vec{r}} + \vec{U} \cdot \dot{\vec{u}},$$

where $T = \partial L/\partial \dot{t}$, $\vec{R} = \partial L/\partial \dot{\vec{r}}$, $\vec{U} = \partial L/\partial \dot{\vec{u}}$. A particular Lagrangian could be, for example

$$L = \frac{m}{2}\frac{\dot{\vec{r}}^2}{\dot{t}} - \frac{m}{2\omega^2}\frac{\dot{\vec{u}}^2}{\dot{t}}, \tag{25}$$

If we consider that the evolution parameter is dimensionless, all terms in the Lagrangian have dimensions of action. The parameter m represents the mass of the particle while the parameter ω, with dimension time^{-1}, represents an internal frequency: it is the frequency of the internal *Zitterbewegung*. In terms of the essential degrees of freedom, which reduce to the three position variables $\vec{r}$, and using the time as the evolution parameter, the Lagrangian can also be written as

$$L = \frac{m}{2}\left(\frac{d\vec{r}}{dt}\right)^2 - \frac{m}{2\omega^2}\left(\frac{d^2\vec{r}}{dt^2}\right)^2. \tag{26}$$

The dynamical equations obtained from the Lagrangian (26) are:

$$\frac{1}{\omega^2}\frac{d^4\vec{r}}{dt^4} + \frac{d^2\vec{r}}{dt^2} = 0, \tag{27}$$

whose general solution is

$$\vec{r}(t) = \vec{A} + \vec{B}t + \vec{C}\cos\omega t + \vec{D}\sin\omega t, \tag{28}$$

in terms of the 12 integration constants $\vec{A}$, $\vec{B}$, $\vec{C}$ and $\vec{D}$.

We see that the kinematical momentum $\vec{K}$ in (22) differs from the point particle case in the term $-\vec{U}$. The definition of the vector $\vec{k} = \vec{U}/m$, implies that $\vec{K} = 0$ leads to the equation $\vec{P} = md(\vec{r} - \vec{k})/dt$, as before, and $\vec{q} = \vec{r} - \vec{k}$ represents the position of the centre of mass of the particle. It is defined in this example as

$$\vec{q} = \vec{r} - \frac{1}{m}\vec{U} = \vec{r} + \frac{1}{\omega^2}\frac{d^2\vec{r}}{dt^2}. \tag{29}$$

In terms of the center of mass, the dynamical equations (27) can be separated into the form

$$\frac{d^2\vec{q}}{dt^2} = 0, \tag{30}$$

$$\frac{d^2\vec{r}}{dt^2} + \omega^2(\vec{r} - \vec{q}) = 0, \tag{31}$$

where (30) is just equation (27) after twice differentiation of (29), and equation (31) is (29) after all terms on the left hand side have been collected.

From (30) we see that the point $\vec{q}$ moves in a straight trajectory at constant velocity while the motion of point $\vec{r}$, given in (31), is an isotropic harmonic motion of angular frequency ω around the point $\vec{q}$.

The spin of the system $\vec{S}$ is defined as

$$\vec{S} = \vec{J} - \vec{q}\times\vec{P} = \vec{J} - \frac{1}{m}\vec{K}\times\vec{P}, \tag{32}$$

and since it is written in terms of constants of motion it is clearly another constant of motion. Its magnitude S^2 is also a Galilei invariant quantity which characterizes the system. From its definition we get

$$\vec{S} = \vec{u}\times\vec{U} + \vec{k}\times\vec{P} = -m(\vec{r} - \vec{q})\times\frac{d}{dt}(\vec{r} - \vec{q}) = -\vec{k}\times m\frac{d\vec{k}}{dt}, \tag{33}$$

which appears as the (anti)orbital angular momentum of the relative motion of the point $\vec{r}$ around the centre of mass position $\vec{q}$ at rest, so that the total angular momentum can be written as

$$\vec{J} = \vec{q} \times \vec{P} + \vec{S} = \vec{L} + \vec{S}. \tag{34}$$

The total angular momentum is the sum of the orbital angular momentum $\vec{L}$, associated with the motion of the centre of mass, and the spin part $\vec{S}$. For a free particle both $\vec{L}$ and $\vec{S}$ are separate constants of motion. We use the term (anti)orbital to suggest that if the vector $\vec{k}$ represents the position of a point of mass m, the angular momentum of its motion is in the opposite direction from what we obtain here for the spin observable. But, as we shall see in a moment, the vector $\vec{k}$ represents not the position of the mass m, but the position of the charge of the particle.

5.6   Interaction with an external electromagnetic field

If the point $\vec{q}$ represents the position of the centre of mass of the particle, then what position does point $\vec{r}$ represent? The point $\vec{r}$ represents the position of the charge of the particle. This can be seen by considering interaction with an external field. The homogeneity condition of the Lagrangian in terms of the derivatives of the kinematical variables suggests an interaction term of the form

$$L_I = -e\phi(t,\vec{r})\dot{t} + e\vec{A}(t,\vec{r})\cdot\dot{\vec{r}}, \tag{35}$$

which is linear in the derivatives of the kinematical variables t and $\vec{r}$, and where the external potentials are only functions of t and $\vec{r}$.

The dynamical equations obtained from the Lagrangian $L + L_I$ are

$$\frac{1}{\omega^2}\frac{d^4\vec{r}}{dt^4} + \frac{d^2\vec{r}}{dt^2} = \frac{e}{m}\left(\vec{E}(t,\vec{r}) + \vec{u}\times\vec{B}(t,\vec{r})\right), \tag{36}$$

where the electric field $\vec{E}$ and magnetic field $\vec{B}$ are expressed in terms of the potentials in the usual form $\vec{E} = -\nabla\phi - \partial\vec{A}/\partial t$, $\vec{B} = \nabla\times\vec{A}$. Because the interaction term does not depend on $\vec{u}$, the function $\vec{U} = m\vec{k}$ has the same expression as in the free particle case. Therefore the spin and the centre of mass definitions, (33) and (29) respectively, remain the same as in the previous free case. Dynamical equations (36) can again be separated into the form

$$\frac{d^2\vec{q}}{dt^2} = \frac{e}{m}\left(\vec{E}(t,\vec{r}) + \vec{u}\times\vec{B}(t,\vec{r})\right), \tag{37}$$

$$\frac{d^2\vec{r}}{dt^2} + \omega^2(\vec{r} - \vec{q}) = 0. \tag{38}$$

The centre of mass $\vec{q}$ satisfies Newton's equations under the action of the total external Lorentz force, while the point $\vec{r}$ still satisfies the isotropic harmonic motion of angular frequency ω around the point $\vec{q}$. But the external force and the fields are defined at the point $\vec{r}$ and not at point $\vec{q}$. It is the velocity $\vec{u}$ of the point $\vec{r}$ which appears in the magnetic term of the Lorentz force. The point $\vec{r}$ clearly represents the position of the charge. In fact, this minimal coupling we have considered is the coupling of the electromagnetic potentials with the particle current, which, in the relativistic case, can be written as $j_\mu A^\mu$. The current j_μ is associated with the motion of a charge e at the point $\vec{r}$.

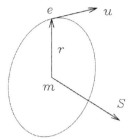

Figure 1: Charge motion in the C.M. frame.

The charge has an oscillatory motion of very high frequency ω, which in the case of the relativistic electron will be $\omega = 2mc^2 / \hbar \approx 1{,}55 \times 10^{21} \text{s}^{-1}$, as shown later. The average position of the charge is the centre of mass, but it is this internal orbital motion which gives rise to the spin structure and also to the magnetic properties of the particle.

When analysed in the centre of mass frame (see Fig. 1), $\vec{q} = 0$, $\vec{r} = \vec{k}$, and the system reduces to a point charge whose motion is in general an ellipse. If we choose $C = D$, and $\vec{C} \cdot \vec{D} = 0$, it reduces to a circle of radius $r = C = D$, orthogonal to the spin. Because the particle has a charge e, it produces a magnetic moment, which according to the usual classical definition is [11]

$$\vec{\mu} = \frac{1}{2} \int \vec{r} \times \vec{j} d^3 \vec{r} = \frac{e}{2} \vec{k} \times \frac{d\vec{k}}{dt} = -\frac{e}{2m} \vec{S}, \tag{39}$$

where $\vec{j} = e\delta^3 (\vec{r} - \vec{k}) d\vec{k} / dt$ is the vector current associated with the motion of a charge e located at the point $\vec{k}$. The magnetic moment is orthogonal to the *Zitterbewegung* plane and opposite to the spin if $e > 0$. The particle also has a non-vanishing electric dipole moment with respect to the centre of mass $\vec{d} = e\vec{k}$. It oscillates and is orthogonal to $\vec{\mu}$, and therefore to $\vec{S}$, in the centre of mass frame. Its time average value vanishes for times larger than the natural period of this internal motion. Although this is a nonrelativistic example, it is interesting to compare this analysis with Dirac's relativistic analysis of the electron, [5] in which both momenta $\vec{\mu}$ and $\vec{d}$ appear, giving rise to two possible interacting terms in Dirac's Hamiltonian.

6. Relativistic elementary particles

The Poincaré group can be parameterised in terms of exactly the same ten parameters $\{b, \vec{a}, \vec{v}, \vec{\alpha}\}$ as the Galilei group and with the same dimensions as before. We therefore maintain the interpretation of these variables respectively as the time, position, velocity and orientation of the particle. The homogeneous spaces of the Poincaré group can be classified in the same manner, but with some minor restrictions. For instance, the kinematical space of the example of the spinning particle with orientation as in section 5.4, $X = \mathcal{G}/G_K$, can no longer be defined in the Poincaré case, because the three dimensional set G_K of Lorentz boosts is not a subgroup of G; but the most general structure of a spinning particle still holds.

The Poincaré group has three different maximal homogeneous spaces spanned by the variables $\{b, \vec{a}, \vec{v}, \vec{\alpha}\}$, which are classified according to the range of the velocity parameter $\vec{v}$. If $v < c$ we have the Poincaré group itself. When $v > c$, this homogeneous space describes particles whose charge is moving faster than light. Finally, if $v = c$, we have a homogeneous space which describes particles whose position $\vec{r}$ is always moving at the speed of light. This is the manifold which defines the kinematical space of photons and electrons [6,7]. The first manifold gives, in the low velocity limit, the same models as in the nonrelativistic case. It is the Poincaré group manifold, which is transformed into the Galilei group by the limiting process $c \to \infty$. But this limit cannot be applied to the other two manifolds. Accordingly, the Poincaré group describes a larger set of spinning objects.

6.1   Spinning relativistic elementary particles

We shall review the main points of the relativistic spinning particles whose kinematical space is the manifold spanned by the variables $\{t, \vec{r}, \vec{u}, \vec{\alpha}\}$, interpreted as the time, position, velocity and orientation of the particle, but with $u = c$. This is a homogeneous space homomorphic to the manifold G/V, where V is the one-dimensional subgroup of pure Lorentz transformations in a fixed arbitrary direction.

For these systems the most general form of the Lagrangian is

$$L = T\dot{t} + \vec{R} \cdot \dot{\vec{r}} + \vec{U} \cdot \dot{\vec{u}} + \vec{W} \cdot \vec{\omega},$$

where $T = \partial L / \partial \dot{t}$, $R_i = \partial L / \partial \dot{r}^i$, $U_i = \partial L / \partial \dot{u}^i$ and $W_i = \partial L / \partial \omega^i$ will be, in general, functions of the ten kinematical variables $\{t, \vec{r}, \vec{u}, \vec{\alpha}\}$ and homogeneous functions of zero degree in terms of the derivatives $\{\dot{t}, \dot{\vec{r}}, \dot{\vec{u}}, \dot{\vec{\alpha}}\}$.

The Noether constants of motion are now the following conserved observables:

$$\text{energy} \quad H = -T - \vec{u} \cdot \frac{d\vec{U}}{dt}, \tag{40}$$

$$\text{linear momentum} \quad \vec{P} = \vec{R} - \frac{d\vec{U}}{dt}, \tag{41}$$

$$\text{kinematical momentum} \quad \vec{K} = H\vec{r}/c^2 - \vec{P}t - \vec{S} \times \vec{u}/c^2, \tag{42}$$

$$\text{angular momentum} \quad \vec{J} = \vec{r} \times \vec{P} + \vec{S}, \tag{43}$$

where

$$\vec{S} = \vec{u} \times \vec{U} + \vec{W}. \tag{44}$$

The difference from the Galilei case comes from the different behaviour of the Lagrangian under the Lorentz boosts when compared with the Galilei boosts. In the nonrelativistic case the Lagrangian is not invariant. However, the relativistic Lagrangian is invariant and the kinematical variables transform in a different way. This gives rise to the term $\vec{S} \times \vec{u}/c^2$ instead of the term $\vec{U}$ which appears in the kinematical momentum (42). The angular momentum observable (44) is not properly speaking the spin of the system, if we define spin

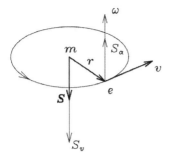

Figure 2. Motion of the centre of charge of the electron around its centre of mass in the C.M. frame.

as the difference between the total angular momentum and the orbital angular momentum associated with the centre of mass. It is the angular momentum of the particle with respect to the point $\vec{r}$, as in the nonrelativistic case. Nevertheless, the observable $\vec{S}$ is the classical equivalent of Dirac's spin observable because in the free particle case it satisfies the same dynamical equation,

$$\frac{d\vec{S}}{dt} = \vec{P} \times \vec{u},$$

as Dirac's spin operator does in the quantum case. It is only a constant of motion for the centre of mass observer. This can be seen by taking the time derivative of the constant total angular momentum $\vec{J}$ given in (43). We shall keep the notation $\vec{S}$ for this angular momentum observable, because when the system is quantized it gives rise to the usual quantum mechanical spin operator in terms of the Pauli spin matrices.

6.2   Dirac's equation

Dirac's equation is the quantum mechanical expression of the Poincaré invariant linear relationship [6,7] between the energy H and the linear momentum $\vec{P}$

$$H - \vec{P} \cdot \vec{u} - \vec{S} \cdot \left(\frac{d\vec{u}}{dt} \times \vec{u} \right) = 0,$$

where $\vec{u}$ is the velocity of the charge ($u = c$), $d\vec{u}/dt$ the acceleration and $\vec{S} = \vec{S}_u + \vec{S}_\alpha$ Dirac's spin observable (see Figure 2). This expression can be obtained from (42) by making the time derivative of that constant observable and a final scalar product with the velocity $\vec{u}$. The Dirac spin has two parts: one $\vec{S}_u = \vec{u} \times \vec{U}$, is related to the orbital motion of the charge, and $\vec{S}_\alpha = \vec{W}$ is due to the rotation of the particle and is directly related to the angular velocity, as it corresponds to a spherically symmetric object.

The centre of mass observer is defined as the observer for whom $\vec{K} = \vec{P} = 0$, because this implies that $\vec{q} = 0$ and $d\vec{q}/dt = 0$. By analysing the observable (42) in the centre of mass frame where $H = mc^2$, we get the dynamical equation of the point $\vec{r}$,

$$\vec{r} = \vec{S} \times \vec{u}/mc^2$$

where $\vec{S}$ is a constant vector in this frame. The solution is the circular motion depicted in Figure 2.

The radius and angular velocity of the internal classical motion of the charge are, respectively, $R = S/mc$, and $\omega = mc^2/S$. The energy of this system is not definite positive. The particle of positive energy has the total spin $\vec{S}$ oriented in the same direction as the $\vec{S}_u$ part while the orientation is the opposite for the negative energy particle. This system corresponds to the time reversed motion of the other. When the system is quantized, the orbital component $\vec{S}_u$, which is directly related to the magnetic moment, quantizes with integer values, while the rotational part $\vec{S}_\alpha$ requires half integer values. For these particles of spin ½, the total spin is half the value of the $\vec{S}_u$ part. When expressing the magnetic moment in terms of the total spin, we thus obtain a pure kinematical interpretation of the $g = 2$ gyromagnetic ratio [12].

For the centre of mass observer this system appears as a system of three degrees of freedom. Two represent the x and y coordinates of the point charge, and the third is the phase of its rotational motion. However this phase is exactly the same as the phase of the orbital motion of the charge. Because the motion is at constant radius at constant speed c, only one independent degree of freedom is left—say the x variable. Therefore the system is reduced to a one-dimensional harmonic oscillator of angular frequency ω. When the system is quantized, the stationary states of a one-dimensional harmonic oscillator have the energy

$$E_n = \left(n + \frac{1}{2}\right)\hbar\omega, \qquad n = 0,1,2,\ldots$$

But if the system is elementary, then it has no excited states, and in the C.M. frame it is reduced to the ground state of energy

$$E_0 = \frac{1}{2}\hbar\omega = mc^2.$$

If we compare this with the classical result $\omega = mc^2/S$ we see that the constant classical parameter S takes the value $S = \hbar/2$ when quantized. The radius of the internal motion is $R = \lambda_C/2$, half Compton's wavelength.

We see that all Lagrangian systems with the same kinematical space as the one considered in this model have exactly the same dynamics for the point r, describe spin ½ particles and satisfy Dirac's equation when quantized. The formalism describes an object whose charge is located at a single point $\vec{r}$, but it is nevertheless moving in a confined region of radius of order λ_C. It has a magnetic moment produced by the motion of the charge, and also an oscillating electric dipole moment, with respect to the centre of mass, of average value zero.

To conclude this section, and with the above model of the electron in mind, it is convenient to remember some of the features that Dirac obtained for the motion of a free electron [5]. Let the point $\vec{r}$ be the position vector in terms of which Dirac's spinor $\psi(t,\vec{r})$ is defined. When computing the velocity of the point $\vec{r}$, Dirac arrives at:

1. The velocity $\vec{u} = i / \hbar [H, \vec{r}] = c\vec{\alpha}$, is expressed in terms of the $\vec{\alpha}$ matrices and he writes, ... "*a measurement of a component of the velocity of a free electron is certain to lead to the result* $\pm c$."

2. The linear momentum does not have the direction of this velocity $\vec{u}$, but must be related to some average value of it: ... "*the* x_1 *component of the velocity,* $c\alpha_1$, *consists of two parts, a constant part* $c^2 p_1 H^{-1}$, *connected with the momentum by the classical relativistic formula, and an oscillatory part, whose frequency is at least* $2mc^2 / h$,....."

3. About the position $\vec{r}$: "*The oscillatory part of* x_1 *is small... which is of order of magnitude* $\hbar / mc$ ...."

And when analyzing the interaction of the electron with an external electromagnetic field in his original 1928 paper [13], after taking the square of Dirac's operator, he obtains two new interaction terms:

$$\frac{e\hbar}{2mc} \vec{\Sigma} \cdot \vec{B} + \frac{ie\hbar}{2mc} \vec{\alpha} \cdot \vec{E},$$

Here Dirac's spin operator is written as $\vec{S} = \hbar \vec{\Sigma} / 2$ where

$$\vec{\Sigma} = \begin{pmatrix} \vec{\sigma} & 0 \\ 0 & \vec{\sigma} \end{pmatrix},$$

in terms of σ-Pauli matrices. $\vec{E}$ and $\vec{B}$ are the external electric and magnetic fields, respectively. He says, "*The electron will therefore behave as though it has a magnetic moment* $(e\hbar / 2mc)\Sigma$ *and an electric moment* $(ie\hbar / 2mc)\vec{\alpha}$. *The magnetic moment is just that assumed in the spinning electron model*" (Pauli model). "*The electric moment, being a pure imaginary, we should not expect to appear in the model.*"

In the last sentence it is difficult to understand why Dirac, who did not reject the negative energy solutions, disliked the existence of this electric dipole, which was obtained from the formalism on an equal footing with the magnetic dipole term. Properly speaking this electric dipole does not represent the existence of a particular positive and negative charge distribution for the electron. The negative charge of the electron is at a single point but because this point is not the centre of mass, there exists a non-vanishing electric dipole moment with respect to the centre of mass even in the centre of mass frame. This is the observable Dirac disliked. It is oscillating at very high frequency and basically plays no role in low energy electron interactions because its average value vanishes, but it is important in high energy processes or in very close electron-electron interactions.

All real experiments to determine very accurately the gyromagnetic ratio are based on the determination of precession frequencies. But these precession frequencies are independent of the spin orientation. However, the difficulty separating electrons in a Stern-Gerlach type experiment suggests polarization experiments have to be done to determine in a direct way whether the spin and magnetic moment for elementary particles are either parallel or antiparallel to each other. One of the predictions of this formalism is that for both particle and the corresponding antiparticle the spin and magnetic moment have to have the same relative orientation, either parallel or antiparallel.

6.3   Dynamical equation of the relativistic spinning electron

We recall from elementary differential geometry some basic properties of any arbitrary three-dimensional curve $\vec{r}(s)$. If it is expressed in parametric form in terms of the arc length s as the parameter, it has associated the three orthogonal unit vectors $\vec{v}_i$, $i = 1,2,3$ called respectively tangent, normal and binormal. These unit vectors satisfy the so called Frenet-Serret differential equations:

$$\dot{\vec{v}}_1(s) = \kappa(s)\vec{v}_2(s)$$
$$\dot{\vec{v}}_2(s) = -\kappa(s)\vec{v}_1(s) \qquad\qquad +\tau(s)\vec{v}_3(s),$$
$$\dot{\vec{v}}_3(s) = \qquad\qquad -\tau(s)\vec{v}_2(s)$$

where κ and τ are respectively the curvature and torsion. Since the unit tangent vector is $\vec{v}_1 = \dot{\vec{r}} \equiv \vec{r}^{(1)}$, when successive derivatives are taken it yields

$$\vec{r}^{(1)} = \vec{v}_1,$$

$$\vec{r}^{(2)} = \kappa\vec{v}_2,$$

$$\vec{r}^{(3)} = \dot{\kappa}\vec{v}_2 + \kappa\dot{\vec{v}}_2 = -\kappa^2\vec{v}_1 + \dot{\kappa}\vec{v}_2 + \kappa\tau\vec{v}_3,$$

$$\vec{r}^{(4)} = -3\kappa\dot{\kappa}\vec{v}_1 + (\ddot{\kappa} - \kappa^3 - \kappa\tau^2)\vec{v}_2 + (2\dot{\kappa}\tau + \kappa\dot{\tau})\vec{v}_3.$$

The elimination of the $\vec{v}_i$ vectors between these equations implies that the most general curve in three-dimensional space satisfies the fourth-order ordinary differential equation:

$$\vec{r}^{(4)} - \left(\frac{2\dot{\kappa}}{\kappa} + \frac{\dot{\tau}}{\tau}\right)\vec{r}^{(3)} + \left(\kappa^2 + \tau^2 + \frac{\dot{\kappa}\dot{\tau}}{\kappa\tau} + \frac{2\dot{\kappa}^2 - \kappa\ddot{\kappa}}{\kappa^2}\right)\vec{r}^{(2)} + \kappa^2\left(\frac{\dot{\kappa}}{\kappa} - \frac{\dot{\tau}}{\tau}\right)\vec{r}^{(1)} = 0.$$

All the coefficients in brackets, in front of the s-derivatives $\vec{r}^{(i)}$, can be expressed in terms of the scalar products $\vec{r}^{(i)} \cdot \vec{r}^{(j)}$, $i,j = 1,2,3$. For helical motions there is a constant relationship $\kappa/\tau = $ constant, and therefore the coefficient of $\vec{r}^{(1)}$ vanishes.

Our example of the nonrelativistic spinning particle also satisfies the fourth order differential equation (27). Similarly, the point $\vec{r}$ of the relativistic spinning electron also satisfies a fourth order ordinary differential equation which has been calculated from invariance principles [14]. It takes the following form for any arbitrary inertial observer:

$$\vec{r}^{(4)} - \frac{3(\vec{r}^{(2)} \cdot \vec{r}^{(3)})}{(\vec{r}^{(2)} \cdot \vec{r}^{(2)})}\vec{r}^{(3)} +$$

$$\left(\frac{2(\vec{r}^{(3)} \cdot \vec{r}^{(3)})}{(\vec{r}^{(2)} \cdot \vec{r}^{(2)})} - \frac{3(\vec{r}^{(2)} \cdot \vec{r}^{(3)})^2}{4(\vec{r}^{(2)} \cdot \vec{r}^{(2)})^2} - (\vec{r}^{(2)} \cdot \vec{r}^{(2)})^{1/2}\right)\vec{r}^{(2)} = 0. \qquad (46)$$

It corresponds to a helical motion since the term in the first derivative $\vec{r}^{(1)}$ is lacking, and it reduces to circular central motion at constant velocity c in the centre of mass frame. Here we use space-time units such that the internal radius $R = 1$ and the *Zitterbewegung* frequency $\omega = 1$.

The centre of mass position is defined by

$$\vec{q} = \vec{r} + \frac{2(\vec{r}^{(2)} \cdot \vec{r}^{(2)})\vec{r}^{(2)}}{(\vec{r}^{(2)} \cdot \vec{r}^{(2)})^{3/2} + (\vec{r}^{(3)} \cdot \vec{r}^{(3)}) - \dfrac{3(\vec{r}^{(2)} \cdot \vec{r}^{(3)})^2}{4(\vec{r}^{(2)} \cdot \vec{r}^{(2)})}}. \tag{47}$$

We can check that both $\vec{q}$ and $\vec{q}^{(1)}$ vanish for the centre of mass observer. The fourth order dynamical equation for the position of the charge (46) can also be rewritten as a system of two second order ordinary differential equations for the positions of the points $\vec{q}$ and $\vec{r}$

$$\vec{q}^{(2)} = 0, \qquad \vec{r}^{(2)} = \frac{1 - \vec{q}^{(1)} \cdot \vec{r}^{(1)}}{(\vec{q} - \vec{r})^2}(\vec{q} - \vec{r}), \tag{48}$$

i.e., a free motion for the centre of mass $\vec{q}$ and a kind of central motion for the charge position $\vec{r}$ around the centre of mass. Equation (46) emerges from (47) after differentiation twice with respect to time. The last equation of (48) is just (47) written in terms of $\vec{q}$ and $\vec{q}^{(1)}$.

For the relativistic electron, when the centre of mass velocity is small, $\vec{q}^{(1)} \to 0$, and because $|\vec{q} - \vec{r}| = 1$ in these units, we obtain the equations of the Galilei case

$$\vec{q}^{(2)} = 0, \qquad \vec{r}^{(2)} = \vec{q} - \vec{r} \tag{49}$$

i.e., a free motion for the centre of mass and a harmonic motion around $\vec{q}$ of angular frequency $\omega = 1$, for the position of the charge, as happened in the nonrelativistic example analysed in (30) and (31).

6.4   Interaction with an external field

The free equation for the centre of mass motion $\vec{q}^{(2)} = 0$ represents the conservation of linear momentum $d\vec{P}/dt = 0$. But the linear momentum is written in terms of centre of mass velocity as $\vec{P} = m\gamma(q^{(1)})\vec{q}^{(1)}$, so that the free dynamical equation (48) in the presence of an external field should be replaced by

$$\vec{P}^{(1)} = \vec{F}, \qquad \vec{r}^{(2)} = \frac{1 - \vec{q}^{(1)} \cdot \vec{r}^{(1)}}{(\vec{q} - \vec{r})^2}(\vec{q} - \vec{r}), \tag{50}$$

where $\vec{F}$ is the external force and the second equation is left unchanged. We consider the same definition of the centre of mass position (47) as in the free particle case, because it corresponds to the fact that the internal structure of an elementary particle is not modified by any external interaction, and the charge moves in the same way around the centre of mass as in the free case. Since

$$\frac{d\vec{P}}{dt} = m\gamma(q^{(1)})\vec{q}^{(2)} + m\gamma(q^{(1)})^3(\vec{q}^{(1)} \cdot \vec{q}^{(2)})\vec{q}^{(1)}$$

it yields

$$m\gamma(q^{(1)})^3(\vec{q}^{(1)} \cdot \vec{q}^{(2)}) = \vec{F} \cdot \vec{q}^{(1)}$$

and by leaving the highest derivative $\vec{q}^{(2)}$ on the left hand side we finally obtain the differential equations that describe the evolution of a relativistic spinning electron in the presence of an external electromagnetic field:

$$m\vec{q}^{(2)} = \frac{e}{\gamma(q^{(1)})}\left[\vec{E} + \vec{r}^{(1)} \times \vec{B} - \vec{q}^{(1)}\left([\vec{E} + \vec{r}^{(1)} \times \vec{B}]\cdot\vec{q}^{(1)}\right)\right], \tag{51}$$

$$\vec{r}^{(2)} = \frac{1 - \vec{q}^{(1)}\cdot\vec{r}^{(1)}}{(\vec{q} - \vec{r})^2}(\vec{q} - \vec{r}). \tag{52}$$

7.   Gyromagnetic ratio

The Hilbert space which describes the wave functions of the spinning electron is a complex vector space of squared integrable functions $\psi(t,\vec{r},\vec{u},\vec{\alpha})$ of the kinematical variables. The general structure of the quantum mechanical angular momentum operator acting on this Hilbert space, in either the relativistic or nonrelativistic approach, is

$$\vec{J} = \vec{r} \times \frac{\hbar}{i}\nabla + \vec{S} = \vec{r} \times \vec{P} + \vec{S}, \tag{53}$$

where the spin operator takes the form $S = Z + W$

$$\vec{S} = \vec{u} \times \frac{\hbar}{i}\nabla_u + \vec{W}. \tag{54}$$

The operator ∇_u is the gradient operator with respect to the velocity variables and $\vec{W}$ is a linear differential operator which depends only on the orientation variables $\vec{\alpha}$; it therefore commutes with ∇_u. For example, in the $\vec{\rho} = \vec{n}\tan(\alpha/2)$ parameterization $\vec{W}$ is written as

$$\vec{W} = \frac{\hbar}{2i}[\nabla_\rho + \vec{\rho} \times \nabla_\rho + \vec{\rho}(\vec{\rho}\cdot\nabla_\rho)], \tag{55}$$

where ∇_ρ is the gradient operator with respect to the $\vec{\rho}$ variables.

The first part Z in (54) is related to the *Zitterbewegung* spin and has only integer eigenvalues. This is because it has the form of an orbital angular momentum operator in terms of the $\vec{u}$ variables. Half-integer eigenvalues come only from the operator (55). This operator takes into account the change of orientation, *i.e.*, the rotation of the particle.

We have seen, in both relativistic and non-relativistic examples, that if the only spin content of the particle $\vec{S}$ is related to the *Zitterbewegung* part $\vec{Z} = \vec{u} \times \vec{U}$, then the relationship between the magnetic moment and *Zitterbewegung* spin is given by

$$\vec{\mu} = \frac{e}{2}\vec{k} \times \frac{d\vec{k}}{dt} = \frac{e}{2m}\vec{Z}, \tag{56}$$

i.e., with a normal gyromagnetic ratio $g = 1$. If the electron has a gyromagnetic ratio $g = 2$, this necessarily implies that another part of the spin arises from the angular velocity of the body, but makes no contribution to the magnetic moment.

For the electron, therefore, both parts $\vec{W}$ and $\vec{Z}$ contribute to the total spin. But the $\vec{W}$ part, which is related to the angular variables that describe its orientation in space, does not contribute to the separation $\vec{k}$ between the centre

Figure 3. Scattering of two spinning electrons with parallel
spins, in their centre of mass
frame. It is also depicted the
scattering of two spinless electrons with the same energy and
linear momentum.

of charge and the centre of mass. It turns out that the magnetic moment of a general particle is still related to the motion of the charge by the expression (56), *i.e.*, in terms of the $\vec{Z}$ part, but not to the $\vec{W}$ part. It is precisely when we express the magnetic moment in terms of the total spin $\vec{S}$ that the concept of gyromagnetic ratio arises.

We now assume that both $\vec{Z}$ and $\vec{W}$ terms contribute to the total spin $\vec{S}$ with their lowest admissible values. In the model of the spinning electron $\vec{Z}$ and $\vec{W}$ have opposite orientation.

For Dirac's particles, the classical *Zitterbewegung* is a circular motion at the speed of light of radius $R = \hbar / 2mc$ and angular frequency $\omega = 2mc^2 / \hbar$, on a plane orthogonal to the total spin. The total spin $\vec{S}$ and the $\vec{Z}$ part are both orthogonal to this plane, and parallel to each other. Let us define the gyromagnetic ratio by $Z = gS$. For the lowest admissible values of the quantized spins $z = 1$ and $w = \frac{1}{2}$ in the opposite direction, this gives rise to a total $s = \frac{1}{2}$ perpendicular to the *Zitterbewegung* plane, and therefore $g = 2$.

8.   Bound motion of two electrons

If we have relativistic and nonrelativistic differential equations satisfied by the spinning electrons we can analyze the interaction between them by assuming, for example, a Coulomb interaction between their charges. This leads to a system of differential equations of the form (37-38) or (51-52) for each particle. For example, the external field acting on the charge e_1 is replaced by the instantaneous Coulomb field created by the other charge e_2 at the position of e_1, and similarly for the other particle. The integration is performed numerically by means of the numerical integration program *Dynamics Solver* [15].

Figure 3 represents the scattering of two spinning electrons analysed in their common centre of mass frame [14]. We send the particles with their spins parallel and with a non vanishing impact parameter. In addition to the helical motion of their charges, we can also depict the trajectories of their centre of mass. If we compare this motion with the Coulomb interaction of two spinless electrons coming from the same initial position and with the same velocity as the centre of mass of the spinning electrons, we obtain the solid trajectories marked with an arrow. Basically, this corresponds to the trajectory of the centre of mass of each spinning particle, provided the two particles do not approach each other below the Compton wavelength. This can be understood because the average position of the centre of charge of each particle approximately coincides with its centre of mass, and if they do not approach each other too closely

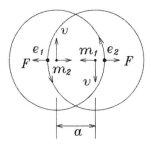

Figure 4: Initial position and velocity of the centre of mass and charges for a bound motion of a two-electron system with parallel spins. The circles would correspond to the trajectories of the charges if considered free. The interacting Coulomb force F is computed in terms of the separation distance between the charges.

the average Coulomb force is the same. The difference comes out when we consider a very deep interaction or very close initial positions.

Figure 4 represents the initial positions of a pair of particles with parallel spins. Recall that the radius of the internal motion is half the Compton wavelength. The initial separation of their centres of mass a is a distance smaller than the Compton wavelength. The centre of mass of each particle is considered to be moving with a velocity $\vec{v}$, as depicted.

That the spins of the two particles are parallel is reflected by the fact that the internal motions of the charges, represented by the oriented circles that surround the corresponding centre of mass, have the same orientation. It must be remarked that the internal motion of the charge around its centre of mass can always be characterised by a phase. The phases of the particles are chosen opposite to one another. We also depict the repulsive Coulomb force F computed in terms of the separation of charges. This interaction force F has also been drawn attached to the corresponding centre of mass, so that the net force acting on the point m_2 is directed toward the point m_1, and conversely. This external force determines the motion of each centre of mass. We thus see that a repulsive force between the charges represents an attractive force between their centres of mass when located at such a short distance.

In Figure 5 we depict the evolution of the charges and masses of this two-electron system for $a = 0,4\lambda_c$ and $v = 0,004c$ during a short time interval. Figure 6 represents only the motions of the centres of mass of both particles for a longer time. It shows that the centre of mass of each particle remains in a bound region.

The evolution of the charges is not shown in this last figure because it

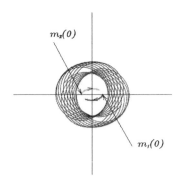

Figure 5: Bound motion of two electrons with parallel spins during a short period of time

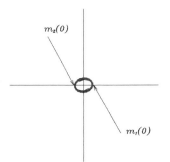

Figure 6. Evolution of the centres of mass of both particles for a longer time

blurs the picture, but it can be inferred from the previous figure. We have found bound motions at least for the range $0 \leq a \leq 0,8\lambda_C$ and velocity $0 \leq v \leq 0,01c$. We can also obtain similar bound motions if the initial velocity v has a component along the OX axis. Bound motion can also be obtained for initial charge positions different from the ones depicted in Figure 4. This range for the relative phase depends on a and v, but in general bound motion is more likely if the initial phases of the charges are opposite to each other. If, instead of the instantaneous Coulomb interaction between the charges, we consider the retarded electromagnetic field of each charge, we obtain a similar behaviour for the bound motion of this electron-electron interaction.

We thus see that if the separation between the centre of mass and centre of charge of a particle (*Zitterbewegung*) is responsible of part of the spin structure, then this attractive effect can be easily interpreted.

A bound motion for classical spinless electrons is not possible. We can conclude that one of the salient features of the present formalism is the existence, from the classical viewpoint, of possible bound states for spinning electron-electron interaction. If the centres of mass of two electrons are separated by a distance greater than the Compton wavelength, they always repel each other as in the spinless case. But if the centres of mass of two electrons are separated by a distance less than the Compton wavelength, then from the classical viewpoint they can form bound states, provided certain initial conditions regarding their relative initial spin orientation, position of charges and centre of mass velocity are fulfilled. The difficulty may be to prepare a pair of electrons in the initial configuration depicted in Figure 4. A high-energy deep scattering can bring electrons to a very close approach. At low energy, if we consider the electrons in the conduction band of a solid, their interaction with the lattice could do this job. If we have a very thin layer under a huge external magnetic field perpendicular to the surface, as in the quantum Hall effect measurements, most of the electrons in this layer will have the spins parallel. If this happens to be true, we have a mechanism associated with the spin structure of the elementary particles for the plausible formation of a spin 1 Bose-Einstein condensate. This is just a classical prediction, not a quantum prediction, associated with a model which satisfies the Dirac equation when quantized. The possible quantum mechanical bound states must be obtained from the corresponding analysis of two interacting quantum Dirac particles, a problem which has not been solved yet. From the classical viewpoint, bound states for a hydrogen atom can exist for any negative energy and any arbitrary angular momentum. The quantum analysis of the atom gives the correct answer for the allowed stationary bound states.

References

[1] J. M. Levy-Leblond, Non-relativistic particles and wave equations, *Comm. Math. Phys.* **6**, 286, (1967).

[2] V.I. Fushchich, A. Nikitin and V.A. Sagolub, On the non-relativistic motion equations in the Hamiltonian form, *Rep. Math. Phys.* **13**, 175 (1978).

[3] M. Rivas, *Kinematical theory of spinning particles*, (Kluwer, Dordrecht) 2001.

[4] V. Bargmann, On unitary ray representations of continuous groups, *Ann. Math.* **5**, 1 (1954).

[5] P.A.M. Dirac, *The Principles of Quantum Mechanics*, 4th ed. (Oxford U. P.), §69, 1958.

[6] M. Rivas, Quantization of generalized spinning particles. New derivation of Dirac's equation, *J. Math. Phys.* **35**, 3380 (1994).

[7] M. Rivas, Classical Relativistic Spinning Particles, *J. Math. Phys.* **30**, 318 (1989).

[8] J.M. Levy-Leblond, Galilei Group and Galilean Invariance, in E.M. Loebl, *Group Theory and its applications, Acad. Press*, NY 1971, vol. 2, p. 221.

[9] M. Rivas, Classical particle systems: I. Galilei free particle, *J. Phys.* A **18**, 1971 (1985).

[10] A.O. Barut and A.J. Bracken, Zitterbewegung and the internal geometry of the electron, *Phys. Rev.* D **23**, 2454 (1981).

[11] J.D. Jackson, *Classical Electrodynamics*, John Wiley & Sons, NY, 3rd. ed. p.186, 1998.

[12] M. Rivas, J.M. Aguirregabiria and A. Hernandez, A pure kinematical explanation of the gyromagnetic ratio g = 2 of leptons and charged bosons, *Phys. Lett.* A **257,** 21 (1999).

[13] P.A.M. Dirac, The quantum theory of the electron, *Proc. Roy. Soc. Lon.* A117, 610 (1928).

[14] M. Rivas, The dynamical equation of the spinning electron, *J. Phys. A,* **36,** 4703 (2003) .

[15] J.M. Aguirregabiria *Dynamics Solver*, 2002. Computer program for solving different kinds of dynamical systems. It is available from his author through the web site <http://tp.lc.ehu.es/jma.html>.

Relativistic Wave Equations, Clifford Algebras and Orthogonal Gauge Groups

Claude Daviau
La Lande, 44522 Pouille-les-coteaux, France
e-mail : daviau.claude@wanadoo.fr
Fondation Louis de Broglie, 23 rue Marsoulan
75012 Paris, France

The Dirac equation is considered the wave equation for a particle with spin ½. We write this equation in the frame of real linear spaces. We present the change resulting from new frames: we can construct new relativistic wave equations, which may not be equivalent to the Dirac equation. One of these new wave equations is proposed for neutrinos. The diversity of relativistic waves is connected to with the diversity of particles with spin ½.

Classical electromagnetism itself was not sufficient to explain the electron, because it predicted neither the electron's stability, nor quantification of energy levels, nor the electron's spin. Louis de Broglie discovered [1] the electron's wave. The non-relativistic equation for the electron was introduced by Erwin Schrödinger. From relativistic considerations the Klein-Gordon relativistic equation was proposed for the electron's wave. However, this equation contains two kinds of defects. It gives a conserved current without the probability density, and it does not give expected quantum numbers and the expected number of states in the case of the hydrogen atom.

In order to obtain a current with the probability density, Dirac introduced a relativistic wave equation [2] based on the Pauli equation, and sought a wave equation with first order derivatives only. This equation lead to interesting results. In the case of the H atom, the Dirac equation gives the expected quantum numbers, the expected number of states, and precise energy levels. Moreover, on the basis of the Pauli principle the Dirac equation (not the Schrödinger equation) leads to the periodic classification of chemical elements. This equation also gives a correct calculation of the Zeeman effect and explains the Lande factors. Consequently, the Dirac equation is the basis for quantum field theories. Further, when experimental physicists discovered new particles with spin ½ (muons, neutrinos and quarks), the Dirac equation was used as the wave equation for all these particles.

Today it is possible to write the Dirac equation differently. Moreover, these real formalisms lead to new relativistic wave equations that are not equivalent to the Dirac equation. The diversity of these wave equations may be linked to the diversity of particles. For instance, it is possible to obtain a chiral relativistic wave equation with a mass term but without the possibility of a charge term, which corresponds to the neutrino's properties. It is possible to derive the different wave equations in triplicate, equivalent, but distinct equations, and this fact may be related to the existence of three and only three generations of particles. Thus, a study of relativistic wave equations provides simple explanations of well-established but not yet understood facts, such as the insensitivity of leptons and sensitivity of quarks to strong interactions, or the presence of charged and uncharged leptons in each generation.

1. Classical framework of the Dirac equation

We use the usual matrices here

$$\psi = \begin{pmatrix} \psi_1 \\ \psi_2 \\ \psi_3 \\ \psi_4 \end{pmatrix}; \quad \gamma_0 = \gamma^0 = \begin{pmatrix} I & 0 \\ 0 & -I \end{pmatrix}; \quad \gamma_j = -\gamma^j = \begin{pmatrix} 0 & -\sigma_j \\ \sigma_j & 0 \end{pmatrix}; \tag{1}$$

$$I = \sigma_0 = \sigma^0 = \begin{pmatrix} 1 & 0 \\ 0 & 1 \end{pmatrix}; \quad \sigma_1 = -\sigma^1 = \begin{pmatrix} 0 & 1 \\ 1 & 0 \end{pmatrix};$$

$$\sigma_2 = -\sigma^2 = \begin{pmatrix} 0 & -i \\ i & 0 \end{pmatrix}; \quad \sigma_3 = -\sigma^3 = \begin{pmatrix} 1 & 0 \\ 0 & -1 \end{pmatrix}; \tag{2}$$

$$\gamma_{ij} = \gamma_i \gamma_j; \qquad \gamma_5 = -i\gamma_{0123}. \tag{3}$$

Thus, the Dirac equation reads

$$[\gamma^\mu(\partial_\mu + iqA_\mu) + im]\psi = 0, \tag{4}$$

$$q = \frac{e}{\hbar c}, \quad m = \frac{m_0 c}{\hbar}, \tag{5}$$

where e is the negative electron's charge and A_μ are the covariant components of the electromagnetic potential vector. The probability vector current is one of the tensorial quantities of this theory. These quantities have the form:

$$\Omega_1 = \overline{\psi}\psi; \qquad \overline{\psi} = \psi^\dagger \gamma_0, \tag{6}$$

$$J^\mu = \overline{\psi}\gamma^\mu \psi, \tag{7}$$

$$S^{\mu\nu} = i\overline{\psi}\gamma^\mu\gamma^\nu\psi, \tag{8}$$

$$K^\mu = -\overline{\psi}\gamma_5\gamma^\mu\psi, \tag{9}$$

$$\Omega_2 = -i\overline{\psi}\gamma_5\psi, \tag{10}$$

where $\psi^\dagger$ is the adjoint, Ω_1 is an invariant scalar; J is the probability current, whose time component

$$J^0 = \overline{\psi}\gamma^0\psi = \sum_{i=1}^{4} |\psi_i|^2 \tag{11}$$

is the probability density. Authors who present the Dirac theory are usually very happy to get 16 tensorial densities without derivatives, because the 4×4

complex matrix algebra is 16-dimensional above the complex field. And we have 16 densities: the Ω_1, scalar, the J, vector, the S, bivector, the K, pseudo-vector, the Ω_2, pseudo-scalar, but these densities are real, not complex. More generally, it is very difficult to adjust the Dirac theory to the quantum theory, because the Dirac matrices are not Hermitian. We have $\gamma_0^\dagger = \gamma_0$, but, on the contrary, we have $\gamma_1^\dagger = -\gamma_1$. Many difficulties arise from this fact in terms of adjusting the Dirac equation to quantum principles derived from non-relativistic quantum mechanics. We shall see that these sixteen tensorial densities are not the only existing densities of the theory.

The γ matrices and the ψ wave are not uniquely defined in the Dirac equation, and this fact has led some to see the wave only as a tool for calculations, without physical reality. It is always possible to replace the γ matrices and ψ by

$$\gamma'^\mu = S\gamma^\mu S^{-1}; \quad \psi' = S\psi; \quad S^{-1} = S^\dagger, \tag{12}$$

where S is any fixed unitary matrix. The gauge transformations with $S = e^{ia I_4}$ are among the gauge transformations (12), where I_n is the unitary $n \times n$ matrix. This gauge transformation is very important, because it is both global and local . The gauge (12) can be made local, but the U(4) gauge group is too small to give an acceptable frame for local gauge invariance of the standard model.

2.   Real mathematical frames for the Dirac equation

Due to the fact that each ψ has a real and an imaginary part, the Dirac spinor is made of eight real components. Clifford algebras present two kinds of eight-dimensional algebras over $\mathbb{R}$ which can be used to obtain the Dirac equation, the Clifford algebra Cl_3 of the three-dimensional physical space, and the even subalgebra of the space-time algebra $Cl_{1,3}$.

A.   Space algebra

This Clifford algebra, isomorphic to the Pauli algebra, is generated by the eight elements 1, σ_1, σ_2, σ_3, σ_{23}, σ_{31}, σ_{12}, σ_{123}. To get the Dirac equation with this frame it is sufficient [3] to associate with each ψ of the Dirac theory the $\phi = f(\psi)$, defined by

$$f(\psi) = \phi = a_1 + a_2\sigma_{32} + a_3\sigma_{31} + a_4\sigma_{12} + a_5\sigma_{123} + a_6\sigma_1 + a_7\sigma_2 + a_8\sigma_3 \tag{13}$$

where a_j are

$$\psi_1 = a_1 + ia_4; \quad \psi_2 = -a_3 - ia_2$$
$$\psi_3 = a_8 + ia_5; \quad \psi_4 = a_6 + ia_7 \tag{14}$$

The space algebra Cl_3 is isomorphic to the Pauli algebra $M_2(\mathbb{C})$, but this isomorphism is not an isomorphism of linear space above $\mathbb{C}$, it is only an isomorphism of linear space above $\mathbb{R}$, because we get

$$f(i\psi) = \phi\sigma_{12} \tag{15}$$

but not $f(i\psi) = if(\psi)$. Therefore, only linear spaces and algebras above $\mathbb{R}$ are convenient here.

We chose the notation $\psi *$ for the complex conjugate of ψ. We also use

$$\tilde{\phi} = \phi^\dagger = a_1 - a_2\sigma_{32} - a_3\sigma_{31} - a_4\sigma_{12} - a_5\sigma_{123} + a_6\sigma_1 + a_7\sigma_2 + a_8\sigma_3 , \qquad (16)$$

$$\hat{\phi} = a_1 + a_2\sigma_{32} + a_3\sigma_{31} + a_4\sigma_{12} - a_5\sigma_{123} - a_6\sigma_1 - a_7\sigma_2 - a_8\sigma_3 , \qquad (17)$$

$$\overline{\phi} = a_1 - a_2\sigma_{32} - a_3\sigma_{31} - a_4\sigma_{12} + a_5\sigma_{123} - a_6\sigma_1 - a_7\sigma_2 - a_8\sigma_3 , \qquad (18)$$

and for each A and B we get

$$(AB)^\dagger = B^\dagger A^\dagger; \quad \overline{AB} = \overline{B}\,\overline{A}; \quad \hat{A} = \overline{A}^\dagger; \quad \hat{\overline{AB}} = \hat{A}\hat{B} . \qquad (19)$$

The f isomorphism yields for each ψ

$$f(\psi^*) = \sigma_2\hat{\phi}\sigma_2; \quad f(\gamma^\mu\psi) = \sigma_\mu\hat{\phi} . \qquad (20)$$

And we get

$$f\left(\gamma^\mu\left[\left(\partial_\mu + iqA_\mu\right) + im\right]\psi\right) = 0 .$$

This is

$$\sigma^\mu\partial_\mu\hat{\phi} + q\sigma^\mu A_\mu\hat{\phi}\sigma_{12} + m\phi\sigma_{12} = 0 \qquad (21)$$

Within space algebra we use

$$\nabla = \sigma^\mu\partial_\mu; \quad A = \sigma^\mu A_\mu , \qquad (22)$$

$$\hat{\nabla} = \partial_0 + \vec{\partial}; \quad \vec{\partial} = \sigma_1\partial_1 + \sigma_2\partial_2 + \sigma_3\partial_3 \\ \hat{A} = A^0 - \vec{A}; \quad \vec{A} = A^1\sigma_1 + A^2\sigma_2 + A^3\sigma_3 . \qquad (23)$$

And the Dirac equation in the space algebra frame is

$$\nabla\hat{\phi} + qA\hat{\phi}\sigma_{12} + m\phi\sigma_{12} = 0 \qquad (24)$$

Tensorial densities without derivatives are

$$J = \phi\phi^\dagger = \sigma^\mu J_\mu , \qquad (25)$$

$$K = \phi\sigma_3\phi^\dagger = \sigma^\mu K_\mu , \qquad (26)$$

$$R = \phi\overline{\phi} = \Omega_1 + \Omega_2\sigma_{123} , \qquad (27)$$

$$S = \phi\sigma_3\overline{\phi} = S^{23}\sigma_1 + S^{31}\sigma_2 + S^{12}\sigma_3 + S^{10}\sigma_{23} + S^{20}\sigma_{31} + S^{30}\sigma_{12} . \qquad (28)$$

If J and R are single, we immediately see that K and S, with σ_3, may be chosen as a case $K = K_{(3)}$, $S = S_{(3)}$ of

$$K_{(j)} = \phi\sigma_j\phi^\dagger; \quad S_{(j)} = \phi\sigma_j\overline{\phi} \qquad (29)$$

We therefore get the old 16 tensorial densities without derivatives, and 20 new tensorial densities without derivatives. More generally, from a spinor with 2^n real components it is possible to construct $(2^n + 1) \times 2^{n-1}$ tensorial densities without derivatives. The 16 densities of the classical theory are electric gauge invariant. Many attempts have been made to reduce the Dirac spinor to its tensorial densities [4] [5]. It is well known that we cannot know the entire wave from the only 16 gauge invariant tensorial densities, because they tell us nothing about the phase, which changes with the gauge. The gauge transformation

$$\psi' = e^{ia}\psi; \quad \phi' = \phi e^{a\sigma_{12}} \qquad (30)$$

induces a rotation between $K_{(1)}$ and $K_{(2)}$ and between $S_{(1)}$ and $S_{(2)}$ as

$$K'_{(1)} = \phi'\sigma_1\phi'^\dagger = \cos(2a)K_{(1)} - \sin(2a)K_{(2)} , \qquad (31)$$

$$K'_{(2)} = \phi'\sigma_2\phi'^\dagger = \sin(2a)K_{(1)} + \cos(2a)K_{(2)}. \tag{32}$$

Space algebra also enables us to describe the chirality of relativistic waves, and that is made, in the complex formalism, by the Weyl spinors. These spinors are defined by

$$U\psi = \begin{pmatrix} \xi \\ \eta \end{pmatrix}; \quad U = U^{-1} = \frac{1}{\sqrt{2}}(\gamma_0 + \gamma_5)$$

$$\xi = \frac{1}{\sqrt{2}}\begin{pmatrix} \psi_1 + \psi_3 \\ \psi_2 + \psi_4 \end{pmatrix}; \quad \eta = \frac{1}{\sqrt{2}}\begin{pmatrix} \psi_1 - \psi_3 \\ \psi_2 - \psi_4 \end{pmatrix}. \tag{33}$$

But we simply have

$$\phi = \sqrt{2}\left(\xi \quad \sigma_{13}\eta^*\right); \quad \hat{\phi} = \sqrt{2}\left(\eta \quad \sigma_{13}\xi^*\right). \tag{34}$$

Thus, using the matrix representation of the space algebra, ξ is simply the left column of ϕ and η is the left column of $\hat{\phi}$. From (21) it is easy to write the wave equation for ξ and η

$$\nabla\eta + iqA\eta + im\xi = 0, \tag{35}$$

$$\hat{\nabla}\xi + iq\hat{A}\xi + im\eta = 0 \tag{36}$$

Chirality in the Dirac theory is linked to the existence of two different representations of the proper Lorentz group. If M is an element of the $SL(2,\mathbb{C})$ group, and if V is a space-time vector

$$V = \sigma^\mu V_\mu \tag{37}$$

the transformation $r: V \mapsto V' = MVM^\dagger$ is a Lorentz rotation, a component of the restricted Lorentz group $\mathcal{L}^\uparrow_+$. With

$$\phi' = M\phi; \quad \nabla' = M\nabla M^\dagger; \quad A' = MAM^\dagger \tag{38}$$

we have $M^\dagger = \hat{M}^{-1}$, and $\bar{M} = M^{-1}$, and we get

$$\nabla'\hat{\phi}'\sigma_{12} = m\phi' + qA'\hat{\phi}' \tag{39}$$

We notice that the linkage between ϕ, ξ and η is invariant under the restricted Lorentz group $\mathcal{L}^\uparrow_+$, because ξ and η are transformed as

$$\xi' = M\xi; \quad \eta' = (M^\dagger)^{-1}\eta = \hat{M}\eta; \quad \eta'^* = \hat{M}^*\eta^* = \sigma_2 M\sigma_2\eta^*$$

$$\sigma_{13}\eta'^* = M\sigma_{13}\eta^* \tag{40}$$

The Ω_1 and the Ω_2 are invariant:

$$\Omega'_1 + \Omega'_2\sigma_{123} = \phi'\bar{\phi}' = M\phi\bar{\phi}\bar{M} = MM^{-1}\phi\bar{\phi} = \Omega_1 + \Omega_2\sigma_{123}, \tag{41}$$

$$S'_{(j)} = \phi'\sigma_j\bar{\phi}' = MS_{(j)}M^{-1}, \tag{42}$$

While the J and $K_{(j)}$ vectors become

$$J' = \phi'\phi'^\dagger = MJM^\dagger; \quad K'_{(j)} = \phi'\sigma_j\phi'^\dagger = MK'_{(j)}M^\dagger. \tag{43}$$

These transformations under a Lorentz rotation are sufficient to prove the tensoriality of $K_{(j)}$ and $S_{(j)}$. Therefore we must regard the complex formalism of the Dirac equation as very deficient.

B.   Space-time algebra

The Clifford algebra $Cl(1,3)$ constructed above the space-time with a signature $+ - - -$ was used by Hestenes [6], Boudet [7], Lasenby [8] to write a complete relativistic physics, and particularly the Dirac equation. It is impossible to summarize these works here. We will simply describe how we can use the matrix representations of space algebra and space-time algebra to go from one to the other. The space-time vector A reads

$$A = \begin{pmatrix} 0 & \hat{A} \\ A & 0 \end{pmatrix} = \gamma^\mu A_\mu \tag{44}$$

$$\gamma^j = \gamma^j; \quad j = 1, 2, 3; \quad \gamma^0 = \gamma_5 \tag{45}$$

The gradient $\partial = \gamma^\mu \partial_\mu$ reads

$$\partial = \begin{pmatrix} 0 & \hat{\nabla} \\ \nabla & 0 \end{pmatrix} \tag{46}$$

With each $\phi = f(\psi)$ we associate the $\Psi = g(\phi)$ defined by

$$\Psi = \begin{pmatrix} \hat{\phi} & 0 \\ 0 & \phi \end{pmatrix} \tag{47}$$

$$= a_1 + a_2 \gamma_{23} + a_3 \gamma_{13} + a_4 \gamma_{21} + a_5 \gamma_{0123} + a_6 \gamma_{10} + a_7 \gamma_{20} + a_8 \gamma_{30}$$

We notice that Ψ has value in the even subalgebra of the space-time algebra, and g is an isomorphism of the space algebra to the even subalgebra. The Dirac equation takes the Hestenes form

$$\partial \Psi \gamma_{12} = m \Psi \gamma_0 + q A \Psi \tag{48}$$

Invariance under the restricted Lorentz group reads

$$\partial \to \partial' = R \partial \tilde{R}; \quad A \to A' = R A \tilde{R}; \tag{49}$$

$$\Psi \to \Psi' = R \Psi, \quad R = \begin{pmatrix} \hat{M} & 0 \\ 0 & M \end{pmatrix}, \quad \tilde{R} = \begin{pmatrix} M^\dagger & 0 \\ 0 & \overline{M} \end{pmatrix} = R^{-1} \tag{50}$$

where $\sim$ "tilde" is reversion, defined by

$$\tilde{\gamma}_\mu = \gamma_\mu; \quad (\widetilde{AB}) = \tilde{B}\tilde{A}. \tag{51}$$

As f and g are isomorphisms, each result in one of these mathematical frames may automatically be translated into another.

But we may also use a third mathematical frame, with real matrices.

C.   The algebra of real matrices

Quantum mechanics gives great importance to hermiticity and unitarity, because quantum theory always uses a Hermitian scalar product, which is, for Dirac spinors

$$\langle \psi | \psi' \rangle = \iiint \left(\sum_{i=1}^{i=4} \psi^*_i \, \psi'_i \right) dv. \tag{52}$$

This Hermitian scalar product is associated with the norm

$$\|\psi\|^2 = \langle \psi | \psi \rangle = \iiint \left(\sum_{i=1}^{i=4} \psi *_i \psi_i \right) dv = \iiint J^0 dv. \tag{53}$$

Transposition of this norm to real algebras is very easy, because

$$J^0 = \sum_{i=1}^{i=8} a_i^2. \tag{54}$$

And we get the norm

$$\|\phi\|^2 = \|\Psi\|^2 = \iiint J^0 dv = \iiint \left(\sum_{i=1}^{i=8} a_i^2 \right) dv. \tag{55}$$

It is always possible to translate something from one formalism to another, so it is possible to transpose this Hermitian scalar product to the real Clifford algebra, by calculating the real and imaginary parts of the scalar product separately. But with a linear space above the real field a Hermitian scalar product is pure nonsense. The only scalar product naturally linked to the norm of ϕ is the Euclidean scalar product

$$\phi \cdot \phi' = \iiint \left(\sum_{i=1}^{i=8} a_i a_i' \right) dv. \tag{56}$$

But we get this scalar product very simply as

$$\phi \cdot \phi' = \iiint \Phi' \Phi' dv, \tag{57}$$

associating with each ϕ of space algebra the real matrix

$$\Phi = \begin{pmatrix} a_1 \\ a_2 \\ \vdots \\ a_8 \end{pmatrix}. \tag{58}$$

With this real matrix the Dirac equation reads

$$\left[\Gamma^\mu \left(\partial_\mu + q A_\mu P \right)_3 + m P_3 \right] \Phi = 0, \tag{59}$$

where the Γ_μ and P_j are

$$\Gamma^0 = \Gamma_0 = \begin{pmatrix} I_4 & 0 \\ 0 & -I_4 \end{pmatrix}, \qquad \Gamma^1 = -\Gamma_1 = \begin{pmatrix} 0 & \gamma_{013} \\ \gamma_{013} & 0 \end{pmatrix}, \tag{60}$$

$$\Gamma^2 = -\Gamma_2 = \begin{pmatrix} 0 & -\gamma_{03} \\ \gamma_{03} & 0 \end{pmatrix}, \qquad \Gamma^3 = -\Gamma_3 = \begin{pmatrix} 0 & -\gamma_{01} \\ \gamma_{01} & 0 \end{pmatrix}, \tag{61}$$

$$P_1 = \begin{pmatrix} \gamma_{013} & 0 \\ 0 & \gamma_{013} \end{pmatrix}, \qquad P_2 = \begin{pmatrix} -\gamma_{05} & 0 \\ 0 & \gamma_{05} \end{pmatrix}, \qquad P_3 = \begin{pmatrix} -\gamma_{135} & 0 \\ 0 & \gamma_{135} \end{pmatrix}. \tag{62}$$

These matrices yield

$$\Gamma^\mu \Gamma^\nu + \Gamma^\nu \Gamma^\mu = 2 g^{\mu\nu} I_8; \qquad \Gamma^\mu P_j = P_j \Gamma^\mu. \tag{63}$$

The P_3 matrix, with square $-I_8$, replaces the "i" of the classical formalism. For example, the gauge invariance of the wave equation now reads

$$A_\mu \to A'_\mu = A_\mu - \frac{1}{q}\partial_\mu a, \tag{64}$$

$$\Phi \to \Phi' = e^{a\Gamma_3}\Phi. \tag{65}$$

The real matrix formalism uses

$$\overline{\Phi} = \Phi'\Gamma_0. \tag{66}$$

Transposing the Dirac equation, and multiplying by Γ_0 by the right, with $P'_j = -P_j$ and $(\Gamma^\mu)'\Gamma_0 = \Gamma_0\Gamma^\mu$ we obtain the equation

$$\overline{\Phi}\left[\left(\partial_\mu - qA_\mu P_3\right)\Gamma^\mu - mP\right]_3 = 0. \tag{67}$$

Tensorial densities without derivatives are

$$\Omega_1 = \overline{\Phi}\Phi, \tag{68}$$

$$J^\mu = \overline{\Phi}\Gamma^\mu\Phi, \tag{69}$$

$$S^{\mu\nu}_{\ (j)} = \overline{\Phi}\Gamma^{\mu\nu}P_{(j)}\Phi, \tag{70}$$

$$K^{\mu\nu\rho}_{\ (j)} = \overline{\Phi}\Gamma^{\mu\nu\rho}P_{(j)}\Phi, \tag{71}$$

$$\Omega_2 = \overline{\Phi}\Gamma^{0123}\Phi. \tag{72}$$

This formalism immediately shows the similarity between $S^{\mu\nu}_{\ (3)}$, $K^{\mu\nu\rho}_{\ (3)}$ of the complex formalism and the new densities $S^{\mu\nu}_{\ (1)}$, $S^{\mu\nu}_{\ (2)}$, $K^{\mu\nu\rho}_{\ (1)}$ and $K^{\mu\nu\rho}_{\ (2)}$. But the real formalism yields much more. While with the complex formalism the algebra generated by the γ^μ and their products is the complete 4×4 matrix algebra, the matrix algebra generated by the Γ^μ and their products is also 16-dimensional, but it is an algebra above $\mathbb{R}$ and the real 8×8 matrix algebra is 64-dimensional above $\mathbb{R}$, so these two algebras are not identical. It is possible to establish that any matrix of $M_8(\mathbb{R})$ may be written in only one way under the form

$$M = M_0 + M_1P_1 + M_2P_2 + M_3P_3, \tag{73}$$

where the M_j are linear combinations of the Γ^μ and their products. The P_j commutes with the Γ^μ, while I_8, P_1, P_2, and P_3 generate an algebra isomorphic to the quaternion field.

From the existence of three matrices with square $-I_8$ which replace the indistinct "i" of quantum mechanics results that we may write, in addition to the Dirac equation, two more equations

$$\left[\Gamma^\mu\left(\partial_\mu + qA_\mu P_1\right) + mP_1\right]\Phi = 0, \tag{74}$$

$$\left[\Gamma^\mu\left(\partial_\mu + qA_\mu P\right)_2 + mP_2\right]\Phi = 0. \tag{75}$$

And so we get three similar wave equations. These equations are equivalent, since an arbitrary solution of one can be associated with a single solution of another. There is a small but very interesting difference between these three equations. In the case of the hydrogen atom, when the Dirac equation is solved, the kinetic momentum operators J^2 and J_3 are diagonalized. The third axis is always used. As early as 1934 Louis de Broglie [9] showed this shocking failure of symmetry, and tried to save the Dirac theory, indicating that with a rotation it is always possible to put the third axis into any direction. But when a rotation is made (with the Dirac equation) a multiplication is done on ψ or on ϕ with a left-acting matrix, which changes nothing on the third preferred axis. This number is the index of one P matrix, corresponding to a multiplication of ϕ by the right. After rotation, the third axis is always used again. And this third axis will always be used when the Zeeman effect is calculated: this calculation requires a magnetic field in the direction of the third axis. What is the result of a magnetic field in any other direction, not orthogonal to the plane of the trajectory? If this calculation is made, the Dirac equation does not give the experimental results.

The first axis is the preferred axis for equation (74). If we solve the equation in the case of the hydrogen atom, it will be J^2 and J_1 which become diagonalized. And to calculate the Zeeman effect, we must take the magnetic field in the direction of the first axis to obtain the experimental results. Evidently it is the same with (75), where the second axis is the preferred axis.

The P_1, P_2, P_3 matrices are matrices from right multiplication of ϕ by σ_{23}, σ_{31} and σ_{12}. They commute with the Γ^μ matrices, which are matrices from left multiplication of ϕ by σ^μ. Therefore they commute with $\Gamma^{\mu\nu}$, which are the generators of the Lorentz rotations. The index of the preferred axis is a relativistic invariant.

But in experimental physics, a very similar situation exists. We know that in addition to electrons there are muons and tauons. A muon acts exactly as an electron, but it is not an electron, does not have the same mass, is not forced by the Pauli principle into an electron cloud. Absolutely nothing explains the existence of three kinds of electrons. Nevertheless these three kinds exist, and now the study of the Z_0 boson indicates that only three kinds exist.

Well, the simplest hypothesis that we can make is to associate each generation with one of the three possible indices, with one of these three possible objects with square -1, which give a Dirac equation. Separately, these objects will act in a very similar way, as these three equations are equivalent. But together the three objects will be different, because if we make a rotation, it is impossible to put a direction into the third and the first axis at the same time. Therefore, the spin of one cannot be added to the spin of the other.

A muon decays into an electron by emitting two neutrinos: one has the muonic preferred direction, while the other carries the preferred direction of the electron family.

From our hypothesis, we cannot know if the muon's family is the first, the second or the third family. But since inversion between a first and a second axis is a spatial symmetry, we should not be astonished if chirality, the difference between left and right, plays a fundamental role.

The 64 matrices ΓP —where Γ is a product of matrices Γ^μ and where P is I_8 or one of the P_j—form a basis for $M_8(\mathbb{R})$. This basis splits into two subsets: 36 have square I_8, yield $M' = M = M^{-1}$ and give the 36 tensorial densities $\Phi'M\Phi$; 28 have square $-I_8$, yield $M' = -M = M^{-1}$, do not give tensorial densities, but generate the Lie algebra of the orthogonal group SO(8).

This orthogonal group is a global gauge group of the Dirac equation, since (59) is invariant under the transformations

$$\Phi \to \Phi' = A\Phi, \qquad A^{-1} = A', \tag{76}$$

$$\Gamma^\mu \to \Gamma'^\mu = A\Gamma^\mu A^{-1}, \tag{77}$$

$$P_3 \to P_3' = AP_3A^{-1}. \tag{78}$$

In quantum mechanics, replacing unitarity by orthogonality is a heresy, which should immediately lead to catastrophic results. For example, solving the Dirac equation in the case of the H atom, we orthonormalize the different solutions corresponding to the different possible quantum states, and we use this orthonormalization when we calculate the Zeeman effect. But there is a particular coincidence, and the orthonormalization for the Hermitian scalar product is exactly identical to the orthonormalization for the Euclidean scalar product (56) in this case [10].

3.   New relativistic wave equations

A.   Chiral wave with mass for the neutrino

One of the most unexpected discoveries of particle physics was parity violation by weak interactions, and even maximal violation of this parity. This violation led some to think that a wave of neutrinos is purely chiral and that charge conjugation reverses chirality. The charge conjugate of the left neutrino is the right antineutrino. The chirality of relativistic waves may be described with Weyl spinors, and we have seen that they transform as

$$\xi' = M\xi; \quad \eta' = \hat{M}\eta. \tag{79}$$

We obtain them in the complex formalism by considering the left part ψ_L and the right part ψ_R of a Dirac spinor ψ:

$$\psi_L = \frac{1}{2}(I_4 - \gamma_5)\psi; \qquad \psi_R = \frac{1}{2}(I_4 + \gamma_5)\psi. \tag{80}$$

Translation into real formalisms is

$$\varphi_L = \phi\frac{1}{2}(1-\sigma_3), \quad \hat{\phi}_L = \sqrt{2}(\eta \quad 0),$$

$$\varphi_R = \phi\frac{1}{2}(1+\sigma_3) = \sqrt{2}(\xi \quad 0), \tag{81}$$

$$\Psi_L = \Psi\frac{1}{2}(1-\gamma_{30}), \qquad \Psi_R = \Psi\frac{1}{2}(1+\gamma_{30}), \tag{82}$$

$$\Phi_L = \frac{1}{2}(I_8 - \Gamma^{0123}P_3)\Phi, \qquad \Psi_R = \frac{1}{2}(I_8 + \Gamma^{0123}P_3)\Phi. \tag{83}$$

The Dirac equation gives (35) and (36); the mass term links ξ with η. Consequently we have only two possibilities: either a mass term exists and the wave has a left and a right part, or the wave is purely chiral, with only a left or a right part, and the mass term must be zero.

It is easy to give a relativistic chiral wave equation, with only a left wave and a mass term [11]:

$$\nabla \hat{\phi}_L = m\phi_L \sigma_2. \tag{84}$$

This equation gives for the left column, which is the Weyl spinor η :

$$\nabla \eta = m\sigma_2 \eta *. \tag{85}$$

Conjugating, we get

$$\hat{\nabla} \phi_L = -m\hat{\phi}_L \sigma_2. \tag{86}$$

And for the second order we get

$$\Box \phi_L = \nabla \hat{\nabla} \phi_L = -m\nabla \phi_L \sigma_2 = -m \left(m\phi_L \sigma_2 \right) \sigma_2 = -m^2 \phi_L, \tag{87}$$

$$\left(\Box + m^2 \right) \phi_L = 0. \tag{88}$$

To obtain the plane wave solutions we let

$$\phi_L = \cos \left(p_\mu x^\mu \right) \phi_{1L} + \sin \left(p_\mu x^\mu \right) \phi_{2L}, \tag{89}$$

where ϕ_{1L} and ϕ_{2L} are fixed left terms. With $p = \sigma^\mu p_\mu$, the wave equation is equivalent to

$$-p\hat{\phi}_{1L} = m\phi_{2L} \sigma_2 \tag{90}$$

$$p\hat{\phi}_{2L} = m\phi_{1L} \sigma_2. \tag{91}$$

The first equation gives

$$\phi_{2L} = -\frac{p}{m} \hat{\phi}_{1L} \sigma_2 \tag{92}$$

and substituting in the second equation we obtain

$$\left(p\hat{p} - m^2 \right) \hat{\phi}_{1L} \sigma_2 = 0. \tag{93}$$

A not identically null wave results only if

$$p\hat{p} = m^2, \quad (p^0)^2 - (\vec{p})^2 = m^2, \tag{94}$$

which is the relativistic condition between mass and impulse. If this condition is found, ϕ_{1L} is anything and ϕ_{2L} is given by (92).

With space-time algebra, the wave equation (84) reads

$$\partial \Psi_L = m\Psi_L \gamma_2; \quad \Psi_L = \begin{pmatrix} \hat{\phi}_L & 0 \\ 0 & \phi_L \end{pmatrix}. \tag{95}$$

It is possible to build a Lagrangian formalism for this wave equation. But to get this Lagrangian, we must consider in the same time a left wave Ψ_L and a right wave Ψ_R. We shall use here the method explained by Lasenby in [12]. $<A>$ is the scalar part of a multivector A. The Lagrangian density is

$$\mathcal{L} = \left\langle \partial \Psi_L \gamma_1 \tilde{\Psi}_R + m\Psi_L \gamma_{12} \tilde{\Psi}_R \right\rangle. \tag{96}$$

Lagrangian equations are

$$\partial_{\tilde{\Psi}_L} \mathcal{L} = \partial \left(\partial_{\partial\tilde{\Psi}_L} \mathcal{L} \right). \tag{97}$$

$$\partial_{\Psi_R} \mathcal{L} = \partial\left(\partial_{\partial\tilde{\Psi}_R} \mathcal{L}\right), \tag{98}$$

Here we have

$$\partial_{\tilde{\Psi}_R} \mathcal{L} = \partial\Psi_L \gamma_1 + m\Psi_L \gamma_{12}, \quad \partial_{\partial\tilde{\Psi}_R} \mathcal{L} = 0, \tag{99}$$

$$\partial_{\tilde{\Psi}_L} \mathcal{L} = m\Psi_R \gamma_{21}, \quad \partial_{\partial\tilde{\Psi}_L} \mathcal{L} = \Psi_L \gamma_1 \tag{100}$$

so the Lagrangian equations give

$$\partial\Psi_L = m\Psi_L \gamma_2 \tag{101}$$

$$\partial\Psi_R = m\Psi_R \gamma_2 \tag{102}$$

We may consider Ψ_R the charge conjugate of Ψ_L. The Lagrangian formalism itself implies equality between particle mass and antiparticle mass.

With Ψ_L alone we can build only ten tensorial densities without derivatives, which are the components of

$$J = \Psi_L \gamma_0 \tilde{\Psi}_L = -\Psi_L \gamma_3 \tilde{\Psi}_L = -K \tag{103}$$

$$S_{(2)} = \Psi_L \gamma_{13} \tilde{\Psi}_L = \gamma_{0123} S_{(1)}. \tag{104}$$

The probability current is at the same time conservative and isotropic

$$J \cdot J = 0, \tag{105}$$

$$\partial \cdot J = 0 \tag{106}$$

B.   Chiral wave with mass and charge terms

When the Dirac equation is transposed to the space-time algebra, a question inevitably arises. Why does the wave has a value only in the even subalgebra of the space-time algebra? Furthermore, for a chiral wave restriction to even subalgebra renders the existence of a wave equation with a mass term and a charge term impossible, because no even term with square -1 commutes with γ_{03} and γ_2. But if we do not restrict the wave to the even subalgebra, we can have electric gauge invariance and a charge term. The wave now reads

$$\Psi_L = \Psi \frac{1}{2}(1 - \gamma_{30}), \quad \Psi_L = \Psi_1 + \Psi_2 \gamma_2, \tag{107}$$

$$\begin{aligned} \Psi_1 = a_1 + a_2\gamma_{23} + a_3\gamma_{13} + a_4\gamma_{21} \\ - a_4\gamma_{0123} + a_3\gamma_{10} + a_2\gamma_{20} - a_1\gamma_{30}, \end{aligned} \tag{108}$$

$$\begin{aligned} \Psi_2 = a_5 + a_6\gamma_{23} + a_7\gamma_{13} + a_8\gamma_{21} \\ - a_8\gamma_{0123} + a_7\gamma_{10} + a_6\gamma_{20} - a_5\gamma_{30}, \end{aligned} \tag{109}$$

The chiral wave equation with charge and mass term is

$$\partial\Psi_L + qA\Psi_L \gamma_2 = m\Psi_L \gamma_2 \tag{110}$$

Relativistic invariance of this equation is again (49)-(50). We obtain gauge invariance under the transformations

$$\Psi_L \to \Psi_L' = \Psi_L e^{a\gamma_2}; \quad A \to A' = A - \frac{1}{q}\partial a \tag{111}$$

We find a Lagrangian formalism for our equation with

$$\mathfrak{L} = \left\langle \partial \Psi_L \gamma_1 \tilde{\Psi}_R + m \Psi_L \gamma_{12} \tilde{\Psi}_R + q A \Psi_L \gamma_{12} \tilde{\Psi}_R \right\rangle, , \tag{112}$$

which, in addition to (110), gives the equation

$$\partial \Psi_R + q A \Psi_R \gamma_2 = m \Psi_R \gamma_2 \tag{113}$$

Thus, if charge conjugation is given as the exchange $L \leftrightarrow R$ or if $C = P$, we must see this equation as

$$\partial \Psi_R + (-q)(-A) \Psi_R \gamma_2 = m \Psi_R \gamma_2 \tag{114}$$

This is equivalent to hypothesizing (with Ziino [13]) that the charge conjugation, which changes the sign of each charge, must also change the sign of the resulting electromagnetic potential vector. Here also, the Lagrangian formalism itself creates equality between the proper mass of the particle and the proper mass of the antiparticle.

It is possible to solve equation (110) in the case of the hydrogen atom, and our equation gives exactly the same results as the Dirac equation. In fact, we can attach a unique solution of the Dirac equation to each solution of our equation, and *vice versa*. To see this equivalence we use in addition to (107):

$$\Psi_1 = \begin{pmatrix} \hat{\phi}_1 & 0 \\ 0 & \hat{\phi}_1 \end{pmatrix}; \quad \Psi_2 = \begin{pmatrix} \hat{\phi}_2 & 0 \\ 0 & \hat{\phi}_2 \end{pmatrix}. \tag{115}$$

The wave equation (110) is equivalent to the system

$$\nabla \hat{\phi}_1 - q A \hat{\phi}_2 = m \hat{\phi}_1 \sigma_2, \quad \hat{\nabla} \hat{\phi}_1 - q \hat{A} \hat{\phi}_2 = -m \hat{\phi}_1 \sigma_2, \tag{116}$$

$$\nabla \hat{\phi}_2 + q A \hat{\phi}_1 = m \hat{\phi}_2 \sigma_2, \quad \hat{\nabla} \hat{\phi}_2 + q \hat{A} \hat{\phi}_1 = -m \hat{\phi}_2 \sigma_2, \tag{117}$$

$\hat{\phi}_1$ and $\hat{\phi}_2$ are two left spinors, and thus have the form

$$\hat{\phi}_1 = \sqrt{2}\begin{pmatrix} \eta_1 & 0 \end{pmatrix}, \quad \hat{\phi}_2 = \sqrt{2}\begin{pmatrix} \eta_2 & 0 \end{pmatrix}, \tag{118}$$

$$\hat{\phi}_1 = \sqrt{2}\begin{pmatrix} 0 & \sigma_{13}\eta_1^* \end{pmatrix}, \quad \hat{\phi}_2 = \sqrt{2}\begin{pmatrix} 0 & \sigma_{13}\eta_2^* \end{pmatrix} \tag{119}$$

where η_1 and η_2 are Weyl spinors. The preceding system is equivalent to

$$\nabla \eta_1 - q A \eta_2 = im\sigma_{13}\eta_1^*, \quad \hat{\nabla}\sigma_{13}\eta_1^* - q \hat{A}\sigma_{13}\eta_2^* = im\eta_1, \tag{120}$$

$$\nabla \eta_2 + q A \eta_1 = im\sigma_{13}\eta_2^*, \quad \hat{\nabla}\sigma_{13}\eta_2^* + q \hat{A}\sigma_{13}\eta_1^* = im\eta_2, \tag{121}$$

If we let

$$\eta = \eta_1 + \eta_2, \quad \xi = -\sigma_{13}\eta_1^* - \sigma_{13}\eta_2^*, \tag{122}$$

the preceding system becomes

$$\nabla \eta + iqA\eta + im\xi = 0, \tag{123}$$

$$\hat{\nabla} \xi + iq\hat{A}\xi + im\eta = 0, \tag{124}$$

which is the system (36)-(37), equivalent to the Dirac equation. From this equivalence and with our equation we obtain exactly the same results as with the Dirac equation. Because the conjugation $\hat{\phi}$ changes the spinor's parity, we can associate a left spinor with any right spinor and *vice versa*. We can therefore also assume either (with the Dirac theory) that the wave is made of a right spinor and a left spinor, or (with equation (110)) that the electron wave is made of two left spinors, *i.e.*, of two neutrino waves. The electrical interaction which links these two spinors is then not of a different nature from the weak interaction linking the left wave of a neutrino to a left wave of an electron.

4.   Space-time algebra and real matrices

If Ψ is a wave with value into the full space-time algebra it reads

$$\Psi = a_1 + a_2\gamma_{23} + a_3\gamma_{13} + a_4\gamma_{21} + a_5\gamma_{0123} + a_6\gamma_{10}$$
$$+ a_7\gamma_{20} + a_8\gamma_{30} + a_9\gamma_0 + a_{10}\gamma_{023} + a_{11}\gamma_{013} \qquad (125)$$
$$+ a_{12}\gamma_{021} + a_{13}\gamma_{132} + a_{14}\gamma_1 + a_{15}\gamma_2 + a_{16}\gamma_3$$

We associate Ψ with the real single-column matrix

$$X = \begin{pmatrix} a_1 \\ a_2 \\ \vdots \\ a_{16} \end{pmatrix}. \qquad (126)$$

The scalar product and associated norm are

$$X \cdot X' = \iiint X'X'dv; \qquad \|X\|^2 = \iiint X'X dv. \qquad (127)$$

This Euclidean scalar product is invariant under the orthogonal transformations

$$X \to X' = MX, \qquad M^{-1} = M', \qquad (128)$$

where M is any orthogonal 16×16 matrix.

The linear space $M_{16}(\mathbb{R})$ formed by the matrices of any linear application from space-time algebra into itself is 256-dimensional above $\mathbb{R}$. It contains the 16 matrices L that are matrices of left multiplication $\Psi \to \gamma\Psi$ and the 16 matrices R, which are matrices of the right multiplication $\Psi \to \Psi\gamma$. We call $I_8, L_\mu, L_{\mu\nu}, L_{\mu\nu\rho}, L_{0123}$ respectively the matrices of left multiplication and $I_8, R_\mu, R_{\mu\nu}, R_{\mu\nu\rho}, R_{0123}$ the matrices of right multiplication by $1, \gamma_\mu, \gamma_{\mu\nu}, \gamma_{\mu\nu\rho}, \gamma_{0123}$. We notice that $L_{\mu\nu} = L_\mu L_\nu$ while $R_{\mu\nu} = R_\nu R_\mu$. It is possible to establish that the 256 matrices $M = LR = RL$ form a basis of $M_{16}(\mathbb{R})$ and as with $M_8(\mathbb{R})$, these 256 matrices split into two subsets: $136 = 16 \times 17/2$ yield $M^2 = I_{16}$, $M' = M$ and give the 136 tensorial densities $X'MX$. $120 = 16 \times 15/2$ yield $M^2 = -I_{16}$, $M' = -M$ and form the basis of the Lie algebra of the orthogonal group $SO(16)$. It is possible to compute the 256 matrices $M = LR$ from:

$$L_\mu = \begin{pmatrix} 0 & \Gamma_\mu \\ \Gamma_\mu & 0 \end{pmatrix}, \qquad \mu = 0,1,2,3, \qquad (129)$$

$$R_0 = \begin{pmatrix} 0 & I_8 \\ I_8 & 0 \end{pmatrix}, \qquad R_j = \begin{pmatrix} 0 & \Gamma_{0123}P_j \\ -\Gamma_{0123}P_j & 0 \end{pmatrix}, \qquad j = 1,2,3. \qquad (130)$$

The left and right parts of the wave read

$$X_L = \frac{1}{2}(I_{16} - R_{03})X; \qquad X_R = \frac{1}{2}(I_{16} + R_{03})X. \qquad (131)$$

The chiral wave equation (110) with real matrices reads

$$L^\mu\left(\partial_\mu + qA_\mu R_2\right)\overline{X}_L = mR_2 X_L. \tag{132}$$

With $\overline{X}_L = X'_L L_0$, transposing the matrices and multiplying by L_0 from the right we get

$$\overline{X}_L\left(\partial_\mu - qA_\mu R_2\right)L^\mu = -m\overline{X}_L R_2. \tag{133}$$

Multiplying (132) by $\overline{X}_L$ by the left, (133) by X_L by the right and adding we obtain the conservation of the probability vector current

$$\partial_\mu J^\mu = 0; J^\mu = \overline{X}_L L^\mu X_L. \tag{134}$$

The wave equation (132) is invariant under a gauge group comprising the transformations

$$X_L \to X'_L = MX_L; \qquad M^{-1} = M', \tag{135}$$

$$X_R \to X'_R = MX_R, \tag{136}$$

$$L^\mu \to L'^\mu = ML^\mu M^{-1}. \tag{137}$$

This orthogonal gauge group G is a subgroup of $SO(16)$ isomorphic to $SO(8)$. The 28 matrices generate the Lie algebra of G:

$R_1 p_3, \ L_0 R_1 p_3, \ L_{01} R_1 p_3, \ L_{02} R_1 p_3, \ L_{03} R_1 p_3, \ L_{123} R_1 p_3, \ R_2 p_3, \ L_0 R_2 p_3, \ L_{01} R_2 p_3, \ L_{02} R_2 p_3,$

$L_{03} R_2 p_3, \ L_{123} R_2 p_3, \ R_3 p_3, \ L_0 R_3 p_3, \ L_{01} R_3 p_3, \ L_{02} R_3 p_3, \ L_{03} R_3 p_3, \ L_{123} R_3 p_3, \ L_1 p_3, \ L_2 p_3,$

$L_3 p_3, \ L_{12} p_3, \ L_{23} p_3, \ L_{31} p_3, \ L_{012} p_3, \ L_{023} p_3, \ L_{031} p_3, \ L_{0123} p_3,$

where

$$p_3 = \frac{1}{2}\left(I_{16} - R_{03}\right); \qquad q_3 = \frac{1}{2}\left(I_{16} + R_{03}\right). \tag{138}$$

If $N = Mp_3$ is one of the 28 generators, we get

$$e^{aN} = I_{16} + \sum_{i=1}^{\infty}\frac{(aMp_3)^n}{n!} = p_3 + q_3 + \left(\sum_{i=1}^{\infty}\frac{a^n M^n}{n!}\right)p_3 = q_3 + e^{aM}p_3, \tag{139}$$

$$p_3 X_R = 0, \qquad q_3 X_R = X_R, \qquad q_3 X_L = 0, \qquad p_3 X_L = X_L, \tag{140}$$

$$e^{aN} X_R = X_R, \qquad e^{aN} X_L = e^{aM} X_L. \tag{141}$$

So G leaves the right part of the wave invariant and acts only on the left part, which remains a left part. We can define an isomorphic group, G, by replacing p_3 by q_3, which leaves invariant the left part of the wave and acts only on the right part, transformed into a right part.

The electric gauge invariance (111) is one of the preceding gauge invariances, with $R_2 p_3$ as generator. It is one of the 3 generators $\{R_1 p_3, \ R_2 p_3, \ R_3 p_3\}$, which generate an algebra isomorphic to the Lie algebra of $SU(2)$. We also notice that $\{R_1 p_3, \ R_2 p_3, \ R_3 p_3, \ L_{0123} p_3\}$ generate an algebra isomorphic to the Lie algebra of $U(1) \times SU(2)$.

Finally we notice that the G' group is isomorphic to $SO(8)$. But the adjoint representation of $SU(3)$ is 8-dimensional, and consequently goes into $SO(8)$. It is therefore easy to find all the parts of the Lie algebra of the $U(1) \times SU(2) \times SU(3)$ group coming from the standard model into the algebra of our orthogonal matrices. But here, contrary to the standard model where the group was built up progressively from experimental results, the structure of the wave itself gives the invariance group. And we obtain isomorphic groups if we replace p_3 by p_1 or by p_2 everywhere. The need to treat the three generations separately comes from the fact that, if the three gauge groups are isomorphic, they are not identical, and do not have the same generators.

Previously we supposed that the signature of space-time is $+ - - -$. It is also possible to use a space-time with signature $- + + +$. We get then the Clifford algebra $Cl_{3,1}$, which is generated by e_0, e_1, e_2, e_3 with $e_\mu e_\nu = -e_\nu e_\mu$ and $e_0{}^2 = -1$, $e_1{}^2 = e_2{}^2 = e_3{}^2 = 1$. Then we write Ψ as

$$\Psi = a_1 + a_2 e_{32} + a_3 e_{31} + a_4 e_{12} + a_5 e_{0123} + a_6 e_{01} + a_7 e_{02} + a_8 e_{03} + a_9 e_0$$
$$+ a_{10} e_{032} + a_{11} e_{031} + a_{12} e_{012} + a_{13} e_{123} + a_{14} e_1 + a_{15} e_2 + a_{16} e_3$$

(142)

And with Ψ we associate the matrix

$$X = \begin{pmatrix} a_1 \\ a_2 \\ \vdots \\ a_{16} \end{pmatrix}.$$

(143)

If L'_μ is the matrix of left multiplication $\Psi \to e_\mu \Psi$ and R'_μ the matrix of right multiplication $\Psi \to \Psi e_\mu$, we obtain

$$L'_\mu = L_\mu S; \qquad R'_\mu = S R_\mu; \qquad S = -L_{0123} R_{0123} = \begin{pmatrix} I_8 & 0 \\ 0 & -I_8 \end{pmatrix}.$$

(144)

Consequently the set of 256 matrices $M' = L'R'$ is identical to the set of 256 matrices $M = LR$. We therefore obtain the same results.

Using this $Cl_{3,1}$ mathematical frame we have studied [14] the wave equation

$$\partial \Psi = m\Psi + qA\Psi e_{0123}.$$

(145)

This wave equation cannot be equivalent to the Dirac equation, because the wave is made up of 16 real components, not just 8. Nevertheless this equation yields results close to the Dirac theory. For example, we obtain the same energy levels in the case of the hydrogen atom.

Using the real matrices, it is possible to associate a wave equation written in $Cl_{1,3}$ with each wave equation written in $Cl_{3,1}$, and *vice versa*. For example, with the real matrices, (145) reads

$$L'^\mu \partial_\mu X = mX + qL'^\mu A_\mu R'_{0123} X,$$

(146)

which is equivalent to

$$L^\mu \partial_\mu X = -mSX + qL^\mu A_\mu R_{0123} X.$$

(147)

And this equation is the matrix translation of the wave equation

$$\partial \Psi + m\gamma_{0123} \Psi \gamma_{0123} + qA\Psi \gamma_{0123} = 0.$$

(148)

Now if we solve this equation in the case of the H atom, we will obtain the same results as with (145). We find the same energy levels as with the Dirac equation, but for each quantum state we obtain a set of solutions presenting an internal $SO(4)$ symmetry [14].

Concluding remarks

Many experimental facts of particle physics can be understood from a study of the different kinds of relativistic waves. The existence of three generations of particles comes from the dimension of the physical space, which gives three kinds of chiral projectors and three equivalent wave equations by permutation of the index of right multiplication. The three different generations must be treated separately in weak interactions, because the gauge groups of the three generations are isomorphic but not identical: they do not have the same generators.

We know that each generation has two leptons, one charged one and the other neutral. The neutral lepton, called the neutrino, has no electric charge. Leptons are insensitive to strong interactions. With weak interactions, parity is maximally violated. We can easily find this result if we suppose that the differences between these objects is due to different kinds of relativistic wave. A neutrino has a wave with value in the even subalgebra, and only left. For the charged lepton we have a left wave, while quarks have value in the full space-time algebra. The neutrino's wave allows only a mass term, but there is no possible charge. With a left wave, which is not restricted to the even subalgebra, we obtain a wave equation for the charged lepton, and this minimal coupling may be extended to the weak interactions. But this wave produces insensitivity to strong interactions, which acts only on the missing part of the wave. Only a wave with value in the full space-time algebra shows the full algebra of the $U(1) \times SU(2) \times SU(3)$ Lie algebra from the standard gauge group. This algebra is a subalgebra of the Lie algebra of the $SO(16)$ gauge group, and arises naturally from the Euclidean scalar product, which is the translation of the Hermitian scalar product of quantum mechanics into Clifford algebra.

For leptons with or without charge, we may identify the P and C symmetries or CP with identity: the wave of the particle is purely left and the wave of the antiparticle is purely right. The gauge group of strong interactions acts only on the missing part of the lepton wave, leading to total leptons insensitivity to strong interactions.

The use, in quantum mechanics and quantum field theories, of a unique and indeterminate "i," and complex linear spaces makes the existence of three and only three generations incomprehensible, while three generations are completely natural if we use real Clifford algebras seriously.

Computers are used intensively in physics today, but computer science has yet to be fully integrated into physicists' brains. Computer science has two main parts, algorithmic and data structures: how to act, what to act on. Quantum theory tells us how to calculate, and (with the Fock space) assumes that the question "what we calculate on" must not be asked. Nevertheless, this question is important, because gauge groups coming from a wave with value in a real Clifford algebra are not the same as when the wave has value in a complex linear space. As with computer science, progress in physics will come if we also ask questions about data structures and properties of objects used in calculations.

References

[1] L. de Broglie: *Recherches sur la théorie des quantas*, thèse, Paris 1924, new edition: *Ann. Fond. Louis de Broglie*, **17** No1 1992.

[2] P. A. M. Dirac: *Proc. Roy. Soc.* (London) **117**, 610 (1928)

[3] C. Daviau: Dirac equation in the Clifford algebra of space, in: *Clifford Algebras and their Application in Mathematical Physics*, Aachen 1996, Kluwer, Dordrecht, C. Daviau: Sur l'équation de Dirac dans l'algèbre de Pauli, *Ann. Fond. Louis de Broglie*, **22** No1 1997. C. Daviau: Sur les tenseurs de la théorie de Dirac en algèbre d'espace, *Ann. Fond. Louis de Broglie*, **23** No 1 1998 C. Daviau: Application a la théorie de la lumière de Louis de Broglie d'une réécriture de l'équation de Dirac, *Ann. Fond. Louis de Broglie*, **23** No 3 - 4, 1998 C. Daviau: Equations de Dirac et fermions fondamentaux, première partie: *Ann. Fond. Louis de Broglie*, **24** No 1 - 4, 1999; deuxième partie: **25** No 1, 2000. C. Daviau: Vers une mécanique quantique sans nombre complexe, *Ann. Fond. Louis de Broglie*, **26** No 1-3 2001.

[4] T. Takabayasi: Relativistic Hydrodynamics of the Dirac Matter, *Prog. Theor. Phys.* Suppl. **4**, 1957.

[5] P. Lounesto: Clifford Algebras and Hestenes Spinors, *Foundations of Physics*, Vol **23**, No. 9, 1993.

[6] D. Hestenes: *Space-Time Algebra* (Gordon & Breach, New York 1966, 1987, and 1992). D. Hestenes: Real Spinor Fields. *J. Math. Phys.*, **8** No 4 1967 D. Hestenes: Local observables in the Dirac theory. J. Math. Phys, **14** No 7 1973 D. Hestenes: Proper particle mechanics. *J. Math. Phys.*, **15** No 10 1974 D. Hestenes: Proper dynamics of a rigid point particle. *J. Math. Phys.*, **15** No 10 1974 D. Hestenes: Observables, operators, and complex numbers in the Dirac theory. *J. Math. Phys.*, **16** No 3 1975 D. Hestenes: A unified language for Mathematics and Physics in: *Clifford algebras and their applications in Mathematics and Physics*. JSR Chisholm & AK Common Eds, (Reidel, Dordrecht, 1986)

[7] R. Boudet: La géométrie des particules du groupe SU (2) et l'algèbre réelle d'espace-temps. *Ann. Fond. Louis de Broglie*, **13** No 1 1988. R. Boudet: The Takabayasi moving Frame, from A Potential to the Z Boson, in *The Present Status of the Quantum Theory of the Light*, S. Jeffers and J.P. Vigier eds., (Kluwer Dordrecht 1995) R. Boudet: The Glashow-Salam-Weinberg Electroweak Theory in the Real Algebra of Spacetime, *Advances in Applied Clifford Algebras* **7** (S) 321-336, 1997

[8] A. Lasenby, C. Doran, and S. Gull: Gravity, *Clifford (Geometric) Algebras*, W. E. Baylis, Editor, (Birkhauser, Boston 1996)

[9] Louis de Broglie: *L'électron magnétique*, Hermann, Paris 1934 page 138.

[10] C. Daviau: Solutions of the Dirac equation and of a nonlinear Dirac equation for the Hydrogen Atom, *Int. Conference on the Theory of the Electron*, Mexico 1995

[11] C. Daviau: Chiral Dirac Equation, in *Clifford Algebras, Applications to Mathematics, Physics and Engineering*, Rafal Ablamowicz Editor, Birkhäuser Boston 2004, p 431-450

[12] A. Lasenby, C. Doran, and S. Gull: A Multivector Derivative Approach to Lagrangian Field Theory. *Found. of Phys.* **23** No 10, 1993

[13] G. Ziino, Massive chiral fermions: a natural account of chiral phenomenology in the framework of Dirac's fermion Theory, *Ann. Fond. L. de Broglie*, **14** No 4, 1989 G. Ziino, On the true meaning of "maximal parity violation": ordinary mirror symmetry regained from "CP symmetry" *Ann. Fond. L. de Broglie*, **16** No 3, 1991

What is the Electron?

H. Sallhofer
Bahnhofstrasse 36, A-5280
Braunau Austria

1.   The hydrogen atom in scalar form

In my youth I often had talks with Erwin Schrödinger (my teacher at that time)
about methods for deriving his famous equation. I tried to simplify the complicated derivation of the time-dependent Schrödinger equation, and showed him
that one only had to insert the Hamilton analogy,

$$N = \frac{c\sqrt{2m(U - \Phi)}}{U},\tag{1}$$

into the classical equation for light

$$\left(\Delta - \frac{N^2}{c^2}\frac{\partial^2}{\partial t^2}\right)\Psi = 0.\tag{2}$$

Then, assuming harmonic solutions, one would obtain the time-dependent
Schrödinger equation

$$\left[\Delta - 2m\left(\frac{\Phi}{\hbar^2} - \frac{i}{\hbar}\frac{\partial}{\partial t}\right)\right]\Psi = 0.\tag{3}$$

Thus, the classical light equation contains Schrödinger's wave mechanics.
On the basis of its derivation, equation (1) is simply an equation for light refraction. If light refraction is introduced into a classical light equation, the solutions necessarily describe the light fields. Therefore, it would be most obvious
to treat harmonic solutions of (2) and (3) as the light fields. In this case the Copenhagen interpretation of (3) can be dropped. The harmonic solutions of (3)
for the Coulomb potential Φ produce the hydrogen spectrum. Since these solutions describe standing waves, they substantiate the interpretation:

Matter is standing light.

Let us visualize this.

Any wave train of light has its energy centre. Let us imagine two wave
trains of light of "equal weight," *e.g.*, photons, which interact with each other.
They may, for instance, orbit each other in such a way that each one is reflected by the field of the other. Their energy centres form a Kepler system that
generates the hydrogen spectrum. This idea thus describes, in general, the light
model of the hydrogen atom. It does not need an electron.

Using Ehrenfest's centre-of-gravity theorem in this conception, we see that the energy centres of the light-hydrogen atom move according to Newtonian mechanics [1].

Until his death, Einstein, and with him Lorentz, von Laue, de Broglie, Dirac, Landé, Hartmann, and others, were of the opinion that Copenhagen interpretation of quantum mechanics is not complete. From that time until recently [2] only Schrödinger changed his mind. In a commentary to a commemorative publication celebrating his seventieth birthday in 1949, Einstein wrote about a quantum mechanics that he could accept. He pointed out that the established part of conventional quantum theory would be found again in the desired new theory, and that the new theory most probably had to be of statistical nature. Then he stated,

> Statistical quantum theory—in case such efforts were successful—would have a status within the framework of classical mechanics. I am rather firmly convinced that the development of theoretical physics will be such, but the way will be lengthy and difficult.

The handicap of Einstein induced some colleagues to look for the "missing link" between statistical mechanics and Schrödinger's wave mechanics. One searcher was the U. Hoyer of Munster, who 20 years ago in his book [3] *Wellenmechanik auf Statistischer Grundlage* (wave mechanics on a statistical basis), put forward a precise derivation of Schrödinger's wave mechanics *via* statistical mechanics. On the particle path, Hoyer goes directly from Boltzmann to Schrödinger, from Vienna to Vienna.

By inserting his statistical theory into the gaping void between Boltzmann and Schrödinger, Hoyer relieves present theoretical physics from the misery of paradoxes and other nuances of the Copenhagen interpretation, which include:

1.1 The indeterminism of microphysics caused by the suspension of the law of causality within its realm.

1.2 The requirement from 1.1 of an extension of classical logic toward one in which the theorem of the excluded third no longer holds.

1.3 The complementarity of contradicting basic conceptions.

The opportunity for a general amelioration has not induced our physics community (thus far) to take advantage of the potential for reform offered by Hoyer. The physicist and philosopher Hoyer remains a voice crying in the wilderness.

2.   The hydrogen atom in vector form

At first, Schrödinger found my proposal for an abridged derivation of his wave mechanics, mentioned in section 1, "interesting." Later on, he retracted de Broglie's position by pointing out that with my derivation the Schrödinger function may be interpreted as light. The latter showed [4] that the "light" interpretation is not complete. "If you know better," Schrödinger meant, "you have to support your ideas vectorially."

Therefore, I tried at first, on an analogy to the steps in section 1, to derive the Dirac equation from Maxwell's equations. That took quite some time. The breakthrough did not come until the end of the seventies [5]. But at last I was able to write down the connection between Maxwell's equation and Dirac's theory with breathtaking brevity [6]

$$\vec{\sigma} \cdot \vec{E} = D.$$

($\vec{E}$: Maxwell's electrodynamics, $\vec{\sigma}$: Pauli vector,   (4)

    D: Dirac equation)

The simplicity of (4) is not accidental. The far-reaching consequences of this relation inspired hope that the Copenhagen interpretation would finally be brought down. This equation, which is known today as the "Maxwell-Dirac isomorphism," constitutes a new relation in natural science. To me it appears both alarming and binding, as well as basic and absolutely necessary for quantum physics and philosophy.

To be succinct, we may say the following. On the assumption of harmonic solutions, source-free electrodynamics may always be put into the amplitude representation

$$\left[\vec{\gamma} \cdot \nabla + i \frac{\omega}{c} \begin{pmatrix} \varepsilon 1 & 0 \\ 0 & \mu 1 \end{pmatrix} \right] \psi^{el} = 0. \tag{5}$$

If we now compare (5) with Dirac's amplitude equation,

$$\left[\vec{\gamma} \cdot \nabla + i \frac{\omega}{c} \begin{pmatrix} \left(1 - \dfrac{\Phi - m_0 c^2}{\hbar \omega} \right) & 0 \\ 0 & \left(1 + \dfrac{\Phi + m_0 c^2}{\hbar \omega} \right) \end{pmatrix} \right] \psi^{D} = 0, \tag{6}$$

we see the following. Just as Schrödinger's theory (3) is contained in the classical light equation (2), Dirac's theory (6) is contained in electrodynamics (5).

Further, some reasons can be suggested why possible radical changes in the foundations of the theory are sometimes accepted nowadays by contemporary physicists, even leading physicists. Among other things, the cause may be that scientists busy at cyclotron physics are not listened to as much as before. This occurred because the Superconducting Super Collider beacon in Waxahachie, Texas was shelved. Nevertheless, scientists sometimes take the view that it is not their job to take notice of changes in the foundations.

I find myself today in a similar position to Hoyer's. Whereas I took the wave path from Maxwell to Dirac, and thereby eliminated the Copenhagen interpretation, Hoyer took the particle route (almost at the same time) from Boltzmann to Schrödinger, thus eliminating the Danish interpretation for a second time. How often does it have to be eliminated?

3.   The electron

3.1   In the Standard Model the electron figures as a basic entity among the leptons. In the neighbouring section, the quarks, the electron appears to be broken. What a new super-paradox!

3.2   From the Maxwell-Dirac isomorphism it becomes evident that the standing electron must have a field equation in electrodynamics.

3.3   The spin of the electron would wind the electron field, normally considered static, around the spin axis.

3.4   The electron will lead to an epiphany for physics. Even though still frequently used in planetary technology, it has remained the most enigmatic.

References

1. H. Sallhofer, *Sackgasse of Quantenphysik*, p. 36, Universitas, Munchen 2000

2. H. Sallhofer, *Der physikerstreit*, p. 189, Universitas, Munchen 2001

3. U. Hoyer, *Wellenmechanik auf Statistischer Grundlage*, p. 11—68, Schmidt & Klaunig, Kiel 1983

4. L. de Broglie, "Une nouvelle conception de la lumière" in *Actualities scientifiques et industrielles*, No. 181, Paris 1934

5. H. Sallhofer, "Elementare Herleitung der Dirac Gleichung 1," *Z. Naturforsch.* **33a**, 1378 (1978)

6. H. Sallhofer, "Elementary Derivation of the Dirac Equation X," *Z. Naturforsch.* **41a**, 468 (1986); or shorter: H. Sallhofer, *Sackgasse of Quantenphysik*, p. 71, Universitas, Munchen 2000.

The Electron as a System of Classical Electromagnetic and Scalar Fields

Volodimir Simulik
Institute of Electron Physics
Ukrainian National Academy of Sciences
21 Universitetska Str., 88000 Uzhgorod, Ukraine
e-mail: sim@iep.uzhgorod.ua

The electron is described as a system of classical electromagnetic and scalar fields, *i.e.*, a coupled system of two bosons (photon and massless boson with spin $s = 0$). A mathematical proof of electron structure is given. The main properties of the electron are explained without reference to quantum mechanics or quantum field theory. The slightly generalized classical Maxwell equations are proved to be the equations for the electron.

1. Introduction

The 19th century conception that matter is electromagnetic in nature is extended here to the idea of more a complete bosonic nature of matter. This means that all bosonic fields (not only photonic or electromagnetic) are treated as sources of fermionic fields, which, therefore, are the consequence of bosonic fields.

The first elementary particle—the electron—is dealt with here in order to demonstrate the possibilities of this idea. It must be stressed that the electron is not an elementary object, but has a structure. The electron is presented as a system of classical electromagnetic and scalar fields, a standing electromagnetic-scalar wave in its stationary states. In other words, the electron is treated as a coupled system of two bosons (photon and massless boson with spin $s = 0$).

The proof of this assertion is presented below on the basis of the following four arguments: the Maxwell-like equation for the electron, its unitary relationship with the Dirac theory, the symmetry principle, derivation of atomic spectra from the new equation.

Our non-quantum-mechanical model of the microworld is a model of the atom based on slightly generalized Maxwell's equations, *i.e.,* in the framework of a moderately extended classical microscopic electrodynamics of media. This model is free of probability interpretation, and can explain many inner-atomic phenomena by means of classical physics. Despite the fact that we construct a classical model, in building it we use essentially an analogy with the Dirac equation and results achieved on the basis of this equation. It should also be noted that electrodynamics is considered here in terms of field strengths (without any reference to vector potentials as the initial variables of the theory).

The first step is to define the unitary relationship (and broad analogy) between the Dirac equation and slightly generalized Maxwell equations [1,2].

The symmetry principle is the second step. On the basis of this principle we introduce in [3,4] the most symmetrical form of generalized Maxwell equations which can now describe both bosons and fermions, because they have [3,4] both spin 1 and spin ½ symmetries. Moreover, these equations are unitary connected with the Dirac equation.

In the third step we refer to Sallhofer, who suggested [5-7] the possibility of introducing interaction with an external field as interaction with specific media in the Maxwell theory (a new way of introducing interaction into the field equations). Nevertheless, our atom (and electron) model [1,2] is essentially different from Sallhofer's. We have used another, *unitary relationship*, with the Dirac theory. On the basis of these three main ideas we are able to construct an electrodynamic model of the atom and atomic electron.

Interest in the problem of the relationship between the Dirac and Maxwell equations emerged immediately after the creation of quantum mechanics [8-18]. However, the authors of these papers considered the simplest example of a free, massless Dirac equation. Interest has grown in recent years thanks to new results [5-7], with investigation of the physically meaningful case (the mass $m_0 \neq 0$, interaction potential $\Phi_0 \neq 0$), and our own research [1-4].

In another approach [19-26], the quadratic relations between the fermionic and bosonic amplitudes were found and used. In our papers [1-4, 27-36], and here we discuss linear relations between the fermionic and bosonic amplitudes.

We have found a relationship between the symmetry properties of the Dirac and Maxwell equations [27-32], the complete set of 8 transformations linking these equations, a relationship between the conservation laws for electromagnetic and spinor fields, a relationship between the Lagrangians for these fields and two possibilities for quantization. We have also laid the foundations for a classical electrodynamic model of the atom. In recent publications [33-36] we add a physical interpretation to these mathematical results [1-4, 27-32]. Here I present a review of our results together with new interpretations.

2.   The Maxwell-like equations for the electron

In the history of theoretical physics the electron appeared within classical electrodynamics as the particle with minimum elementary electric charge. Yet there was no place for the electron in the framework of the classical electrodynamics of the atom; the difficulties of Rutherford's purely electrodynamic model of the atom are well known, and the properties of the electron could not be described in the framework of classical electrodynamics. A quest for another theory seemed necessary. Quantum mechanics and quantum electrodynamics were much more successful in the microregion. Strangely though, the theory (classical electrodynamics) could not explain the physical object to which it gave birth. We must therefore answer the questions: "Why can classical electrodynamics not describe its principal fundamental object—the electron? What could be done better in classical theory in order to bring it into line with the experimental facts of atomic and electron physics?"

The aim of our investigations [1-4, 33-36], and our starting point, is as follows: "The electron as the fundamental object of classical electrodynamics must be described in the framework of this theory without recourse to quantum mechanics or quantum electrodynamics. The solution should exist within classical electrodynamics." The result of our investigations has opened up this possibility. However, the equations we used were *slightly generalized Maxwell equations*. The inneratomic electrodynamics of the microworld appears to require more general Maxwell-like equations than the ordinary Maxwell equations of the macroworld. Further, *our electron is a wave object*, not a particle.

The slightly generalized classical Maxwell equations [1,2] are considered in a specific medium that models the relativistic atom. It is easily seen that they differ from the standard Maxwell electrodynamics by the presence of magnetic sources (in one interpretation), or scalar fields (in another interpretation).

The slightly generalized Maxwell equations in a medium representing a system of coupled electromagnetic $(\vec{E}, \vec{H})$ and scalar fields (E^0, H^0)

$$\text{curl}\vec{H} - \varepsilon \partial_0 \vec{E} = \text{grad}E^0, \quad \text{curl}\vec{E} + \mu \partial_0 \vec{H} = -\text{grad}H^0,$$

$$\text{div}\vec{E} = -\mu \partial_0 E^0, \quad \text{div}\vec{H} = -\varepsilon \partial_0 H^0, \quad \partial_0 \equiv \frac{\partial}{\partial x^0}, \tag{1}$$

are postulated. We emphasize that equations (1) are not proposed *ad hoc*. In the stationary case these Maxwell equations are unitarily connected [1, 2, 33, 34] with the Dirac equation for a massive particle in an external field $\Phi(\vec{x})$ if the electric ε and magnetic μ permeabilities are of the form [5-7]

$$\varepsilon(\vec{x}) = 1 - \frac{\Phi(\vec{x}) + m_0}{\omega}, \quad \mu(\vec{x}) = 1 - \frac{\Phi(\vec{x}) + m_0}{\omega}, \tag{2}$$

i.e., permeabilities are defined by the parameters m_0, ω, and the given function $\Phi(\vec{x})$. (Below we will demonstrate this relation in detail, and step by step, all the reasons for our choice of the form (1) will be explained.) Here the system of units $\hbar = c = 1$ is used, and transition to the standard system is fulfilled by the substitution $\omega \to \hbar\omega$, $m_0 \to m_0 c^2$, etc.

Due to the time independence of ε and μ, equations (1) may be rewritten in the equivalent form

$$\text{curl}\vec{H} - \partial_0 \varepsilon \vec{E} = \vec{j}_{el}, \quad \text{curl}\vec{E} + \partial_0 \mu \vec{H} = -\vec{j}_{mag},$$

$$\text{div}\varepsilon\vec{E} = \rho_{el}, \quad \text{div}\mu\vec{H} = \rho_{mag}, \tag{3}$$

where electric and magnetic current and charge densities have the form

$$\vec{j}_{el} = \text{grad}E^0, \quad \vec{j}_{mag} = \text{grad}H^0,$$

$$\rho_{el} = -\varepsilon\mu\partial_0 E^0 + \vec{E}\text{grad}\varepsilon, \quad \rho_{mag} = -\varepsilon\mu\partial_0 H^0 + \vec{H}\text{grad}\mu. \tag{4}$$

Due to the presence in equations $(1) = (3)$ of both electric and magnetic sources, we called them "slightly generalized Maxwell equations."

For a compact description of the system $(\vec{E},\vec{H},E^0,H^0)$ of electromagnetic and scalar fields, it is useful to introduce the following complex vector and tensor functions:

$$\mathcal{E} \equiv (\mathcal{E}^\mu) = \begin{vmatrix} \mathcal{E}^j \\ \mathcal{E}^0 \end{vmatrix} = \text{column} \left| E^1 - iH^1, E^2 - iH^2, E^3 - iH^3, E^0 - iH^0 \right|, \tag{5}$$

$$\mathcal{E} \equiv (\mathcal{E}^{\mu\nu}): \quad \mathcal{E}^{0j} = \mathcal{E}^{j0} = \mathcal{E}^j, \quad \mathcal{E}^{mn} = i\varepsilon^{mnj}\mathcal{E}^j, \quad \mu = 0,1,2,3, \; j = 1,2,3, \tag{6}$$

where $\vec{\mathcal{E}} = (\mathcal{E}^j) = (E^j - iH^j) = \vec{E} - i\vec{H}$ is the well-known form for the electromagnetic field used by Majorana as early as 1930 (see, e.g., [9]), and $\mathcal{E}^0 = E^0 - iH^0$ is a complex scalar field.

To illustrate the essence of our generalization of the Maxwell equations and the group-theoretical foundations of a description of fermions in terms of bosons, we consider the simplest version of equations (1), i.e., the case $\varepsilon = \mu = 1$ (no medium):

$$\partial_0 \vec{E} = \text{curl}\vec{H} - \text{grad}E^0, \quad \partial_0 \vec{H} = -\text{curl}\vec{E} - \text{grad}H^0,$$
$$\text{div}\vec{E} = -\partial_0 E^0, \quad \text{div}\vec{H} = -\partial_0 H^0. \tag{7}$$

In terms of functions (5), (6) equations (7) can be rewritten in the following equivalent forms:

$$\partial_0 \vec{\mathcal{E}} = i\text{curl}\vec{\mathcal{E}} - \text{grad}\mathcal{E}^0, \quad \text{div}\vec{\mathcal{E}} = -\partial_0 \mathcal{E}^0, \tag{8}$$

$$\partial_\mu \mathcal{E}_\nu - \partial_\nu \mathcal{E}_\mu + i\varepsilon_{\mu\nu\rho\sigma}\partial^\rho \mathcal{E}^\sigma = 0, \quad \partial_\mu \mathcal{E}^\mu = 0, \tag{9}$$

$$\partial_\nu \mathcal{E}^{\mu\nu} = -\partial^\mu \mathcal{E}^0, \quad \left(j_\mu = -\partial_\mu \mathcal{E}^0 \right), \tag{10}$$

$$\left(i\partial_0 + \vec{s}\cdot\vec{p} \right)\vec{\mathcal{E}} + i\text{grad}\mathcal{E}^0 = 0, \quad \partial_\mu \mathcal{E}^\mu = 0, \tag{11}$$

$$\tilde{\gamma}^\mu \partial_\mu \mathcal{E}(x) = 0, \tag{12}$$

where $\vec{s} \equiv (s^j)$ are the Hermitian generators of irreducible representation D(1) of the group SU(2), $\vec{p} = (p^j) = -i\partial_j$, matrices $\tilde{\gamma}$ contain the operator C of complex conjugation, $C\mathcal{E} = \mathcal{E}*$:

$$\tilde{\gamma}^0 = \begin{vmatrix} 1 & 0 & 0 & 0 \\ 0 & 1 & 0 & 0 \\ 0 & 0 & 1 & 0 \\ 0 & 0 & 0 & -1 \end{vmatrix}C, \quad \tilde{\gamma}^1 = \begin{vmatrix} 0 & 0 & 0 & 1 \\ 0 & 0 & -i & 0 \\ 0 & i & 0 & 0 \\ -1 & 0 & 0 & 0 \end{vmatrix}C,$$

$$\tilde{\gamma}^2 = \begin{vmatrix} 0 & 0 & i & 0 \\ 0 & 0 & 0 & 1 \\ -i & 0 & 0 & 0 \\ 0 & -1 & 0 & 0 \end{vmatrix}C, \quad \tilde{\gamma}^3 = \begin{vmatrix} 0 & -i & 0 & 0 \\ i & 0 & 0 & 0 \\ 0 & 0 & 0 & 1 \\ 0 & 0 & -1 & 0 \end{vmatrix}C, \tag{13}$$

and satisfy the relations of the Clifford-Dirac algebra: $\tilde{\gamma}^\mu \tilde{\gamma}^\nu + \tilde{\gamma}^\nu \tilde{\gamma}^\mu = 2g^{\mu\nu}$.

The general solution of equations (8) = (9) = (10) = (11) = (12) was found [31,32] in the manifold $(S(\mathbb{R}^4)\otimes C^4)*$ of Schwartz's generalized functions directly by the Fourier method. In terms of helicity amplitudes $c^\mu(\vec{k})$ the solution has the form

$$\mathcal{E}(x) = \int d^3k \sqrt{\frac{2\omega}{(2\pi)^3}} \left\{ \left[c^1 \mathbf{e}_1 + c^3 \left(\mathbf{e}_3 + \mathbf{e}_4 \right) \right] e^{-ikx} + \left[c *^2 \mathbf{e}_1 + c *^4 \left(\mathbf{e}_3 + \mathbf{e}_4 \right) \right] e^{ikx} \right\},$$
(14)

$$\omega \equiv \sqrt{\vec{k}^2},$$

where 4-component basis vectors $\mathbf{e}_\alpha$ are taken in the form

$$\mathbf{e}_j = \text{column } (\vec{\mathbf{e}}_j, 0), \qquad \mathbf{e}_4 = \text{column } (0,1).$$
(15)

Here the 3-component basis vectors $\vec{\mathbf{e}}_j$ are the eigenvectors of the quantum-mechanical helicity operator for the spin $s = 1$,

$$\left(\hat{h} \equiv \frac{\vec{s} \cdot \vec{k}}{\omega} \right) \vec{\mathbf{e}}_j = \lambda_j \vec{\mathbf{e}}_j \quad \text{with} \quad \lambda_j = \mp 1, 0 \quad \text{for} \quad j = 1,2,3.$$
(16)

Without loss of generality these vectors can be taken as

$$\vec{\mathbf{e}}_1 = \frac{1}{\omega \sqrt{2 \left(k^1 k^1 + k^2 k^2 \right)}} \begin{vmatrix} \omega k^2 - k^1 k^3 \\ -\omega k^1 - ik^2 k^3 \\ i(k^1 k^1 + k^2 k^2) \end{vmatrix}, \quad \vec{\mathbf{e}}_2 = \vec{\mathbf{e}}_1 *, \quad \vec{\mathbf{e}}_3 = \frac{\vec{k}}{\omega}.$$
(17)

It should be noted that if the quantities E^0, H^0 in equations (7) are some given functions for which the representation

$$E^0 - iH^0 = \int d^3k \sqrt{\frac{2\omega}{(2\pi)^3}} \left(c^3 e^{-ikx} + c^4 e^{ikx} \right),$$
(18)

is valid, then equations (7) are the Maxwell equations with the sources $j_\mu^{el} = -\partial_\mu E^0$, $j_\mu^{mag} = -\partial_\mu H^0$. (We call these 4 currents gradient-like sources). In this case the general solution of the Maxwell equations $(7) = (8) = (9) = (10) = (11) = (12)$ with the given sources, as follows from the solution (14), has the form

$$\vec{E}(x) = \int d^3k \sqrt{\frac{\omega}{2(2\pi)^3}} (c^1 \vec{\mathbf{e}}_1 + c^2 \vec{\mathbf{e}}_2 + \alpha \vec{\mathbf{e}}_3) + c.c,$$

$$\vec{H}(x) = i \int d^3k \sqrt{\frac{\omega}{2(2\pi)^3}} (c^1 \vec{\mathbf{e}}_1 - c^2 \vec{\mathbf{e}}_2 + \beta \vec{\mathbf{e}}_3) + c.c,$$
(19)

where the amplitudes of longitudinal waves $\vec{\mathbf{e}}_3 \exp(-ikx)$ are $\alpha = c^3 + c^4$, $\beta = c^3 - c^4$ and c^3, c^4 are determined by functions E^0, H^0 from equation (18).

Longitudinal electromagnetic waves were investigated by Hvorostenko [37]. Now we are able: (i) to add to his results the exact solution of the Maxwell equations with gradient-like sources, which contains the longitudinal waves, and (ii) to identify the location of these waves in the same space-time domain where the gradient-like sources are located (since the amplitudes c^3, c^4, which define the waves and the gradient-like sources, are the same).

Note that in the procedure to find the solutions (19), as an arbitrary step we can make $H^0 = 0$, or $c^4 = 0$, and easily treat the partial case with only one scalar field $E^0 \neq 0$, corresponding to electric sources.

The procedure by which we have generalized the standard Maxwell equations involves two steps. We first add the magnetic currents and charges (generalization). Second, we suppose that electric and magnetic sources are gradients of two scalar fields (E^0, H^0), *i.e.*, we consider the partial case of sources. In this second step we simplify (rather than generalize) standard Maxwell electrodynamics. Lastly, we deal with the slightly generalized Maxwell equations for the system $(\vec{E}, \vec{H}, E^0, H^0)$ of interacting electromagnetic and scalar fields.

Generalization of the standard Maxwell equations was not undertaken *ad hoc*, *i.e.*, without motivation. There is no doubt that the classical Maxwell electrodynamics of the macroworld (without generalization) is sufficient to describe electrodynamic phenomena in the macroregion. Yet it is well known that for micro-phenomena (inneratomic region), classical Maxwell electrodynamics and classical mechanics do not work and must be replaced by quantum theory. In attempting to extend classical electrodynamics into the inner-atomic region, we concluded that this could be done by generalizing standard Maxwell classical electrodynamics *via* an extension of its symmetry.

We have four reasons for introducing generalized equations (1) to describe micro phenomena. (i) These equations are directly connected with the Dirac equation, whose application in atomic and nuclear physics is well known. (ii) These equations are the maximally symmetrical form among the possible forms of the Maxwell equations, *i.e.*, they are introduced due to the symmetry principle (recall the first use of the symmetry principle by Maxwell). (iii) We show below that these equations describe the spectra of atoms on the same level as the Dirac equation does. (iiii) The relationship of these equations with standard Maxwell theory is evident.

3.   Unitary relationship with the Dirac theory

We briefly show the connection between the stationary Maxwell equations

$$\text{curl}\vec{H} - \omega\varepsilon\vec{E} = \text{grad}E^0, \quad \text{curl}\vec{E} - \omega\mu\vec{H} = -\text{grad}H^0,$$
$$\text{div}\vec{E} = \omega\mu E^0, \quad \text{div}\vec{H} = -\omega\varepsilon H^0, \tag{20}$$

[1,2] that follow from Maxwell-like system (1)—below we shall derive (20)— and the stationary Dirac equation obtained from the ordinary Dirac equation

$$\left(i\gamma^\mu\partial_\mu - m_0 + \gamma^0\Phi\right)\Psi = 0, \qquad \Psi \equiv \left(\Psi^\alpha\right), \tag{21}$$

with $m_0 \neq 0$ and the interaction potential $\Phi \neq 0$.

Assuming the ordinary time dependence

$$\Psi(x) = \Psi(\vec{x})e^{-i\omega t} \Rightarrow \partial_0\Psi(x) = -i\omega\Psi(x), \tag{22}$$

for the stationary states, and using the standard Pauli-Dirac representation for the γ matrices, we obtain the following system of equations for the components $\Psi^\alpha(\vec{x})$ of the spinor $\Psi(\vec{x})$:

$$-i\omega\varepsilon\Psi^1 + (\partial_1 - i\partial_2)\Psi^4 + \partial_3\Psi^3 = 0,$$
$$-i\omega\varepsilon\Psi^2 + (\partial_1 + i\partial_2)\Psi^3 - \partial_3\Psi^4 = 0,$$
$$-i\omega\mu\Psi^3 + (\partial_1 - i\partial_2)\Psi^2 + \partial_3\Psi^1 = 0,$$
$$-i\omega\mu\Psi^4 + (\partial_1 + i\partial_2)\Psi^1 - \partial_3\Psi^2 = 0, \tag{23}$$

where ε and μ are the same as in (2). After substitution in equations (23) of the following column for Ψ

$$\Psi = \text{column}\left|-H^0 + iE^3, -E^2 + iE^1, E^0 + iH^3, -H^2 + iH^1\right| \tag{24}$$

we obtain equations (20). A complete set of 8 transformations with the same properties was obtained in our papers [27,28] with the help of the Pauli-Gursey symmetry operators [38].

The relationship (24) may be written in terms of a unitary operator. It is useful to represent the right-hand side of (24) in terms of components of the complex function (5). In these representations the connection between the spinor and electromagnetic (together with the scalar) fields has the form

$$\mathcal{E} = U_{st}\Psi, \qquad \Psi = U_{st}^{\dagger}\mathcal{E}, \tag{25}$$

where the unitary operator U_{st} is the following:

$$U_{st} = \begin{vmatrix} 0 & iC_- & 0 & C_- \\ 0 & -C_+ & 0 & iC_+ \\ iC_- & 0 & C_- & 0 \\ iC_+ & 0 & C_+ & 0 \end{vmatrix}; \quad C_{\mp} \equiv \frac{1}{2}(C \mp 1), \quad C\Psi = \Psi*, \quad C\mathcal{E} = \mathcal{E}*. \tag{26}$$

The unitarity of operator U_{st} (26) can easily be verified by noting that

$$(AC)^{\dagger} = CA^{\dagger}, \quad aC = Ca*, \quad (aC)* = Ca, \tag{27}$$

hold for an arbitrary matrix A and a complex number a. We underline that in the real algebra (i.e., the algebra over the field of real numbers) and in the Hilbert space of quantum mechanical amplitudes, this operator has all the properties of linearity and $U_{st}U_{st}^{-1} = U_{st}^{-1}U_{st} = 1$, $U_{st}^{-1} = U_{st}^{\dagger}$.

The operator (26) transforms the stationary Dirac equation

$$\left[(\omega - \Phi)\gamma^0 + i\gamma^k \partial_k - m_0\right]\Psi(\vec{x}) = 0 \tag{28}$$

from the standard representation (the Pauli-Dirac representation) into the bosonic representation

$$\left[(\omega - \Phi)\bar{\gamma}^0 + i\,\bar{\gamma}^k \partial_k - m_0\right]\mathcal{E}(\vec{x}) = 0 \tag{29}$$

Here the $\bar{\gamma}^{\mu}$ matrices have the following unusual explicit form

$$\bar{\gamma}^0 = \begin{vmatrix} 1 & 0 & 0 & 0 \\ 0 & 1 & 0 & 0 \\ 0 & 0 & 1 & 0 \\ 0 & 0 & 0 & -1 \end{vmatrix} C, \quad \bar{\gamma}^1 = \begin{vmatrix} 0 & 0 & i & 0 \\ 0 & 0 & 0 & -1 \\ i & 0 & 0 & 0 \\ 0 & 1 & 0 & 0 \end{vmatrix},$$

$$\bar{\gamma}^2 = \begin{vmatrix} 0 & 0 & 0 & 1 \\ 0 & 0 & i & 0 \\ 0 & i & 0 & 0 \\ -1 & 0 & 0 & 0 \end{vmatrix}, \quad \bar{\gamma}^3 = \begin{vmatrix} -i & 0 & 0 & 0 \\ 0 & -i & 0 & 0 \\ 0 & 0 & i & 0 \\ 0 & 0 & 0 & i \end{vmatrix}, \tag{30}$$

in which, in comparison with (13) only the $\bar{\gamma}^0$ matrix explicitly contains complex conjugation operator C. We call the representations (13), (30) the bosonic representations of γ matrices. Matrices (13) and (30) are related to one another by the unitary transformation. Due to the presence of operator C these bosonic representations are essentially different from ordinary Pauli-Dirac, Weyl and other standard representations of γ matrices. For example, in bosonic representation (30), imaginary unit i is represented by the 4×4 matrix operator:

$$\bar{i} = \begin{vmatrix} 0 & -1 & 0 & 0 \\ 1 & 0 & 0 & 0 \\ 0 & 0 & 0 & -i \\ 0 & 0 & -i & 0 \end{vmatrix}. \tag{31}$$

Due to the unitarity of the operator U_{st} (26) the $\bar{\gamma}^{\mu}$ matrices (30) (as well as the matrices (13)) still obey the Clifford-Dirac algebra

$$\bar{\gamma}^{\mu}\bar{\gamma}^{\nu} + \bar{\gamma}^{\nu}\bar{\gamma}^{\mu} = 2g^{\mu\nu} \tag{32}$$

and have (together with the matrices (13)) the same Hermitian properties as the Pauli-Dirac γ^{μ} matrices:

$$\bar{\gamma}^{0\dagger} = \bar{\gamma}^{0}, \qquad \bar{\gamma}^{k\dagger} = \bar{\gamma}^{k}. \tag{33}$$

Formulae (13) and (30) thus give two exotic representations of γ matrices.

In vector-scalar form, the equation (29) is as follows

$$-i\text{curl}\vec{\mathcal{E}} + \left[(\omega-\Phi)C - m_0\right]\vec{\mathcal{E}} = -\text{grad}\mathcal{E}^0,$$
$$\text{div}\vec{\mathcal{E}} = \left[(\omega-\Phi)C + m_0\right]\mathcal{E}^0. \tag{34}$$

Completing the transition to common real field strengths according to formula $\mathcal{E} = E - iH$ and separating real and imaginary parts, we obtain equations (20), which are mathematically equivalent to equations (1) in the stationary case.

The mathematical facts considered here prove the one-to-one correspondence between the solutions of the stationary Dirac and the stationary Maxwell equations with gradient-like 4-currents. Hence, using (24), one can write the hydrogen solutions of the Maxwell equations (1) starting from the well-known hydrogen solutions of the Dirac equation (21), *i.e.*, without the special procedure of finding solutions of the Maxwell equations [1]. Moreover, *all successfully solved stationary Dirac problems of atomic physics can easily be reformulated and solved equally well in terms of Maxwell-like equations* (1). Yet we now work in the framework of slightly generalized classical electrodynamics.

We now consider the relationship between the Dirac and Maxwell equations in the simplest case when $m_0 = 0$ and $\varepsilon = \mu = 1$. Equations (8) = (9) = (10) = (11) = (12) are directly connected with the free massless Dirac equation

$$i\gamma^{\mu}\partial_{\mu}\Psi(x) = 0. \tag{35}$$

Substitutiing (the notations are the same as in (26))

$$\Psi = \begin{vmatrix} E^3 + iH^0 \\ E^1 + iE^2 \\ iH^3 + E^0 \\ -H^2 + iH^1 \end{vmatrix} = U\mathcal{E}, \qquad U = \begin{vmatrix} 0 & 0 & C_+ & C_- \\ C_+ & iC_+ & 0 & 0 \\ 0 & 0 & C_- & C_+ \\ C_- & iC_- & 0 & 0 \end{vmatrix}, \tag{36}$$

into Dirac equation (35) with γ matrices in standard Pauli-Dirac representation transforms it into the slightly generalized Maxwell equations (7) = (8) = (9) = (10) = (11) = (12) (the complete set of transformations as in (36) in [27,28]).

Thus, equation (35) with Ψ in the form (36), as well as (12), may be interpreted as the electrodynamic representation of the massless Dirac equation.

The unitarity of the operator (36) can easily be verified with reference to relations (27). Here, as in the case of operator (26), in the real algebra (*i.e.*, the algebra over the field of real numbers) and in the Hilbert space of quantum mechanical amplitudes, this operator has all the properties of unitarity.

We emphasize that equation (12) has the form of the massless Dirac equation for the fermionic field. Hence the $\tilde{\gamma}^\mu$ matrices may be chosen in arbitrary representation (*e.g.*, in each of the Pauli-Dirac, Majorana, Weyl, *etc.* representations). However, only in the exotic representation (13) is equation (12) the Maxwell equation for the system of interacting electromagnetic $\vec{\mathcal{E}} = \vec{E} - i\vec{H}$ and scalar $\mathcal{E}^0 = E^0 - iH^0$ fields. (We thus call the representation (13) "bosonic.") If equation (12) is treated as bosonic, the representation of the $\tilde{\gamma}^\mu$ matrices and their explicit form *must be fixed* in the form (13). In the bosonic interpretation of equation (35) one must fix the explicit form of γ^μ in standard Pauli-Dirac representation, and the form of Ψ must be fixed as column (36). *Thus, we introduce our generalization of the Maxwell equations on the basis of the Dirac equation; or more prescisely, on the basis of the Maxwell-Dirac unitary relationship (the first motivation of our generalization).*

The Maxwell-Dirac relationship presented here may be seen as the derivation of our generalized form of the Maxwell equations from the well-defined Dirac equation. This central conclusion is presented in all our publications. This relationship may be seen as a derivation of the Dirac equation from classical electrodynamics, as demonstrated in other work [35].

Finally, knowing the operator U (36), it is easy to obtain the relationship between the Bose amplitudes $c^\mu(\vec{k})$ (determining the general solution (14) of equations (9)) and the Fermi amplitudes $a^r(\vec{k})$, $b^r(\vec{k})$, $r = 1, 2$, (determining the well-known general solution of the massless Dirac equation (35), we explore the Pauli-Dirac representation). This solution has the form:

$$\Psi(x) = \frac{1}{(2\pi)^{3/2}} \int d^3k \left[a^r\left(\vec{k}\right) v_r^-(\vec{k}) e^{-ikx} + b^{*r}\left(\vec{k}\right) v_r^+\left(\vec{k}\right) e^{ikx} \right];$$

$$r = 1, 2, \quad kx \equiv \omega t - \vec{k}\vec{x}, \quad \omega \equiv \sqrt{\vec{k}^2}, \quad k \in R^3,$$

(37)

where

$$v_1^-(\vec{k}) = \frac{1}{\sqrt{2}} \begin{vmatrix} 1 \\ 0 \\ k^3/\omega \\ (k^1 + ik^2)/\omega \end{vmatrix}, \quad v_2^-(\vec{k}) = \frac{1}{\sqrt{2}} \begin{vmatrix} 0 \\ 1 \\ (k^1 - ik^2)/\omega \\ -k^3/\omega \end{vmatrix},$$

$$v_1^+(\vec{k}) = \frac{1}{\sqrt{2}} \begin{vmatrix} k^3/\omega \\ (k^1 + ik^2)/\omega \\ 1 \\ 0 \end{vmatrix}, \quad v_2^+(\vec{k}) = \frac{1}{\sqrt{2}} \begin{vmatrix} (k^1 - ik^2)/\omega \\ -k^3/\omega \\ 0 \\ 1 \end{vmatrix}.$$

(38)

Corresponding formulae [31, 32, 36] connecting fermionic and bosonic amplitudes have the form:

$$a^1 = \frac{1}{2\omega}\left[i\sqrt{(\omega-k^3)(\omega+k^3)}(c^1-c^2)-(\omega-k^3)c^3+(\omega+k^3)c^4\right],$$

$$a^2 = \frac{1}{2\omega}\left[-i(k^1+ik^2)\left(\sqrt{\frac{\omega+k^3}{\omega-k^3}}c^1+\sqrt{\frac{\omega-k^3}{\omega+k^3}}c^2\right)+(k^1+ik^2)(c^3+c^4)\right],$$

$$b^1 = \frac{1}{2\omega}\left[i\sqrt{(\omega-k^3)(\omega+k^3)}(c^1+c^2)+(\omega+k^3)c^3+(\omega-k^3)c^4\right],$$ (39)

$$b^2 = \frac{1}{2\omega}\left[i(k^1+ik^2)\left(\sqrt{\frac{\omega-k^3}{\omega+k^3}}c^1-\sqrt{\frac{\omega+k^3}{\omega-k^3}}c^2\right)+(k^1+ik^2)(c^3-c^4)\right].$$

In terms of unitary operator V these formulae have the form:

$$\hat{a} \equiv \begin{vmatrix} a^1 \\ a^2 \\ b^1 \\ b^2 \end{vmatrix} = \frac{1}{2\omega}\begin{vmatrix} i\sqrt{pq} & -p & -i\sqrt{pq} & q \\ -iz*\sqrt{\frac{q}{p}} & z* & -iz*\sqrt{\frac{p}{q}} & z* \\ i\sqrt{pq} & q & i\sqrt{pq} & p \\ iz\sqrt{\frac{p}{q}} & z & -iz\sqrt{\frac{q}{p}} & -z \end{vmatrix} \cdot \begin{vmatrix} c^1 \\ c^3 \\ c^2 \\ c^4 \end{vmatrix} = V\cdot\hat{c},$$ (40)

where $p = \omega - k^3$, $q = \omega + k^3$, $z = k^1 - ik^2$, $z* = k^1 + ik^2$, $\omega \equiv \sqrt{\vec{k}^2}$. The operator V (the image of operator U (36) in the space of quantum-mechanical amplitudes $\hat{c}$ and $\hat{a}$, i.e., in the rigged Hilbert space $S_3^4 \subset H \subset S_3^{*4}$, where $S_3^{*4} \equiv (S(R^3)\otimes C^4)*$ is the space of 4-component generalized Schwartz functions) is linear and $VV^{-1} = V^{-1}V = 1$, $V^{-1} = V^\dagger$.

Hence, the fermionic states may be constructed as linear combinations of bosonic states, i.e., states of the coupled electromagnetic $\vec{\mathcal{E}} = \vec{E} - i\vec{H}$ and scalar $\mathcal{E}^0 = E^0 - iH^0$ fields. The inverse relationship between the bosonic and fermionic states is also valid. We prefer the first possibility which is a new (bosonic) realization of the old idea (Thomson, Abraham, etc. [39]) of the electromagnetic nature of mass and the material world. Consequently, today on the basis of (24), (26), (36) (and (70) below) we may speak of the more general idea of the *bosonic field nature of the material world*.

On the basis of this relationship, a connection between the quantized scalar-electromagnetic and massless spinor fields has been obtained [31, 32, 36]. The possibility of both Bose and Fermi quantization types for the electromagnetic-scalar field (and, inversely, for the Dirac spinor field) has been proved. This is interesting for the development of quantum field theory in general.

We will not touch on the problems of quantization. As in most of our publications on this subject, we describe the atom and electron without quantization. Quantization of the electromagnetic-scalar field is addressed elsewhere [36].

4.   Derivation of the slightly generalized Maxwell equations from the symmetry principle

Equation $(7) = (8) = (9) = (10) = (11) = (12)$ is the maximally symmetrical form of all generalized and non-generalized forms of Maxwell equations. Due to the fact that equations (9), $(10) = (11)$, (12) are manifestly covariant vector, tensor-scalar and spinor forms of one and the same equation (7), respectively, the following theorem is valid.

THEOREM 1. The slightly generalized Maxwell equations $(8) = (9) = (10) = (11) = (12)$ are invariant with respect to the three different transformations, which are generated by three different representations P^V, P^{TS}, P^S of the Poincaré group $P(1,3)$ given by the formulae

$$\mathcal{E}(x) \to \mathcal{E}^V(x) = \Lambda \mathcal{E}\left[\Lambda^{-1}(x-a)\right],$$

$$\mathcal{E}(x) \to \mathcal{E}^{TS}(x) = F\Lambda \mathcal{E}\left[\Lambda^{-1}(x-a)\right], \tag{41}$$

$$\mathcal{E}(x) \to \mathcal{E}^S(x) = S\Lambda \mathcal{E}\left[\Lambda^{-1}(x-a)\right],$$

where Λ is a vector (*i.e.*, $\left(\frac{1}{2}, \frac{1}{2}\right)$), $F(\Lambda)$ is a tensor-scalar ($(0,1) \otimes (0,0)$) and $S(\Lambda)$ is a spinor representation ($\left(0, \frac{1}{2}\right) \otimes \left(\frac{1}{2}, 0\right)$) of SL(2,C) group. This means that the equations $(8) = (9) = (10) = (11) = (12)$ have both spin 1 and spin 1/2 symmetries.

Proof. Let us write the infinitesimal transformations, following from (41), in the form

$$\mathcal{E}^{V,TS,S}(x) = \left(1 - a^\rho \partial_\rho - \tfrac{1}{2}\omega^{\rho\sigma} j^{V,TS,S}_{\rho\sigma}\right)\mathcal{E}(x). \tag{42}$$

Then the generators of the transformations (42) have the form

$$\partial_\rho = \frac{\partial}{\partial x^\rho}, \quad j^{V,TS,S}_{\rho\sigma} = x_\rho \partial_\sigma - x_\sigma \partial_\rho + s^{V,TS,S}_{\rho\sigma}, \tag{43}$$

where

$$\left(s^V_{\rho\sigma}\right)^\mu_\nu = \delta^\mu_\rho g_{\sigma\nu} - \delta^\mu_\sigma g_{\rho\nu}, \quad s^V_{\rho\sigma} \in \left(\tfrac{1}{2}, \tfrac{1}{2}\right), \tag{44}$$

$$s^{TS}_{\rho\sigma} = \begin{vmatrix} s^T_{\rho\sigma} & 0 \\ 0 & 0 \end{vmatrix} \in (0,1) \oplus (0,0), \quad s^T_{\rho\sigma} = -s^T_{\sigma\rho}, \quad s^T_{mn} = -i\varepsilon^{mnj}s^j, \quad s^T_{0j} = s^j, \tag{45}$$

($\left(s^1, s^2, s^3\right)^3 \equiv \vec{s}$ are the same as in (11)), and

$$s^S_{\rho\sigma} = \tfrac{1}{4}\left[\tilde{\gamma}_\rho, \tilde{\gamma}_\sigma\right], \quad s^S_{\rho\sigma} \in \left(0, \tfrac{1}{2}\right) \oplus \left(\tfrac{1}{2}, 0\right), \tag{46}$$

where the $\tilde{\gamma}$ matrices in specific bosonic representation are given in (13) and satisfy standard Clifford-Dirac algebra. The proof of the theorem is now reduced to verifying that all generators (43) obey the commutation relations of the $P(1,3)$ group and commute with the operator of the generalized Maxwell equations (12) [3,4].

COROLLARY 1. The transition inverse to (36) transforms the equation $(8) = (9) = (10) = (11) = (12)$ into the massless Dirac equation (35) with matrices $\tilde{\gamma}^\mu$ in standard Pauli-Dirac representation. This means that the massless

Dirac equation has the same three different P^V, P^{TS}, P^S Poincaré symmetries as the slightly generalized Maxwell equations (9).

This result for the slightly generalized Maxwell equations (8) = (9) means that, from a group theoretical point of view, these equations can describe both bosons and fermions. As a result, there are direct group-theoretical grounds for applying these equations to describe the electron, as presented below.

A distinctive feature of equation (9) for the system $\mathcal{E} = \left(\vec{\mathcal{E}}, \mathcal{E}^0 \right)$ (i.e., for the system of interacting irreducible (0,1) and (0,0) fields) is that it is the manifestly covariant equation with a minimum number of components, i.e., the equation without redundant components for this system.

Note that each of the three representations (41) of the P(1,3) group is a local one, because each matrix part of transformations (41) (matrices Λ, $F(\Lambda)$ and $S(\Lambda)$) does not depend on coordinates $x \in R^4$, and, consequently, the generators (43) belong to the Lie class of operators. Each of the transformations in (41) may be understood as connected with special relativity transformations in the space-time $R^4 = (x)$, i.e., with transformations in the manifold of inertial frames of reference.

It follows from equations (9) = (12) that the field $\mathcal{E} = \left(\vec{\mathcal{E}}, \mathcal{E}^0 \right)$ is massless, i.e., $\partial^\nu \partial_\nu \mathcal{E}^\mu = 0$. Therefore it is interesting to note that neither P^V, nor P^{TS} symmetries can be extended to the local conformal C(1,3) symmetry. Only the spinor C^S representation of C(1,3) group, obtained from the local P^S representation, is the symmetry group for the slightly generalized Maxwell equations (9). This fact is understandable: the electromagnetic field $\vec{\mathcal{E}} = \vec{E} - i\vec{H}$ obeying equations (9) is not free; it interacts with the scalar field $\mathcal{E}^0$.

Consider the particular case of standard (non-generalized) Maxwell equations, i.e., the case of equations (8) = (9) without magnetic charge and current densities (when $H^0 = 0$ but $E^0 \neq 0$). The symmetry properties of these standard electrodynamic equations are tightly restricted in comparison with the generalized equations (9): they are invariant only with respect to tensor-scalar (spin 1 or 0) representation of the Poincaré group defined by the corresponding representation $(0,1) \otimes (0,0)$ of the proper orthochronous Lorentz group SL(2,C). Other symmetries mentioned in the theorem are lost for this case too. The proof of this assertion follows from the fact that the vector $(\frac{1}{2}, \frac{1}{2})$ and the spinor $(0, \frac{1}{2}) \otimes (\frac{1}{2}, 0)$ transformations of $\mathcal{E} = \left(\vec{\mathcal{E}}, \mathcal{E}^0 \right)$ mix the $\mathcal{E}^0$ and $\vec{\mathcal{E}}$ components of the field $\mathcal{E}$, and only the tensor-scalar $(0,1) \otimes (0,0)$ transformations do not mix them.

For the free Maxwell equation in vacuum without sources (the case $E^0 = H^0 = 0$) the loss of the symmetries mentioned above is evident, for the same reasons. Moreover, it is well known that these equations are invariant only with respect to the tensor (spin 1) representations of the Poincaré and conformal groups and with respect to the dual transformation: $\vec{E} \to \vec{H}$, $\vec{H} \to -\vec{E}$. We have obtained the extended 32-dimensional Lie algebra [40] (and the corresponding group) of invariance of free Maxwell equations, which is isomorphic to $C(1,3) \oplus C(1,3) \oplus$ dual algebra. We have proved this by a method obtained from the Lie class of symmetry operators. The proof also held in a more general class, namely, in the simplest Lie-Bäcklund class of operators. The corresponding generalization of symmetries of equations (9) presented in the above theorem leads to a wide 246-dimensional Lie algebra in the class of first order Lie-Bäcklund operators.

The Maxwell equations (9) with electric and magnetic gradient-like sources have the maximum possible symmetry properties of all standard and generalized equations of classical electrodynamics!!! We therefore introduce our generalization of the Maxwell equations on the basis of the symmetry principle (the second motivation for our generalization).

5. Derivation of atomic spectra from the Maxwell-like equations

The consideration presented above for the simplest case $\varepsilon = \mu = 1$ furnishes the group-theoretical basis for the classical electrodynamic (non-quantum-mechanical) model of the electron and atom based on the Maxwell equations $(1) = (3)$ in medium with $(\varepsilon, \mu) \neq 1$.

Now we return to the input equations (1) and consider the stationary case. For the stationary solutions with positive energy ω

$$E^0(t, \vec{x}) = E_A^0(\vec{x}) \cos \omega t + E_B^0(\vec{x}) \sin \omega t,$$
$$H^0(t, \vec{x}) = H_A^0(\vec{x}) \cos \omega t + H_B^0(\vec{x}) \sin \omega t,$$
(47)

$$\vec{E}(t, \vec{x}) = \vec{E}_A(\vec{x}) \cos \omega t + \vec{E}_B(\vec{x}) \sin \omega t,$$
$$\vec{H}(t, \vec{x}) = \vec{H}_A(\vec{x}) \cos \omega t + \vec{H}_B(\vec{x}) \sin \omega t,$$
(48)

the slightly generalized Maxwell equations (1) in an electrodynamic medium (2) (which assumes here the role of nuclear field) have the form (20).

Strictly speaking for the 16 time-independent amplitudes, two non-linked subsystems like (20) [1] are obtained:

$$\text{curl}\vec{H}_A - \omega\varepsilon\vec{E}_B = \text{grad}E_A^0, \quad \text{curl}\vec{E}_B - \omega\mu\vec{H}_A = -\text{grad}H_B^0,$$
$$\text{div}\vec{E}_B = \omega\mu E_A^0, \quad \text{div}\vec{H}_A = -\omega\varepsilon H_B^0,$$
(49)

$$\text{curl}\vec{H}_B + \omega\varepsilon\vec{E}_A = \text{grad}E_B^0, \quad \text{curl}\vec{E}_A + \omega\mu\vec{H}_B = -\text{grad}H_A^0,$$
$$\text{div}\vec{E}_A = -\omega\mu E_B^0, \quad \text{div}\vec{H}_B = \omega\varepsilon H_A^0,$$
(50)

We consider only the first of these, because these subsystems are connected by the transformations

$$E \to H, \quad H \to -E, \quad \varepsilon E \to \mu H, \quad \mu H \to -\varepsilon E,$$
$$\varepsilon \to \mu, \quad \mu \to \varepsilon.$$
(51)

It is useful to separate equations (49) into the following subsystems:

$$\begin{cases} \omega\varepsilon E_B^3 - \partial_1 H_A^2 + \partial_2 H_A^1 + \partial_3 E_A^0 = 0, \\ \omega\varepsilon H_B^0 + \partial_1 H_A^1 + \partial_2 H_A^2 + \partial_3 H_A^3 = 0, \\ -\omega\mu E_A^0 + \partial_1 E_B^1 + \partial_2 E_B^2 + \partial_3 E_B^3 = 0, \\ \omega\mu H_A^3 - \partial_1 E_B^2 + \partial_2 E_B^1 - \partial_3 H_B^0 = 0, \end{cases}$$
(52)

$$\begin{cases} \omega\varepsilon E_B^1 - \partial_2 H_A^3 + \partial_3 H_A^2 + \partial_1 E_A^0 = 0, \\ \omega\varepsilon E_B^2 - \partial_3 H_A^1 + \partial_1 H_A^3 + \partial_2 E_A^0 = 0, \\ \omega\mu H_A^1 - \partial_2 E_B^3 + \partial_3 E_B^2 - \partial_1 H_B^0 = 0, \\ \omega\mu H_A^2 - \partial_3 E_B^1 + \partial_1 E_B^3 - \partial_2 H_B^0 = 0. \end{cases} \tag{53}$$

Assuming the spherical symmetry case, when $\Phi(\vec{x}) = \Phi(r)$, $r \equiv |\vec{x}|$, we make the transition into the spherical coordinate system and look for solutions in the spherical coordinates in the form

$$(E,H)(\vec{r}) = R_{(E,H)}(r) f_{(E,H)}(\theta,\phi), \tag{54}$$

where $E \equiv (\vec{E}, E^0)$, $H \equiv (\vec{H}, H^0)$. We choose for the subsystem (52) the d'Alembert *Ansatz* in the form

$$\begin{aligned} \bar{E}_A^0 &= \bar{C}_{E_4} R_{H_4} P_{l_{H_4}}^{\bar{m}_4} e^{-i\bar{m}_4\phi}, \\ \bar{E}_B^k &= \bar{C}_{E_k} R_{E_k} P_{l_{E_k}}^{\bar{m}_k} e^{-i\bar{m}_k\phi}, \\ \bar{H}_B^0 &= \bar{C}_{H_4} R_{E_4} P_{l_{E_4}}^{\bar{m}_4} e^{-i\bar{m}_4\phi}, \qquad k = 1,2,3. \\ \bar{H}_A^k &= \bar{C}_{H_k} R_{H_k} P_{l_{H_k}}^{\bar{m}_k} e^{-i\bar{m}_k\phi}, \end{aligned} \tag{55}$$

We use the following representation for the $\partial_1, \partial_2, \partial_3$ operators in spherical coordinates

$$\partial_1 CRP_l^m e^{\mp im\phi} = \frac{e^{\mp im\phi}C}{2l+1}\cos\phi\left(R_{,l+1}P_{l-1}^{m+1} - R_{,-l}P_{l+1}^{m+1}\right) + e^{\mp i(m-1)\phi}C\frac{m}{\sin\theta}P_l^m\frac{R}{r},$$

$$\partial_2 CRP_l^m e^{\mp im\phi} = \frac{e^{\mp im\phi}C}{2l+1}\sin\phi\left(R_{,l+1}P_{l-1}^{m+1} - R_{,-l}P_{l+1}^{m+1}\right) \mp e^{\mp i(m-1)\phi}C\frac{im}{\sin\theta}P_l^m\frac{R}{r}, \tag{56}$$

$$\partial_3 CRP_l^m e^{\mp im\phi} = \frac{e^{\mp im\phi}C}{2l+1}\left[R_{,l+1}(l+m)P_{l-1}^m + R_{,-l}(l-m+1)P_{l+1}^m\right].$$

Substituting (55) and (56) into subsystem (52), and adopting assumptions

$$\begin{aligned} & R_{E_\alpha} = R_E, \qquad l_{E_\alpha} = l_E, \qquad R_{H_\alpha} = R_H, \qquad l_{H_\alpha} = l_H, \\ & \bar{m}_1 = \bar{m}_2 = \bar{m}_3 - 1 = \bar{m}_4 - 1 = m, \\ & \bar{C}_{H_1} = i\bar{C}_{H_2}, \qquad \bar{C}_{E_2} = -i\bar{C}_{E_1}, \qquad \bar{C}_{H_4} = -i\bar{C}_{E_3}, \qquad \bar{C}_{H_3} = -i\bar{C}_{E_4}, \\ & \bar{C}_{H_2}^I = \bar{C}_{E_4}^I(l_H^I + m + 1), \qquad \bar{C}_{E_3}^I = -C_{E_4}^I \equiv C^I, \qquad \bar{C}_{E_1}^I = \bar{C}_{E_3}^I(l_E^I - m), \\ & \bar{C}_{H_2}^{II} = -\bar{C}_{E_4}^{II}(l_H^{II} - m), \qquad \bar{C}_{E_3}^{II} = -\bar{C}_{E_4}^{II} \equiv \bar{C}^{II}, \qquad \bar{C}_{E_1}^{II} = -\bar{C}_{E_3}^{II}(l_E^{II} + m + 1), \\ & l_H^I = l_E^I - 1 \equiv l^I, \qquad l_H^{II} = l_E^{II} + 1 \equiv l^{II}, \end{aligned} \tag{57}$$

ensures the separation of variables in these equations and leads to a pair of equations for two radial functions R_E, R_H:

$$\varepsilon\omega R_E^I - R_{H,-l}^I = 0, \qquad \mu\omega R_H^I + R_{E,l+2}^I = 0, \tag{58}$$

$$\varepsilon\omega R_E^{II} - R_{H,l+1}^{II} = 0, \qquad \mu\omega R_H^{II} + R_{E,-l+1}^{II} = 0, \qquad R_{,a} \equiv \left(\frac{d}{dr} + \frac{a}{r}\right)R. \tag{59}$$

In the case $\Phi = -ze^2/r$ and for the energy region $0 < \omega < m_0c^2$ the solutions (54) of equations (1) rapidly decrease at the limit $|x| \equiv r \to \infty$, and the possible values of the energy are discrete and coincide with the Sommerfeld-Dirac formula

$$\omega = \omega_{nj}^{hyd} = \frac{m_0 c^2}{\hbar \sqrt{1 + \dfrac{\alpha^2}{\left(n_r + \sqrt{k^2 - \alpha^2}\right)^2}}} \qquad (60)$$

with the notations $n_r = n - k$, $k = j + 1/2$, $\alpha = e^2/\hbar c$ [41]. The reason for this is the coincidence of the radial functions $R_{(E,H)}(r)$ in (58), (59) with those for the stationary Dirac equation (28) for the electron with mass m_0 in the external field $\Phi = -Ze^2/r$. Furthermore, the standard relativistic electron states (the solutions of the Dirac equation (28)) can be obtained from the solutions $(E, H)(\vec{r})$ of the slightly generalized Maxwell equations (1) *via* the unitary operator (24)-(26).

Nevertheless, (and this is the main result!) here we are only working with classical Maxwell equations and do not utilize quantum-mechanical equations.

For the subsystem (53) the d'Alembert *Ansatz* has the form

$$\overset{+}{E_A^0} = \overset{+}{C}_{E_4} R_{H_4} P_{l_{H_4}}^{\overset{+}{m_4}} e^{i \overset{+}{m_4} \phi},$$

$$\overset{+}{E_B^k} = \overset{+}{C}_{E_k} R_{E_k} P_{l_{E_k}}^{\overset{+}{m_k}} e^{i \overset{+}{m_k} \phi},$$

$$\overset{+}{H_B^0} = \overset{+}{C}_{H_4} R_{E_4} P_{l_{E_4}}^{\overset{+}{m_4}} e^{i \overset{+}{m_4} \phi}, \qquad (61)$$

$$\overset{+}{H_A^k} = \overset{+}{C}_{H_k} R_{H_k} P_{l_{H_k}}^{\overset{+}{m_k}} e^{i \overset{+}{m_k} \phi},$$

and the corresponding assumptions are the following:

$$R_{E_\alpha} = R_E, \quad l_{E_\alpha} = l_E, \quad R_{H_\alpha} = R_H, \quad l_{H_\alpha} = l_H,$$

$$\overset{+}{m_1} - 1 = \overset{+}{m_2} - 1 = \overset{+}{m_3} = \overset{+}{m_4} = m,$$

$$\overset{+}{C}_{E_4} = i \overset{+}{C}_{H_3}, \quad \overset{+}{C}_{E_1} = i \overset{+}{C}_{H_4}, \quad \overset{+}{C}_{E_1} = i \overset{+}{C}_{E_2}, \quad \overset{+}{C}_{H_2} = -i \overset{+}{C}_{H_1},$$

$$\overset{+}{C}_{H_3}^I = i \overset{+}{C}_{H_2}^I \left(l_H^I + m + 1 \right), \quad \overset{+}{C}_{E_3}^I = -\overset{+}{C}_{H_2}^I \equiv \overset{+}{C}^I, \quad \overset{+}{C}_{H_4}^I = -i \overset{+}{C}_{E_2}^I \left(l_E^I - m \right), \qquad (62)$$

$$\overset{+}{C}_{H_3}^{II} = -i \overset{+}{C}_{H_2}^{II} \left(l_H^{II} - m \right), \quad \overset{+}{C}_{E_3}^{II} = -\overset{+}{C}_{H_2}^{II} \equiv \overset{+}{C}^{II}, \quad \overset{+}{C}_{H_4}^{II} = -i \overset{+}{C}_{E_2}^{II} \left(l_E^{II} + m + 1 \right),$$

$$l_H^I = l_E^I - 1 \equiv l^I, \quad l_H^{II} = l_E^{II} + 1 \equiv l^{II}.$$

Again we derive the equations (58), (59) and formula (60).

The complete set of solutions of the equations (1) has the form:

$$\bar{E}^{I0} = -\bar{C}^I R_H^I P_l^{m+1} \cos(m+1)\phi, \qquad \bar{H}^{I0} = -\bar{C}^I R_E^I P_{l+1}^{m+1} \sin(m+1)\phi,$$

$$\bar{E}^{I1} = \bar{C}^I R_E^I \left(l - m + 1 \right) P_{l+1}^m \cos m\phi, \qquad \bar{H}^{I1} = -\bar{C}^I R_H^I \left(l + m + 1 \right) P_l^m \sin m\phi,$$

$$\bar{E}^{I2} = -\bar{C}^I R_E^I \left(l - m + 1 \right) P_{l+1}^m \sin m\phi, \qquad \bar{H}^{I2} = -\bar{C}^I R_H^I \left(l + m + 1 \right) P_l^m \cos m\phi, \qquad (63)$$

$$\bar{E}^{I3} = \bar{C}^I R_E^I P_{l+1}^{m+1} \cos(m+1)\phi, \qquad \bar{H}^{I3} = \bar{C}^I R_H^I P_l^{m+1} \sin(m+1)\phi,$$

$$\bar{E}^{II0} = -\bar{C}^{II} R_H^{II} P_l^{m+1} \cos(m+1)\phi, \qquad \bar{H}^{II0} = -\bar{C}^{II} R_E^{II} P_{l-1}^{m+1} \sin(m+1)\phi,$$

$$\bar{E}^{II1} = -\bar{C}^{II} R_E^{II} (l+m) P_{l-1}^{m} \cos m\phi, \qquad \bar{H}^{II1} = \bar{C}^{II} R_H^{II} (l-m) P_l^{m} \sin m\phi, \qquad (64)$$

$$\bar{E}^{II2} = \bar{C}^{II} R_E^{II} (l+m) P_{l-1}^{m} \sin m\phi, \qquad \bar{H}^{II2} = \bar{C}^{II} R_H^{II} (l-m) P_l^{m} \cos m\phi,$$

$$\bar{E}^{II3} = \bar{C}^{II} R_E^{II} P_{l-1}^{m+1} \cos(m+1)\phi, \qquad \bar{H}^{II3} = \bar{C}^{II} R_H^{II} P_l^{m+1} \sin(m+1)\phi,$$

$$\overset{+}{E}{}^{I0} = \overset{+}{C}{}^{I} \overset{+}{R}_H^{I} (l+m+1) P_l^{m} \cos m\phi, \qquad \overset{+}{H}{}^{I0} = -\overset{+}{C}{}^{I} \overset{+}{R}_E^{I} (l-m+1) P_{l+1}^{m} \sin m\phi,$$

$$\overset{+}{E}{}^{I1} = \overset{+}{C}{}^{I} \overset{+}{R}_E^{I} P_{l+1}^{m+1} \cos(m+1)\phi, \qquad \overset{+}{H}{}^{I1} = \overset{+}{C}{}^{I} \overset{+}{R}_H^{I} P_l^{m+1} \sin(m+1)\phi,$$

$$\overset{+}{E}{}^{I2} = \overset{+}{C}{}^{I} \overset{+}{R}_E^{I} P_{l+1}^{m+1} \sin(m+1)\phi, \qquad \overset{+}{H}{}^{I2} = -\overset{+}{C}{}^{I} \overset{+}{R}_H^{I} P_l^{m+1} \cos(m+1)\phi, \qquad (65)$$

$$\overset{+}{E}{}^{I3} = -\overset{+}{C}{}^{I} \overset{+}{R}_E^{I} (l-m+1) P_{l+1}^{m} \cos m\phi, \qquad \overset{+}{H}{}^{I3} = \overset{+}{C}{}^{I} \overset{+}{R}_H^{I} (l+m+1) P_l^{m} \sin m\phi,$$

$$\overset{+}{E}{}^{II0} = -\overset{+}{C}{}^{II} \overset{+}{R}_H^{II} (l-m) P_l^{m} \cos m\phi, \qquad \overset{+}{H}{}^{II0} = \overset{+}{C}{}^{II} \overset{+}{R}_E^{II} (l+m) P_{l-1}^{m} \sin m\phi,$$

$$\overset{+}{E}{}^{II1} = \overset{+}{C}{}^{II} \overset{+}{R}_E^{II} P_{l-1}^{m+1} \cos(m+1)\phi, \qquad \overset{+}{H}{}^{II1} = \overset{+}{C}{}^{II} \overset{+}{R}_H^{II} P_l^{m+1} \sin(m+1)\phi,$$

$$\overset{+}{E}{}^{II2} = \overset{+}{C}{}^{II} \overset{+}{R}_E^{II} P_{l-1}^{m+1} \sin(m+1)\phi, \qquad \overset{+}{H}{}^{II2} = -\overset{+}{C}{}^{II} \overset{+}{R}_H^{II} P_l^{m+1} \cos(m+1)\phi, \qquad (66)$$

$$\overset{+}{E}{}^{II3} = \overset{+}{C}{}^{II} \overset{+}{R}_E^{II} (l+m) P_{l-1}^{m} \cos m\phi, \qquad \overset{+}{H}{}^{II3} = -\overset{+}{C}{}^{II} \overset{+}{R}_H^{II} (l-m) P_l^{m} \sin m\phi.$$

In the first possible interpretation the states of the hydrogen atom are described by these field strength functions $(E^0, H^0, \vec{E}, \vec{H})$ of electromagnetic and scalar fields.

It is evident from (1) that scalar fields (E^0, H^0) generate densities of currents and charges. Therefore the solutions (63)-(66) may be represented in another form, in which (E^0, H^0) are replaced by the corresponding densities of currents and charges:

$$\bar{\rho}_{el}^{I} = \bar{C}^{I} P_l^{m+1} \cos(m+1)\phi \left(\varepsilon R_E^{I} \right)_{,l+2},$$

$$\bar{\rho}_{el}^{II} = \bar{C}^{II} P_l^{m+1} \cos(m+1)\phi \left(\varepsilon R_E^{II} \right)_{,-l+1},$$

$$\overset{+}{\rho}_{el}^{I} = -\overset{+}{C}{}^{I} (l+m+1) P_l^{m} \cos m\phi \left(\varepsilon R_E^{I} \right)_{,l+2},$$

$$\overset{+}{\rho}_{el}^{II} = \overset{+}{C}{}^{II} (l-m) P_l^{m} \cos m\phi \left(\varepsilon R_E^{II} \right)_{,-l+1},$$

$$\bar{\rho}_{mag}^{I} = \bar{C}^{I} P_{l+1}^{m+1} \sin(m+1)\phi \left(\mu R_H^{I} \right)_{,-l}, \qquad (67)$$

$$\bar{\rho}_{mag}^{II} = \bar{C}^{II} P_{l-1}^{m+1} \sin(m+1)\phi \left(\mu R_H^{II} \right)_{,l+1},$$

$$\overset{+}{\rho}_{mag}^{I} = \overset{+}{C}{}^{I} (l-m+1) P_{l+1}^{m+1} \sin(m+1)\phi \left(\mu R_H^{I} \right)_{,-l},$$

$$\overset{+}{\rho}_{mag}^{II} = -\overset{+}{C}{}^{II} (l+m) P_{l-1}^{m} \sin m\phi \left(\mu R_H^{II} \right)_{,l+1},$$

where the following notations are used:

$$\left(\varepsilon R_\beta^\alpha \right)_{,l+2} \equiv \varepsilon \left(\frac{d}{dr} + \frac{l+2}{r} \right) R_\beta^\alpha + R_\beta^\alpha \frac{d\varepsilon}{dr}, \quad etc. \qquad (68)$$

In the second possible interpretation the states of the hydrogen atom are described by the field strength functions $(\vec{E}, \vec{H})$ generated by the corresponding currents and charge densities (67).

The solutions of the second subsystem (50) follow from (63)-(66), or (67), after the application of transformation (51).

As in quantum theory, the numbers $n = 0,1,2,...$ $j = k - \frac{1}{2} = l \mp \frac{1}{2}$ $(k = 1,2,...,n)$ and $m = -l, -l+1,...l$ mark both the terms (60) and the corresponding exponentially decreasing field functions $\vec{E}, \vec{H}$ (and E^0, H^0) in (63)-(66), i.e., they mark the different discrete states of the classical electrodynamic (and scalar) field, which by definition describes the corresponding states of hydrogen atom in the model under consideration.

It is evident from this example that the discreteness of the physical system states (and its characteristics such as energy, etc.) may be a consequence of both quantum systems (Schrödinger, Dirac) and the classical (Maxwell) equations for the given system. In the present case, this discreteness is caused by the properties of the medium, which are given by the electric and magnetic permeabilities (2).

Note that the radial equations (58), (59) cannot be obtained if one neglects the sources in equations (1), or one (electric or magnetic) of these sources. Moreover, in this case there is no solution, which is effectively concentrated in the atomic region.

Bohr's postulates. Now we can show on the basis of this model that the assertions known as *Bohr's postulates are consequences of equations (1) and of their classical interpretation*: i.e., these assertions can be derived from the model, and there is no need to postulate them from beyond the framework of classical physics, as is done in Bohr's theory. To derive Bohr's postulates one can calculate the generalized Poynting vector (and generalized expression for the energy) for the hydrogen solutions (63)-(66), i.e., for the compound system of stationary electromagnetic and scalar fields $\left(\vec{E}, \vec{H}, E^0, H^0\right)$,

$$\vec{P}_{gen} = \int d^3 x \left(\vec{E} \times \vec{H} - \vec{E} E^0 - \vec{H} H^0 \right), \tag{69}$$

$$P^0_{gen} = \frac{1}{2} \int d^3 x \left(\vec{E}^2 + \vec{H}^2 + E_0^2 + H_0^2 \right) = \omega_{nj}^{hyd}. \tag{70}$$

The straightforward calculations of $\vec{P}_{gen}$ show that not only is vector (69) identically equal to zero, but the Pointing vector itself and the term with scalar fields E^0, H^0 are also identically equal to zero. This means that in stationary states the hydrogen atom does not emit any Pointing radiation, neither due to the electromagnetic $\vec{E}, \vec{H}$ field, nor to the scalar E^0, H^0 field. This is the mathematical proof of the first Bohr postulate.

Similar calculations of the energy (70) for the same system give a constant W_{nl}, depending on n, l (or n, j) and independent of m. In our model this constant is to be identified with the parameter ω in equations (1), which in the stationary states of $\left(\vec{E}, \vec{H}, E^0, H^0\right)$ field appears to be equal to the Sommerfeld-

Dirac value ω_{nj}^{hyd} (60). It is very interesting to consider also the analogy of formula (70) in medium $\varepsilon \neq \mu \neq 1$.

By abandoning the $\hbar = c = 1$ system and putting arbitrary "A" in equations (1) instead of $\hbar$ we obtain final ω_{nj}^{hyd} with "A" instead of $\hbar$. The numerical value of $\hbar$ can then be obtained by comparison of ω_{nj}^{hyd} containing "A" with experiment. These facts complete the proof of the second Bohr postulate.

This result means that in this model the Bohr postulates are no longer postulates, but the direct consequences of the classical electrodynamic equation (1). Moreover, together with the Dirac or Schrödinger equations we now have a new equation, which can be used to find the solutions to atomic spectroscopy problems. In contrast to the well-known equations of quantum mechanics, our equation is classical. *Thus, we have verified equations (1) introduced by us in the test case of the hydrogen atom (the third motivation of our generalization).*

Lamb shift. It is very useful to consider the Lamb shift in the approach presented here. This specific quantum electrodynamic effect (as modern theory asserts) can be described here in the framework of the classical electrodynamics of media. In order to obtain the Lamb shift one must add to $\Phi(\bar{x}) = -Ze^2 / r$ in (2) the quasipotential (known, *e.g.*, from [42], which follows, of course, from quantum electrodynamics)

$$-\frac{Ze^4}{60\pi^2 m_0^2} \delta(r), \tag{71}$$

and solve the equations (1) = (3) for this medium, similar to the procedure presented above. Finally one obtains the Lamb shift correction to the Sommerfeld-Dirac formula (60). Therefore, the Lamb shift can be interpreted as a pure classical electrodynamic effect. It may be considered a *consequence of the polarization of the medium (2), and not a polarization of some abstract concept, such as the vacuum in quantum electrodynamics.* This brief example demonstrates that our proposition can essentially extend the limits of application of classical theory in the microworld, which was the main purpose of our investigations.

The electric charge. Due to the unitary connection to the Dirac theory (considered above) the electric charge is still a conserved quantity here in the same sense as in the Dirac model. It may be defined similarly to the Dirac theory, or be derived from it on the basis of the unitary relationship (24)-(26).

Transition to ordinary Maxwell theory. The limiting transition to the ordinary Maxwell theory is fulfilled by assuming that $H^0 = 0$ in the macroworld. This assumption is sufficiently motivated because the field $H^0 \neq 0$ generates (see (4)) the magnetic charge and current densities. And the non-existence of the magnetic monopole in the macroworld is a well-defined experimental fact. In this case, when $H^0 = 0$, one immediately obtains the simplified partial case of the standard Maxwell electrodynamics with the partial case of electric charge and current densities when electric sources are the gradients of the scalar function $E^0 \neq 0$.

The inverse assumption, that in the microworld $H^0 \neq 0$, may be motivated too. In that region the magnetic monopole may exist, much as quarks have "existence," though they are also not observed in free states in the macroworld. We emphasize that the slightly generalized Maxwell equations are not free (even if $\varepsilon = \mu = 1$). They are equations for the interacting coupled system of the electromagnetic and scalar fields $(7) = (8) = (9)$. Above, we considered the three main reasons for this possibility ($H^0 \neq 0$): the relationship with the Dirac theory, the symmetry principle, and derivation of the observed spectra of hydrogen-like atoms.

Brief hypothesis on gravity. A unified theory of electromagnetic and gravitational phenomena may be constructed in the approach under consideration in the following way. The main primary equations again are written as (1) and gravity is treated as a medium in these equations, *i.e.,* the electric ε and magnetic μ permeabilities of the medium are some functions of the gravitational potential Φ_{grav}:

$$\varepsilon = \varepsilon(\Phi_{grav}), \qquad \mu = \mu(\Phi_{grav}). \tag{72}$$

Gravity as a medium may generate all the phenomena that in standard Einstein gravity are generated by Riemann geometry. For example, the refraction of the light beam near a massive star is a typical medium effect in a unified model of electromagnetic and gravitational phenomena. The main idea is as follows. The gravitational interaction between massive objects may be represented as an interaction with some medium, much as the electromagnetic interaction between charged particles is considered in equations (1) here.

6.   Conclusions

Symmetry

One of the general conclusions of this investigation is that a field equation itself does not tell us what kind of particle (Bose or Fermi) it describes. To answer this question one needs to find all the representations of the Poincaré group under which the equation is invariant. If more than one such Poincaré representation is found [3, 4], including representations with integer and half-integer spins, then the equation describes both Bose and Fermi particles, and both quantization types (Bose and Fermi) [3, 4, 31, 32, 36] of the field function, obeying this equation, satisfy the microcausality condition. The strict group-theoretical grounds for this assertion are presented in theorem 1 above.

Interpretation

The above-mentioned conclusion, which follows from our results for symmetries, has direct applications to theoretical physics for the interpretation of theories and models. Now it is clear that only the pair of notions "equation" plus "fixed Bose or Fermi representation of Poincaré group" tells us what kind of particle, boson or fermion, is described. In the example of the electron, this means the following.

The pair "Dirac equation plus reducible, spins 1 and 0, representation" may describe a double bosonic system (photon plus boson).

The pair "Dirac equation plus spin ½ representation" may describe fermions (electron, neutrino, *etc.*).

The pair "generalized Maxwell equation plus spin 1 and 0 representation" may describe a double bosonic system (photon plus boson).

Finally, the pair "generalized Maxwell equation plus spin ½ representation" may describe fermions, *e.g.*, the electron. This latter possibility is considered in this paper.

Using the slightly generalized Maxwell classical electrodynamics (equation (1) for the system of electromagnetic and scalar fields) and taking a spin ½ representation, we arrive at a model of the electron. The electron can be interpreted as a system of electromagnetic and scalar fields (waves) in a medium (2) (compound system of photon plus massless boson with spin equal to zero). The electron is a standing wave in the stationary case. Because it is a system of electromagnetic and scalar waves (not a charged corpuscle), it is free from the radiation difficulties of Rutherford's electron in electrodynamics. The charge here is a secondary quality, generated by interacting electromagnetic-scalar fields. The limit $m_0 \neq 0$, $\Phi \to 0$ produces the free electron. Thus, the electron can be constructed from bosons.

The simplest case $m_0 = 0$, $\Phi = 0$ is treated in detail in formulae (39), where it is shown that amplitudes of fermionic states (or their creation-annihilation operators) are the linear combinations of amplitudes (or of creation-annihilation operators) of bosonic states. In this sense our model, where the electron is considered a compound system of photon plus mass-less spin-less boson—*i.e.,* the electron's states are linear combinations of states of the electromagnetic-scalar field—has an analogy in modern quark models of hadrons. On the basis of (39), together with (70) and discussion after (39) and (70), we are able to construct fermionic states from bosonic states. Moreover, *formula (70) expresses the mass of the atom in terms of bosonic (electromagnetic and scalar) field strengths!* This is the basis for our hypothesis: *the material world is bosonic in nature* (more general than simple electromagnetic).

Furthermore, changing the field strengths in (70) may cause a change in the mass of a material object like an atom. Can a new flying machine, with mass going to zero, be constructed on the basis of this phenomenon?

It is evident from the example of the hydrogen atom presented in Section 5 that the discreetness of the physical system's states (and its characteristics, such as energy, *etc.*) may be a consequence of both quantum systems (Schrödinger, Dirac) and the classical (Maxwell) equations for the given system. In the present case, discreetness is caused by the properties of the medium, which are given by the electric and magnetic permeabilities (2).

The main conclusion from Sec. 3 is the following. The unitary equivalence between the stationary Dirac equation and the stationary Maxwell equations with gradient-like currents and charges in a medium (2) offers the possibility of reformulating all the problems of atomic and nuclear physics (not just the problem of describing the hydrogen atom, which is only one example), which can be solved on the basis of the stationary Dirac equation, in the language of the classical electrodynamic stationary Maxwell equations. This means that our model for the stationary case is just as successful as conventional relativistic quantum mechanics. In the approach based on equations (1), it is possible to solve other stationary problems of atomic physics without appealing to the Dirac equation or the probabilistic or Copenhagen interpretation.

Some non-stationary problems, *e.g.*, the problem of transitions between stationary states caused by external perturbation, can probably be solved in terms of this electrodynamic model, just as this problem is now solved in terms of the stationary Schrödinger equation with corresponding perturbation.

A few words may be said about the interpretation of the Dirac Ψ function. As a result of the consideration presented here, *e.g.*, from the relationships (24) and (36), a new interpretation of the Dirac Ψ function can be put forward: the Ψ function is the combination of electromagnetic field strengths $(\vec{E}, \vec{H})$ and two scalar fields (E^0, H^0) generating electromagnetic sources; *i.e.*, in this case, the probabilistic or Copenhagen interpretation of the Ψ function is not necessary.

Given that many interpretations of quantum mechanics (*e.g.*, Copenhagen, statistical, Feynman's, Everett's, transactional [43-46]) exist, we are under no illusion that our interpretation should be the only one. (Different models of the atom have been proposed in this book and elsewhere [47].) But the main point is that a classical interpretation (without probabilities) is now possible.

In the majority of our publications [1-4, 27-34] we have tried to develop the classical electrodynamic interpretation of the above facts, which is the main purpose of our investigation. Nevertheless, we have also emphasized [35] that a standard quantum mechanical Dirac (or spinor classical field-theoretical) interpretation is certainly also possible here. In this case the above facts only demonstrate in explicit form the classical electrodynamic aspect of the Dirac equation [35]. In other words, our equations may be considered (interpreted) as the Dirac equation for the classical (not quantum) spinor field Ψ in a specific electromagnetic representation. We have written one special paper [35] to admit this possibility, which may be more suitable for readers who are beyond the influence of Standard Model. Magnetic monopole enthusiasts may attempt to develop the monopole interpretation [48]: we note that there are few specific possibilities for interpretation. Thus, the new features that follow from our approach are:

(i)   the classical interpretation,
(ii)   a new equation and method in atomic and nuclear physics based on classical electrodynamics in an inner-atomic medium as in (2),
(iii) hypothesis of the bosonic nature of matter (bosonic structure of fermions),
(iv)   application of classical theory extended further into the microworld,

(v)   foundations of a unified model of electromagnetic and gravitational phenomena, in which gravitation is considered a medium in generalized equations,

(vi)   the electron is described as a classical electromagnetic-scalar wave and is related to the equation of motion.

Acknowledgement

The author is very grateful to Profs. Hans Sallhofer and Boris Struminskiy for many discussions of details and essential support of the main idea. This work is supported by the Ukraine National Fund for Fundamental Research, grant F7/458-2001.

References

[1] V.M. Simulik, Solutions of the Maxwell equations describing the spectrum of hydrogen, *Ukrainian Mathematical Journal*, Vol.49, No7, 1997, pp.1075-1088.

[2] V.M. Simulik, I.Yu. Krivsky, Clifford algebra in classical electrodynamical hydrogen atom model, *Advances in Applied Clifford Algebras*, Vol.7, No1, 1997, pp.25-34.

[3] V.M. Simulik, I.Yu. Krivsky, Fermionic symmetries of the Maxwell equations with gradient-like sources, *Proceedings of the International conference "Symmetry in nonlinear mathematical physics."* - Kiev: 7-13 July, 1997, pp.475-482.

[4] V.M. Simulik, I.Yu. Krivsky, Bosonic symmetries of the massless Dirac equation, *Advances in Applied Clifford Algebras*, Vol.8, No1, 1998, pp.69-82.

[5] H. Sallhofer, Elementary derivation of the Dirac equation, I., *Zeitschrift fur Naturforschung*, Vol. A33, 1978, pp.1379-1381.

[6] H. Sallhofer, Hydrogen in electrodynamics. VI. The general solution, *Zeitschrift fur Naturforschung*, Vol. A45, 1990, pp.1361-1366.

[7] H. Sallhofer, D. Radharose, *Here Erred Einstein*, World Scientific, London, 2001.

[8] C.G. Darwin, The wave equations of the electron, *Proceedings of the Royal Society of London*, Vol.A118, 1928, pp.654-680.

[9] R. Mignani, E. Recami, M. Baldo, About a Dirac-like equation for the photon according to Ettore Majorana, *Letters in Nuovo Cimento*, Vol.11, No12, 1974, pp.572-586.

[10] O. Laporte, G.E. Uhlenbeck, Application of spinor analysis to the Maxwell and Dirac equations, *Physical Review*, Vol.37, 1931, pp.1380-1397.

[11] J.R. Oppenheimer, Note on light quanta and the electromagnetic field, *Physical Review*, Vol.38, 1931, pp.725-746.

[12] R.H. Good, Particle aspect of the electromagnetic field equations, *Physical Review*, Vol.105, No6, 1957, pp.1914-1919.

[13] A.A. Borhgardt, Wave equations for the photon, *Sov. Phys.: Journal of Experimental and Theoretical Physics*, Vol.34, No2, 1958, pp.334-341.

[14] H. E. Moses, A spinor representation of Maxwell's equations, *Nuovo Cimento Supplemento*, Vol.7, No1, 1958, pp.1-18.

[15] J.S. Lomont, Dirac-like wave equations for particles of zero rest mass and their quantization, *Physical Review*, Vol.111, No6, 1958, pp.1710-1716.

[16] A. Da Silveira, Dirac-like equation for the photon, *Zeitschrift fur Naturforschung*, Vol.A34, 1979, pp.646-647.

[17] E. Giannetto, A Majorana - Oppenheimer formulation of quantum electrodynamics, *Letters in. Nuovo Cimento*, Vol.44, No3, 1985, pp.140-144.

[18] K. Ljolje, Some remarks on variational formulations of physical fields, *Fortschritt für Physik*, Vol.36, No1, 1988, pp.9-32.

[19] A. Campolattaro, New spinor representation of Maxwell equations, *International Journal of Theoretical Physics*, Vol.19, No2, 1980, pp.99-126.

[20] A. Campolattaro, Generalized Maxwell equations and quantum mechanics, *International Journal of Theoretical Physics*, Vol.29, No2, 1990, pp.141-155.

[21] C. Daviau, Électromagnetisme, monopoles magnétiques et ondes de matière dans l'algèbre d'espace-temps, *Annales de la Fondation Louis de Broglie*, Vol.14, No3, 1989, pp.273-390.

[22] C. Daviau, G. Lochak, Sur un modèle d'équation spinorielle non linéaire, *Annales de la Fondation Louis de Broglie*, Vol.16, No1, 1991, pp.43-71.

[23] W. Rodrigues Jr., E.C. de Oliveira, Dirac and Maxwell equations in the Clifford and Spin-Clifford Bundles, *International Journal of Theoretical Physics*, Vol29, No4, 1990, pp.397-412.

[24] W. Rodrigues Jr., J. Vaz Jr., From electromagnetism to relativistic quantum mechanics, *Foundations of Physics*, Vol.28, No5, 1998, pp.789-814.

[25] J. Keller, On the electron theory, *Proceedings of the International Conference "The theory of electron"* - Mexico. 24-27 September 1995, *Advances in Applied Clifford Algebras*, Vol.7 (Special), 1997, pp.3-26.

[26] J. Keller, The geometric content of the electron theory, *Advances in Applied Clifford Algebras*, Vol.9, No2, 1999, pp.309-395.

[27] V.M. Simulik, Relationship between the symmetry properties of the Dirac and Maxwell equations. Conservation laws, *Theoretical and Mathematical Physics*, Vol.87, No1, 1991, pp.76-85.

[28] V. M. Simulik, Some algebraic properties of Maxwell-Dirac isomorphism, *Zeitschrift für Naturforschung*, Vol.A49, 1994, pp.1074-1076.

[29] I.Yu. Krivsky, V.M. Simulik, *Foundations of quantum electrodynamics in field strengths terms*, Naukova Dumka, Kiev, 1992.

[30] I.Yu. Krivsky, V.M. Simulik, The Dirac equation and spin 1 representation. Relationship with the symmetries of the Maxwell equation, *Theoretical and Mathematical Physics*, Vol.90, No3, 1992, pp.388-406.

[31] I.Yu. Krivsky, V. M. Simulik, Unitary connection in Maxwell - Dirac isomorphism and the Clifford algebra, *Advances in Applied Clifford Algebras*, Vol.6, No2, 1996, pp.249-259.

[32] I.Yu. Krivsky, V. M. Simulik, The Maxwell equations with gradient-type currents and their relationship with the Dirac equation, *Ukrainian Physical Journal*, Vol.44, No5, 1999, pp.661-665.

[33] V.M. Simulik, I.Yu. Krivsky, Slightly generalized Maxwell classical electrodynamics can be applied to inneratomic phenomena, *Annals of Foundations of Louis de Broglie*, Vol.27, No2 (Special issue: *Contemporary Electrodynamics*), 2002, pp.303-328.

[34] V.M. Simulik, I.Yu. Krivsky, Relationship between the Maxwell and Dirac equation: symmetries, quantization, models of atom, *Reports on Mathematical Physics*, Vol.50, No3, 2002, pp.315-328.

[35] V.M. Simulik, I.Yu. Krivsky, Classical electrodynamical aspect of the Dirac equation, *Electromagnetic Phenomena*, Vol.3, No1(9) (Special issue: *100th anniversary of Dirac's birth*), 2003, pp.103-114.

[36] V.M. Simulik, I.Yu. Krivsky, The Maxwell equations with Gradient-Type Sources, there Applications and Quantization, *In: Developments in quantum physics*, Nova Science, New York, 2004, pp.143-165.

[37] N.P. Hvorostenko, Longitudinal electromagnetic waves, *Sov. Phys. Journ.: Izvestiya Vuzov*, Ser. Physics., No7, 1992, pp.24-29.

[38] N.H. Ibragimov, Invariant variation problems and conservation laws, *Theoretical and Mathematical Physics*, Vol.1, No3, 1969, pp.350-359.

[39] M. Jammer, *Concepts of mass in classical and modern physics*, Harvard University Press, Cambridge - Massachusetts, 1961.

[40] I.Yu. Krivsky, V. M. Simulik, Lagrangian for electromagnetic field in the terms of field strengths and conservation laws, *Ukrainian Physical Journal*, Vol.30, No10, 1985, pp.1457-1459.

[41] H.A. Bethe. and E. E. Salpeter, *Quantum mechanics of one- and two-electron atoms*, Springer-verlag, Berlin, 1957.

[42] F. Halzen, A. Martin, *Quarks and leptons*, John Wiley & Sons, New York, 1984.

[43] M. Jammer, *The philosophy of quantum mechanics. The interpretations of quantum mechanics in historical perspective*, Wiley, New York, 1974.

[44] J.G. Cramer, The transactional interpretation of quantum mechanics, *Review of Modern Physics*, Vol.58, No3, 1986, pp.647-687.

[45] A. Sudbery, *Quantum mechanics and the particles of nature*, Cambridge University, Press. Cambridge, 1986.

[46] M. Paty, Are quantum systems physical objects with physical properties, *European Journal of Physics*, Vol.20, 1999, pp.373-388.

[47] A. Lakhtakia, *Models and modelers of hydrogen*, World Scientific, London, 1996.

[48] J. Lochak, The symmetry between electricity and magnetism and the problem of the existence of the magnetic monopole, *In: Advanced electromagnetism*, World Scientific, London, 1995, pp.105-148.

What Causes the Electron to Weigh?

Malcolm H. Mac Gregor
130 Handley Street
Santa Cruz, CA 95060, USA
e-mail: mhmacgreg@aol.com

In his book on Albert Einstein entitled '*Subtle is the Lord...*', which was published in 1982, Abraham Pais has a section called *Electromagnetic Mass: The First Century*. The last two sentences of this section summarize his conclusions: [1]

> *Recently, unified field theories have taught us that the mass of the electron is certainly not purely electromagnetic in nature.*
>
> *But we still do not know what causes the electron to weigh.*

In the present article we discuss various ideas that bear on this problem [2]. Recent discoveries in physics have sharpened our interest, but have not yet provided answers.

The concept of mass in physics

Our ideas about the concept of *mass* have gone through several stages of development in the past few centuries, and in particular in the past few decades. During this period, our ability to observe masses, or at least the effect of masses, has dramatically increased, both in the very small scale of the atom and the very large scale of the universe. In the seventeenth through the nineteenth centuries, our knowledge of masses seemed to be in pretty good shape. The Newtonian laws of motion delineated the inertial properties of massive objects, and the experiments of Eötvos demonstrated that the gravitational mass is the same as the inertial mass. Physicists didn't worry about the way that mass is distributed in elementary particles or atoms, because there was no way of observing them, and there was no clear proof that they even existed. Astrophysicists didn't worry about the distribution of mass in an expanding universe, because that was not one of their concepts. The discovery of the electron in 1897 ushered in the modern age of observations at the atomic level, and the development of large telescopes early in the twentieth century started our exploration of space and time on an extra-galactic scale. These new observational skills have led us into domains where the behaviour of masses is not at all what we were expecting, both in the very small and the very large. And these new discoveries have, as of now, raised more questions than they have answered, even in the case of the ubiquitous electron.

The concept of electric charge is as mysterious in its fundamental reality as is the concept of mass. We have known since Benjamin Franklin's time that charge comes in two matching forms, which we label positive and negative. A stationary charge gives rise to an electrostatic field that acts on other charges, and perhaps also on itself. A moving charge gives rise to both an electrostatic field and a magnetic field. A moving observer looking at a (to us) stationary charge sees both fields where we see just the one. The nature of these electromagnetic fields, which can exist in a medium or in the vacuum state, has never been clearly revealed, nor has the nature of the electric charge that generates them. We might think that there should also be a "magnetic" charge, but persistent searches have failed to find it. Chemical experiments in the 19th century made it clear that electric charge exists in a quantized form rather than as a continuous "fluid" of some kind. The identification of the electron by Thomson and others showed that it is a carrier of the unit charge $-e$, and that it might in fact simply *be* the charge $-e$. Electron-electron scattering experiments carried out in mid-20th century indicate that the charge on the electron is concentrated in a very small area. In fact, it has no measurable size at all, down to length scales of about 10^{-16} cm. If this point charge is all that there is to the electron, then the large values observed for its spin angular momentum ($J = 1/2\hbar$) and magnetic moment ($\mu_e = e\hbar/2m_e c$) are results for which we have no conventional explanations, since these values, if calculated from the standard formulas of mechanics and electrodynamics, require a length scale of roughly 10^{-11} cm.

The puzzle about the actual size of the electron is compounded when we move to the other basic massive particle—the proton. The proton carries the charge $+e$, where the absolute value of e is precisely the same as that of the electron. The proton also has the same spin as the electron, and it has a magnetic moment $\mu_p \cong 2.8 e\hbar/2m_p c$. But, unlike the electron, the proton has a finite electromagnetic size of about a fermi (10^{-13} cm). The crucial point here is that the measured size of the proton is in fact roughly what we would expect theoretically from the values of its spin and magnetic moment. Why should the proton, with its classically scaled spectroscopy, be so different from the supposedly point-like electron, with its inexplicable spectroscopy? This is really a question about theory more than experiment. The experimentally observed elementary particles have mass values that were not anticipated and are not understood. But we might expect that the theory which gives us some guidance about the spectroscopy of the proton should do the same for the electron. And perhaps, as we suggest below, it might do just that, but with a required extension of our concept of mass.

When we move from protons to quarks, the substates that make up protons and neutrons, the mass mystery becomes even deeper. When quarks were first postulated by Gell-Mann and Ne'eman, it was assumed that, with enough energy, we could knock a proton apart into its basic components. When this failed to happen, it was attributed to the fact that proton quarks are very massive and have large binding energies [3]. When collision energies got very high and quarks still didn't appear, it became apparent that this was not the answer. A different type of binding mechanism based on "gluons" was proposed, and the proton quarks themselves were now assumed to have very small masses, with almost all of the proton mass arising from the quark-gluon plasma.

Further mass mysteries awaited us in outer space. Modern telescopes enabled us to examine large spiral galaxies in detail, and it became apparent that these galaxies do not contain enough visible matter to hold themselves together gravitationally at the observed rotation rates. They must have some kind of "dark matter" that is clumped in the outer regions of the galaxies [4]. Also, the existence of "black holes" is required in order to explain the tremendous energy output of quasars. A black hole is a massive burnt-out star that has collapsed to a small size under the action of its own gravity, and whose gravitational field is so strong that no light or energy can escape from it. As still another result, the Casimir effect [5] demonstrated that if we hold two plates close together in a vacuum, there is an attraction between them that is generated by the ceaseless production and annihilation of charge pairs in the vacuum—the so-called "zero point energy." Thus empty space is not really empty.

As we enter into the 21st century, further refinements in our concepts of mass are being forced upon us. Neutrinos and antineutrinos are electrically neutral leptons which are emitted in (e.g.) neutron, muon or tau decay. Postulated and then identified in the 20th century, neutrinos carry energy, spin ($J = 1/2\ \hbar$) and lepton number. Their interactions are so small that they can penetrate through a hundred light years worth of lead bricks before they interact with the lead. It was originally believed that neutrinos are massless and travel at the speed of light. However, measurements of electron neutrino fluxes from the sun indicated that only 1/3 of the expected events were occurring. Recent experiments have confirmed [6] that electron neutrinos are being transformed into muon or tau neutrinos as they travel through space, which requires that they have finite masses. Since the universe is flooded with neutrinos, even a small neutrino mass causes them to make a substantial (but not decisive) contribution to the overall mass of the universe.

Perhaps the strangest modification in our concept of mass comes from modern cosmology. In the "big bang" theory, the universe expanded from a point-like beginning to its present size, and observations indicate that it is still expanding. Space is filled with low-energy electromagnetic radiation from the original expansion. In order to explain the gravitational attraction that has created galaxies and clusters of galaxies, we had to assume the existence of large amounts of dark matter in space, as mentioned above, whose composition is a mystery. But in order to simultaneously account for the large expansion rate of the universe, we may also have to assume the existence of vast amounts of unobservable negative-pressure "dark energy," sometimes denoted as "quintessence" (the fifth essence) [7]. This postulated dark energy, whose composition is also a mystery, is thinly spread throughout the universe because it has the unique property that it repels itself gravitationally. Current estimates are that dark energy constitutes 73% of the mass in the universe, and exotic dark matter constitutes another 23%. The ordinary baryonic matter with which we are familiar—protons and neutrons—is assigned just 4% of the matter of the universe, with only 1/8 of that amount appearing as visible matter. Thus, from this viewpoint, most of the space in the universe contains matter—energy—in forms that are completely unfamiliar to us.

One familiar object which also has puzzling mass properties is the photon. The photon has a spin of 1 ($J = \hbar$), and a circularly polarized photon beam can rotate a quarter-wave plate. But the photon, unlike the neutrino, has no measurable rest mass, even though it carries energy and momentum. Since an energetic spin 1 photon can be converted into an electron-positron pair of spin ½ particles, the spin of the photon must in some sense be related to the spin of the electron. Thus the models we make for these objects must be interrelated. A particle spin is in principle a spin angular momentum, which involves an *extended mass* rotating around a center. Hence the mass problem emerges in a very crucial way in the dual concepts of the massless photon and the point-like electron.

In the present paper we bring to bear all of the information we have about the *spectroscopy* of the electron to see what it can tell us about the *structure* of the electron. Many years ago Albert Einstein, who spent years worrying about the structure of the photon and the structure of gravitational space, made the following observation: [8]

> You know, it would be sufficient to really understand the electron.

Maybe he was, as usual, correct.

The spectroscopic and bulk sizes of the electron

The estimates of the spectroscopic size of the electron come from its magnetic moment $\mu = e\hbar/2mc$ and spin angular momentum $J = 1/2\, \hbar$. The electron has an electric charge $-e = 4.8 \times 10^{-10}$ esu, and it has a mass $m = 0.511$ MeV/c^2. Measurements on the size of the charge in the electron show point-like behaviour down to at least 10^{-16} cm. But the magnetic moment and spin tell a different story. Magnetic moments arise from current loops, as given by the equation $\mu = \pi R^2 \cdot i = eRv/2c$, where R is the radius of the loop and v is the velocity of the rotating charge e in the current i. The minimum possible value of R corresponds to the maximum possible value of v, which is $v = c$. Inserting this value, we see that $R_{\min} = \hbar/mc$, which is the Compton radius $R_C = 3.86 \times 10^{-11}$ cm, where the Compton wavelength $\lambda = h/mc$ matches and quantizes the de Broglie path length of the rotating charge in the electron current loop. R_C is the smallest radius we can use to represent the magnetic moment μ of the electron as a current loop.

The spin angular momentum of the electron is $J = I\omega$, where I is the moment of inertia of the mass m. As a first-order estimate, we represent m as a solid sphere of radius R. This gives $J = 2/5\, mRv$, where v is the velocity at the equator R. Setting $v = c$ as the limiting value, we obtain $R_{min} = 5/4\ R_C$. This is not the same value as we obtained from the magnetic moment, but it indicates the same Compton-like size for a spinning electron. When we use a relativistically spinning sphere (next section), we get exact agreement.

Other particles have measured electrical sizes which are roughly comparable to their Compton radii [9]. The proton has an rms electric radius of 0.87 fermi, as compared to its Compton radius of 0.21 fermi (which can be taken as indirect evidence for a quark substructure), and the electrically neutral neutron has a magnetic radius of about the same value. The charged pion has an electric radius of 0.67 fermi and a Compton radius of 1.44 fermi. The charged kaon has an electric radius of 0.56 fermi and a Compton radius of 0.40 fermi. Thus all of these rather diverse particles have electric radii that correspond in magnitude to the Compton radii specified by their mass values.

The electric size measurements cited above are for measurements on individual particles. Another way to obtain a size estimate is from the overall bulk density of a collection of particles. A large atomic nucleus, composed of closely-packed nucleons, has a density ρ of about 2×10^{14} g/cm^3 [10]. If we relate ρ to the size R of the nucleon by means of the formula $\rho = m/(2R)^3$, where m is the nucleon mass, we obtain $R = 1.0$ fermi, which is close to the measured rms radius of 0.87 fermi. A neutron star, composed of closely-packed neutrons, has a density in the range of 10^{11} to 10^{15} g/cm^3 [11], which overlaps the atomic nucleon density cited above, and which gives a neutron radius in the range of 0.6 to 13 fermi, as compared to its measured magnetic radius of about 0.9 fermi. A normal star, composed of closely-packed hydrogen atoms, has a density in the range of 10^{-4} to 10 g/cm^3, which gives an atomic radius of 0.3 to 13 angstroms (10^{-8} cm), in agreement with atomic sizes. A white dwarf star, composed of electrons and atomic nuclei (*i.e.*, collapsed atoms), has a density of 10^4 to 10^8 gm/cm^3, and calculated radii of 1.3 to 28×10^{-11} cm. These radii bracket the 4×10^{-11} cm Compton radius that we obtained above for the electron from the values of its magnetic moment and spin, and they suggest that Compton-sized electrons could be providing stability against further collapse in a white dwarf star, in the same manner as neutrons do in a neutron star.

The relativistically spinning sphere (RSS) electron model

We demonstrated above that the magnetic moment of the electron can be reproduced by a current loop of radius R_C, where the charge e is rotating at velocity c. We also showed that if the mass of the electron is in the form of a solid sphere of radius $5/4\ R_C$, then we can reproduce the spin of the electron by the non-relativistic rotation of the sphere, where its periphery is moving at the limiting velocity $v = c$. However, if the sphere is rotating this fast, its rotation will be relativistic, and the mass distribution will be non-uniform. The sphere becomes heavier near the periphery, which increases its moment of inertia and decreases the required radius of the sphere, thus bringing the spin radius more in line with the radius required for the magnetic moment. We can make this result quantitative. The relativistic equation for a spinning ring of matter of radius r is

$$m(r) = m_0(r)/\sqrt{1 - \omega^2 r^2/c^2},$$

where ω is the angular velocity of the ring. This equation follows either from special relativity, where $v = \omega r$ is the instantaneous velocity, or from general relativity, where the mass increase is attributed to the increase in gravitational potential of the rotating ring [12]. Integrating over the volume of the sphere, using cylindrical coordinates, gives

$$M_s = \frac{3M_0}{R^3} \int_0^R \sqrt{\frac{R^2 - r^2}{1 - \omega^2 r^2/c^2}}\, r\, dr.$$

By going to the limiting angular velocity $\omega = c/R$, we obtain $M_s = 3/2\, M_0$. The relativistically spinning sphere (RSS) is half again as massive as its nonspinning counterpart. The spinning mass remains finite because the vanishing of the volume element near the periphery cancels out the increase in the mass element. The relativistic moment of inertia is

$$I = \frac{3M_0}{R^3} \int_0^R \sqrt{\frac{R^2 - r^2}{1 - \omega^2 r^2/c^2}}\, r^3\, dr.$$

In the limit $\omega = c/R$, this becomes $I = \tfrac{3}{4}M_0 R^2 = \tfrac{1}{2}M_s R^2$. Setting $J = I\omega = \tfrac{1}{2}\hbar$ for the spin, we obtain $R = R_C$ as the radius of the relativistically spinning sphere. Thus the magnetic moment and the spin of the electron both lead to the same spectroscopic size for the electron—the Compton radius $R_C = \hbar/mc$.

It is sometimes asserted that the gyromagnetic ratio of the electron—the ratio of its magnetic moment to its spin—is a quantum mechanical result that has no classical explanation [13]. The RSS model presented here—a spinning sphere with a point charge on the equator—stands as a counterexample to this assertion. (It should be noted that the RSS model was not discovered [14] until after the publication of ref. [13].) The gyromagnetic ratio g of the electron in units of $e/2mc$ is $g = \mu/J = 2$, where $e/2mc$ is the $g = 1$ value that applies to electron orbitals in atoms. The fact that the RSS model [2] reproduces g makes it of at least heuristic significance. But can it be in accord with reality? There are several issues to be addressed, which we deal with in the discussions below.

One problem that arises with respect to the RSS model is the question of relativistic stresses in the material. Spinning the sphere relativistically distorts its geometry so that it is no longer Euclidean, as Ehrenfest pointed out long ago [15]. If we fasten a circumferential string at points A and B on a disk and then set it into rotation, an observer in the disk frame sees A and B as moving apart and concludes that the string has become stretched. An observer in the inertial frame sees the distance between A and B as remaining constant (since they had identical acceleration histories), but sees the string as having been relativistically contracted in length, so he also concludes that the string has become stretched. This is the Dewan-Beran stress [16] that has been discussed in the literature. However, a point that was not noted in these discussions was that the string relativistically changes mass in direct proportion to the distance it is stretched [17]. Thus its linear density remains constant, and it does not regard itself as being stretched. We can extend this result to encompass the relativistically spinning sphere, whose volume and mass are relativistically increased in direct proportion to one another, so that the local density in the sphere remains unchanged. If we regard stresses as forces that *change densities*, then the relativistic motion in the sphere does not produce stresses.

Another way of looking at rotational stresses is to study the centrifugal force that arises from the centripetal acceleration. This force is $m\omega^2 r$, where $\omega = d\theta/dt$. As the radius r approaches the peripheral radius R of the sphere, the time dilation slows down the rotation, so that the centrifugal force vanishes in the limit where $r = R$. Another consequence of this result is that, since the effective curvature of the motion vanishes at $r = R$, an equatorial charge e thinks it is going in a straight line and does not radiate.

The quantization of the electron spin at the value $s = \frac{1}{2}\hbar$ follows in a very direct manner from the RSS model: the sphere has the Compton radius (as required for the correct magnetic moment), and it is spinning as fast as it can. From the standpoint of energetics, the fact that a relativistically spinning sphere has $M_s = 3/2\, M_0$ means that three nonspinning mass quanta M_0 can isoergically transform into two spin $\frac{1}{2}\hbar$ quanta M_s (pair formation) that are spinning at the full relativistic limit.

A massive sphere that rotates with its equator traveling at (or infinitesimally below) the velocity c may seem to be an unphysical object to use for representing an electron, but calculations show that the energy of this sphere is not divergent, and it correctly correlates the main spectroscopic features of the electron.

The relativistic transformation properties of the relativistically spinning sphere

Is a spatially-extended RSS electron model consistent with the postulates of special relativity? Specifically, do the mass, spin angular momentum, and magnetic moment of the electron transform properly? In order to answer this question, we must first establish the correct transformation properties in going from the (non-spinning) center-of-mass frame of reference to the laboratory frame. The special-relativistic transformation parameter is

$$\gamma = 1/\sqrt{1 - v^2/c^2},$$

where v is the RSS translational velocity in the laboratory frame. The relationships between the spectroscopic values of the electron in the two frames of reference are [18]

$$m_{lab} = \gamma m_{cm}, \quad J_{lab} = J_{cm}, \quad \mu_{lab} = \mu_{cm}/\gamma, \quad g_{lab} = g_{cm}/\gamma \qquad (1)$$

The mass m increases as γ, the spin J is invariant, and the magnetic moment μ and gyromagnetic ratio g decrease as $1/\gamma$. The Lorentz transformation equations were formulated with respect to point-like particles, so it isn't immediately apparent how to analytically apply them to particles with extended internal structures. However, the problem can be approached numerically by dividing the RSS into small unit cells, transforming each cell as a point-like object, and recombining the results at the end. Empirically, it was found that using about 37,000 individual cells gave accurate results [18]. The two transformations that had to be taken into account were the relativistic contraction of length for a moving electron and the relativistic addition of the rotational and translational velocity of each cell. The magnetic moment that results from the rotational motion of an equatorial point charge was handled by dividing the charge into a series of fractional charges spread equally around the equator to represent an average value over the rotation, and then transforming each fractional charge separately. The calculation was carried out for various orientations of the spin axis of the electron with respect to the translational velocity of the electron. The calculations [18] showed accurate agreement with Eq. (1), thus demonstrating that the RSS has the required Lorentz transformation properties.

In addition to simply calculating the agreement with Eq. (1), it was possible to examine the transformation properties in more detail [18]. When the relativistic corrections were divided into coordinate C (contraction of length) and velocity V components, it was discovered that the relativistic mass increase in Eq. (1) came solely from the V component, the invariance of the spin came from C and V components with opposite signs that cancelled out, and the magnetic moment decrease came from C and V components of the same sign. Adding a small Larmor precessional motion of the spin axis about the axis of spin quantization (see the next section) had no significant effect.

One final RSS calculation is important to consider here. Problems were run in which the sphere was rotated with its equator moving at the reduced velocity $c/2$, instead of at the full limiting velocity c [18]. This gave a spin angular momentum that was too small. The correct spin was then obtained by increasing the value of the radius R of the sphere. But now the calculated magnetic moment was too large. Furthermore, the Lorentz transformation properties of this slowly-rotating sphere were correct for the mass and spin values, but had the wrong angular dependence for the magnetic moment values as a function of the spin orientation with respect to the translational velocity. Thus we have two significant RSS conclusions: (1) The RSS gives the correct *gyromagnetic ratio* for the electron only when the equator is moving at the full relativistic limit c (or infinitesimally below that value); (2) the RSS gives the correct *Lorentz transformation equations* only when the equator is moving at the velocity c.

The vanishing electric quadrupole moment and spin quantization

The spectroscopic RSS model of the electron that we described above features a spinning sphere of uniform matter of some kind (note the title of this paper), and it has an electric charge $-e$ placed on the equator. This model correctly and uniquely reproduces the spin and magnetic moment, and hence the gyromagnetic ratio, of the electron. From the standpoint of the magnetic moment, the charge on the equator could be either a point charge or a continuous charge distributed around the equator, since each of these reproduces the current loop that generates the magnetic field. (Electron-electron scattering mandates a point charge.) But this current loop presents a problem, since it is a spatially extended entity that has an electric quadrupole moment which is large enough to affect electron orbitals in an atom. Thus the electric size of the loop should be observable. However, the electric quadrupole moment can be made to vanish, and in a very intriguing manner. The electrostatic potential V of the current loop in (r, θ, z) cylindrical coordinates is [19]

$$V = e/r - 1/4\,(e/r)(a/r)^2\,(3\cos^2\theta - 1) + \dots\,,$$

where e is the electric charge, a is the radius of the current loop, r is the distance from the center of the ring to a point on the z axis of quantization, and $r > a$. The first term is the coulomb potential V_0 and the second is the electric quadrupole moment V_2. If we choose the quantization angle $\theta = \arccos(1/\sqrt{3})$, then V_2 vanishes identically along the z axis, and it can also be shown [19] to vanish along the x and y axes when averaged over a cycle of precessional motion. The V_4 term in this expansion is smaller than the V_0 term by a factor of order α^4 [19]. Thus, at this prescribed angle, the current loop on the electron appears point-like in nature.

The interesting thing about this result is the manner in which it relates to the spin of the electron. The observed spin of the electron is $J_z = \frac{1}{2}\hbar$, which from a quantum mechanical viewpoint is the projection of the total spin $J = \sqrt{1/2\,(1/2 + 1)}\,\hbar$ onto the z-axis of quantization. The projection angle required to accomplish this is $\theta = \arccos(1/\sqrt{3})$, which is the same angle that is required for the vanishing of the electric quadrupole moment V_2. Thus the vanishing of V_2 in the RSS electron model and the quantum mechanical orientation of its spin angular momentum are directly related effects. In terms of the RSS model itself, tipping the spin axis by $\theta = \arccos(1/\sqrt{3})$ with respect to the z axis means that we must increase the RSS radius to the value $R = \sqrt{3}\,R_C$, so that the observed values of $\frac{1}{2}\hbar$ for the spin and $e\hbar/2mc$ for the magnetic moment are the z-projections of quantities which are each $\sqrt{3}$ larger in absolute magnitude.

It is instructive to examine the electric quadrupole effect of an electron in an atomic orbital in more detail [20]. The orbital frequency of an electron in an atom is typically about 10^{15} Hz (rps). In comparison, the frequency of rotation of the electron about its spin axis is greater than 10^{20} Hz. Thus a point charge on the electron appears as a current loop with respect to its effect on atomic motion. The magnetic field of the electron current loop interacts with the magnetic field produced by the electron orbital motion to cause a Larmor precession of the electron spin axis. The Larmor frequency is roughly 10^{10} Hz, so that the electron current loop has essentially a fixed orientation as it makes a single revolution around the atom. Thus the precessional motion of the current loop about the z axis that is required in order to produce a vanishing V_2 potential in the x and y spatial directions comes from the orbital motion of the electron, and not from its very slow Larmor precession.

The spectroscopic properties of the electron are direct indicators of its "overall size." When combined with its "electric size" (and hence its electric self-energy), they also provide clues as to the nature of its mass, as we now discuss.

The electric "size" and "mechanical" mass of the electron

The RSS model of the electron is of interest in that it correlates the main spectroscopic properties of the electron—its mass, charge, spin angular momentum, and magnetic moment. However, of equal interest is the information it can supply us as to the nature of the "mass" of the electron. What is it that makes the electron weigh? In discussing this problem, we will use the terms "mass" and "energy" interchangeably.

When the mass m and charge e on the electron were first determined, the question immediately arose as to the relationship between them. Is m just the self-energy of the charge e, or is it something else? Suppose that the charge e is spatially distributed, with each element of charge acting on each other element through the laws of classical electrodynamics. Then the self-energy of this configuration is [21]

$$W_E = 1/2 \iint \frac{\rho(\vec{r}_1)\rho(\vec{r}_2)}{|\vec{r}_1 - \vec{r}_2|} d\tau_1 d\tau_2 .$$

Now assume that the charge distribution ρ is spherically symmetric inside a sphere of radius R_E. We then have $W_E = Ae^2/R_E$, where $A = \frac{1}{2}$ for a spherical shell, and $A = 3/5$ for a uniform volume distribution. Let us for simplicity set $A = 1$, and then set $W_E = mc^2 =$ the total mass of the electron. This gives $R_E = e^2/mc^2 = 2.82 \times 10^{-13}$ cm, which is denoted as the *classical electron radius*. Thus R_E is much smaller than the $R_C = 3.86 \times 10^{-11}$ cm Compton radius that we require in the RSS model. Hence if W_E represents the actual mass of the electron, and if R_E represents its actual size, then we have no conventional explanation for its spin and magnetic moment.

This conceptual dilemma about the limitations imposed by W_E and R_E is resolved by the experimental results on electron-electron (Møller) and electron-positron (Bhabha) scattering [21]. The Dirac equations for this scattering accurately reproduce the data, and they indicate point-like scattering down to distances of less than 10^{-16} cm. Furthermore, since they give the absolute values as well as the angular distributions for the scattering, they indicate that the scattering is purely electromagnetic: electrons interact with one another only through their charges. These experiments tell us that $R_E < 10^{-16}$ cm. Since this would give a calculated value for W_E that is much larger than the total mass of the electron, it is apparent that the charge e on the electron is not a distributed entity which interacts with itself. Given this fact, we can in principle choose any value we like for the electrostatic mass W_E. However, in order to preserve the RSS spectroscopic results presented above, where the magnetic moment arises from an equatorial charge distribution and the spin angular momentum results from the rotation of a uniform sphere of matter, we must set $W_E = 0$. Other theoretical approaches—Fokker (1929), Wheeler and Feynman (1945,1949), Rohrlich (1964)—have led to this same conclusion [22].

If $W_E = 0$, then we are thrown back to the title of the present paper: *What causes the electron to weigh?* There is in fact another electromagnetic component in the electron, the *magnetic* energy W_H that is associated with the magnetic moment of the electron. This is discussed in the next section, where it is shown to represent about 0.1% of the total electron energy. But this leaves 99.9% unaccounted-for. There must be a *non-electromagnetic* mass—a new state of matter that is not observed in our familiar macroscopic world. We label it here as *mechanical* matter, just to give it a name. This mechanical matter is required to have several distinctive properties, which we enumerate here:

(1) It forms 99.9% of the mass of the electron.

(2) It furnishes the stability of the electron structure, including the confinement of the electric charge. (It should be noted that, by Earnshaw's theorem [23], no purely electromagnetic configuration is stable.)

(3) It is responsible for the inertial properties of the electron, including its spin.

(4) It forms a uniform continuum, at least within the dictates of the relativistically spinning sphere model and the gyromagnetic ratio of the electron. (Note that the *ordinary* matter with which we are familiar is mostly empty space that is filled with occupying particles which are held together electromagnetically.)

(5) It is non-interacting; that is, its interactions (in Møller and Bhabha scattering) are many orders of magnitude smaller than the electromagnetic interactions of the charge e.

(6) It functions as a *rigid body*, in the sense that it does not distort if internal or external forces are applied. *Internal* forces occur in the RSS rotation, which gives the correct gyromagnetic ratio for the electron only if the rotating sphere is not distorted by the rotational forces. *External* forces occur in Møller or Bhabha scattering, which appear point-like (central) in nature. Since this scattering, from the viewpoint of the RSS model, comes from *equatorial* (non-central) electric charges, the apparent point-like angular distributions of the scattering require Chasle's theorem [24], which states that an external force applied to a rigid body can be separated into two components: *(a)* a translational force that acts through the mass center; and *(b)* a torque that acts around the mass center. With this assumption, we can mimic point-like scattering with spatially extended (but locally point-like) charge distributions [25].

(7) It carries the lepton quantum number.

If we were to start with the electron and then remove its charge, leaving just its mechanical mass, we would have a non-interacting object that has a spin of $\frac{1}{2}\hbar$ and carries the lepton quantum number; these are the salient characteristics of the neutrino.

The magnetic "size" and "magnetic" mass of the electron

Some of the most important information about the electron comes from its magnetic properties [26]. Immediately after the studies of Uhlenbeck and Goudsmit established the existence of the magnetic moment of the electron, Rasetti and Fermi pointed out that the magnetic energy associated with this magnetic moment indicates a larger size for the electron than does the size of the electric charge that creates the magnetic field. It is of interest to reproduce their calculation. The asymptotic magnetic field components in $(r,\ \theta, z)$ polar coordinates are

$$H_r = 2\mu\cos\theta/r^3, \qquad H_\theta = \mu\sin\theta/r^3,$$

where μ is the magnetic moment of the electron. Assuming that these asymptotic forms apply all the way in to a magnetic radius R_H, we obtain an external magnetic field energy

$$W_H^{\text{ext}} = \left(\frac{\mu^2}{8\pi}\right)\int_{R_H}^{\infty}\int_0^{\pi}\left(\frac{1}{r^6}\right)\left(3\cos^2\theta+1\right)2\pi r^2\sin\theta\,d\theta\,dr = \frac{\mu^2}{3R_H^3},\ (r > R_H).$$

This calculation assumes that the magnetic moment μ is spread uniformly over the spherical volume inside R_H. A similar calculation by Born and Schrödinger, assuming a spherical shell distribution for μ, gave $W_H^{\text{ext}} = \mu^2/2R_H^3$, in close agreement with Rasetti and Fermi [26]. If we extend the Rasetti and Fermi integration in to the origin by assuming that the magnetic field stays constant at its R_H value, we obtain $W_H^{\text{int}} = W_H^{\text{ext}}$ as a lower bound on the interior energy. Thus the lower bound on the total magnetic energy is $W_H^{\text{tot}} \geq 2\mu^2/3R_H^3$. To see what lower bound this equation gives for R_H, we put $\mu = e\hbar/2mc$, and we set the magnetic energy equal to the total energy $m_e c^2$ of the electron. This yields

$$R_H^3 \geq (\alpha/6)R_C^3 \quad \Rightarrow \quad W_H(R = R_C) \geq (\alpha/6)m_e c^2, \qquad (2)$$

where $\alpha = e^2/\hbar c \approx 1/137$ is the fine structure constant and R_C is the Compton wavelength. Thus we obtain $R_H \geq 0.106\ R_c = 4.09 \times 10^{-12}$ cm as the lower bound for R_H. This is more than a factor of 10 larger than the classical electron radius $R_E = 2.82 \times 10^{-13}$ cm, as Fermi had foreseen. If we set $R_H = R_C$, where the magnetic field is assumed to arise from the equatorial current loop on the electron, this decreases W_H by a factor of $\alpha/6$, so that it becomes equal to the 0.1% mass value mentioned in the preceding section.

How does this 0.1% magnetic energy component fit into the spectroscopy of the electron? As we now demonstrate, it logically gives rise to the anomalous magnetic moment of the electron. The magnetic moment equation $\mu_e = e\hbar/2m_e c$ contains only one factor that is specific to the electron—its mass m_e. This equation shows that the magnetic moment varies inversely with the mass. The magnetic moment of the muon obeys the same basic equation, $\mu_\mu = e\hbar/2m_\mu c$. The equation $\mu_e/\mu_\mu = m_\mu/m_e$ has an experimental accuracy of better than one part in 10^5. Thus both of these quite different particles have magnetic moments that depend on just their masses. About two decades after the work of Uhlenbeck and Goudsmit, it was discovered experimentally that these magnetic moment equations are not completely accurate, and that μ_e and μ_μ actually obey the equation

$$\mu \cong (e\hbar/2mc)(1 + \alpha/2\pi). \tag{3}$$

Since μ_e and μ_μ depend solely on m_e and m_μ, respectively, their identical anomalies suggest that these anomalies may arise from an additional mass component. Furthermore, these anomalies occur in both the magnetic moments and the gyromagnetic ratios of the electron and muon, and therefore *do not occur in the spins*. This indicates that the anomalous mass component must be *irrotational*, since it does not contribute to the spin. The four different types of energies (masses) we can envision for the electron and muon are: (1) electrostatic self-energy; (2) magnetic self-energy; (3) mechanical mass; (4) gravitational self-energy. We concluded above that the electrostatic self-energy of the charge on the electron (and hence also the muon) is zero. Furthermore, the gravitational self-energy is negligible until we get down to length scales of about 10^{-30} cm. And the spherical mechanical mass was introduced precisely to account for the spin. Thus we are left with the magnetic self-energy as the only candidate for the magnetic moment anomaly, and since it is produced by the current loop of a rotating charge, it is in fact irrotational in nature. We can write Eq. (3) for the electron in the form

$$\mu_e = e\hbar/2(m_e - m_H)c \quad \Rightarrow \quad m_H = m_e \cdot \alpha/2\pi. \tag{4}$$

The experimental (and theoretical) expression $m_H/m_e = \alpha/2\pi$ shown in Eq. (4) is very close to theoretical estimate $m_H/m_e \sim \alpha/6$ given in Eq. (2). Hence the anomalous magnetic moments of the electron and muon are logically attributed to the self-energies of their respective magnetic fields. In the RSS model, this correction factor is reproduced by *(1)* using the Compton radius $R_C = \hbar/(m_e - m_H)c = (1 + \alpha/2\pi)R_C$, as suggested in Eq. (4), which increases the spin and magnetic moment values by that amount, and *(2)* omitting the magnetic mass m_H from the calculation of the spin angular momentum, which restores the spin to its original value and inserts the $\alpha/2\pi$ anomaly into the gyromagnetic ratio g.

The calculation of the magnetic moment of the electron or muon in quantum electrodynamics (QED) involves the spin as a factor, and the spin is not amenable to direct quantitative measurement. Thus what is actually measured is the gyromagnetic ratio g—the ratio of the magnetic moment to the spin. The agreement between the calculated and measured g values of the electron and muon represents one of the greatest triumphs of modern physics. The $\alpha/2\pi$ term shown in Eq. (3) is the first-order QED correction term. When the QED calculation is continued to higher terms, agreement for g is found to an accuracy of about 1 part in 10^{11} for the electron, and 1 part in 10^8 for the muon [27], and the slightly different higher-order QED diagrams for the electron and muon are clearly in evidence. This astounding accuracy far transcends any results that are obtainable with the RSS model, and the QED calculations include effects such as vacuum polarization that are outside of the scope of the present RSS calculations. However, it is important to consider the relationship between these two approaches to electron or muon structure, and to ascertain if they are in any sense complementary. QED is a very formal approach, and the integrals involved in its calculations contain divergences that require very careful cancellations. It clearly provides the correct answers about the anomalous magnetic moments of the electron and muon, but it provides very little information about the structures of the particles themselves. Richard Feynman has delineated this situation very clearly: [28]

> It seems that very little physical intuition has yet been developed in this subject. In nearly every case we are reduced to computing exactly the coefficient of some specific term. We have no way to get a general idea of the result to be expected. To make my view clear, consider, for example, the anomalous electron moment ... . We have no physical picture by which we can easily see that the correction is roughly $\alpha/2\pi$, in fact, we do not even know why the sign is positive (other than by computing it). ... We have been computing terms like a blind man exploring a new room, but soon we must develop some concept of this room as a whole, and to have some general idea of what is contained in it. As a specific challenge, is there any method of computing the anomalous moment of the electron which, on first rough approximation, gives a fair approximation to the α term ... ?

The present RSS electron model, which was discovered [14] after the above quote was published, gives some useful first-order information in response to this challenge.

The relationship between the RSS model and QED can be pursued further by an examination of the RSS current loop that generates the magnetic field of the electron. The full details, including references, are presented elsewhere [29], and we sketch the salient results here. We can represent the current loop as a thin wire in the form of a circle of radius R_C and cross sectional radius $R_E \ll R_C$ The self-inductance L of the loop is

$$L = 4\pi R_C \left\{ \eta(\ln 8 R_C / R_E - 2) + 1/4\eta' \right\} \equiv 4\pi R_C \cdot B,$$

where η and η' are the permeabilities inside of and outside of the wire, and where B denotes the {} bracket term. The self-energy of this current loop is

$$W_H = 1/2 L i^2 = \alpha/2\pi \cdot mc^2 \cdot B.$$

If we simply set $B = 1$, we immediately obtain the correct first-order QED term. Unfortunately, we have no apparent justification for doing this. But we can nevertheless obtain some interesting information from B. The dominant term in B is the divergent logarithm $\ln R_C/R_E$. Expanding B in this term gives

$$W_H \sim \alpha \cdot mc^2 \cdot \ln R_C / R_E,$$

which is the same logarithmic singularity that occurs in the QED calculation of the electromagnetic self-energy of a particle, as was first demonstrated by

Weisskopf [30]. Using the value of m_H shown in Eq. (4), and assuming that $B = 1$, we can work backward and obtain the value $R_E \sim 1 \times 10^{-11}$ cm [29]. This is much larger than the intrinsic size of the rotating charge e, which is $< 10^{-16}$ cm, and it logically corresponds to the vacuum polarization effects that prevent the build-up of intense magnetic fields in the vicinity of the point charge. The present analysis indicates that the QED logarithmic singularity in the electromagnetic self-energy comes from the size R_E of the rotating *electric* charge, but its effect is on the magnitude of the *magnetic* energy W_H (which it has to be since $W_E = 0$).

It is useful to collect together the various aspects of the spectroscopy of the electron that we have discussed above. Figure 1 shows the RSS electron model, and the caption to the figure describes its parameters. This model reproduces the main spectroscopic properties of the electron to first order in α.

Measurement of the size of the electron

The electron interacts with other particles through its electric charge. The charge itself appears to be point-like, but the spin and magnetic moment of the electron suggest a much larger size. This situation was summarized by Asim Barut: [31]

> If a spinning particle is not quite a point particle, nor a solid three dimensional top, what can it be? What is the structure which can appear under probing with electromagnetic fields as a point charge, yet as far as spin and wave properties are concerned exhibits a size of the order of the Compton wave length?

The RSS model of Fig. 1 serves as a response to Barut's queries. However, the RSS model leads to another important question: are there any experiments that reveal the size of the RSS equatorial current loop? If there are no experiments that show this size effect, there is little point in even discussing it. Electrons moving in atomic orbitals with their spins oriented at the quantum-mechanically prescribed angle have vanishing electric quadrupole moments, so they do not show appreciable charge size effects. Also, as we described above, electron-electron and electron-positron elastic scattering angular distributions indicate point-like behaviour, so we cannot use them for size measurements.

One type of experiment that could reveal an extended size for the RSS charge loop on the electron or positron is Mott scattering off atomic nuclei [32, 33]. Atomic nuclei have radii of a few fermis, and thus are much smaller than the Compton-sized current loops on the electron and positron. Mott scattering is the quantum mechanical counterpart of Rutherford scattering, which it closely resembles. In the RSS model, the equatorial charge traces out a helix as the electron or positron approaches the nucleus. If the incident particle is aimed right at the nucleus, then the nucleus lies inside the asymptotically projected area of the helix. As a result, the plane of the coulomb scattering rotates with the helical motion, which serves to enhance the forward scattering relative to the wide-angle scattering. This can be referred to as *Mott helical channelling*,

Figure 1. The relativistically spinning sphere (RSS) model of the electron. The sphere radius is $R_{RSS} = \sqrt{3} \cdot (\hbar/m_e c) \cdot (1 + \alpha/2\pi)$, the quantum mechanical tilt angle is 54.7°, the spinning RSS 'mechanical" mass is $m_m = m_e \cdot (1 - \alpha/2\pi)$, and the irrotational magnetic mass is $m_H = m_e \cdot \alpha/2\pi$. This RSS model reproduces the spin $J_z = 1/2\hbar$, magnetic moment $\mu_z = (e\hbar/2m_e c) \cdot (1 + \alpha/2\pi)$, and gyromagnetic ratio $g = 2 \cdot (e/2m_e c) \cdot (1 + \alpha/2\pi)$ of the electron. The quadrupole moment of the rotating-charge current loop vanishes identically along the z axis of quantization, and it vanishes on the average along the x and y axes. The equator of the RSS is moving at (or infinitesimally below) the limiting velocity c. The relativistically spinning mass, spin angular momentum, and magnetic moment have the proper Lorentz transformations. The rigidity of the mechanical mass serves to produce point-like Møller and Bhabha scattering, as suggested by Chasle's Theorem.

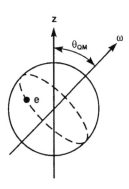

and it only happens for a range of incident energies in the KeV range. If the incident energy is too low, the impact parameters are very large, and the incident particle doesn't get near the nucleus. If the incident energy is too high, there is no time for a helical cycle to be completed.

Extensive computer calculations were carried out for both point-like and helical scattering on several atomic nuclei [32]. These calculations involved screening effects, nuclear size effects, and relativistic corrections to the motion. In all, they required more than 10,000 hours of computer time on a battery of workstation computers. Comparison to experimental data on aluminium, copper, tin and gold [33] showed matching dips in the large-angle cross sections at about the expected energies, which were at the lowest energies in the experiments. Unfortunately, these are the energies at which multiple-scattering effects become most important, so the agreement cannot be regarded as conclusive. Thus further experimental work is indicated. To date, there seem to be no well-established experimental data that confirm the large spectroscopic size of the electron.

Zero-rest-mass photons and electron waves as particle-hole ("zeron") excitations

The difficulties in understanding the nature of the electron *mass* are paralleled by even more difficulties in understanding the nature of the de Broglie electron *wave*. Moving electrons are accompanied by electron wave packets that clearly affect the motion of the electrons. What are these wave packets? Conventionally, they are denoted as mere probability distributions—mathematical constructs. But a physical system such as a wave packet that *does* something ought to *be* something, at least in the opinion of the present author. The electron wave packet is much larger in size than the associated localized electron that emerges in the detection process. When the electron is observed, its wave packet collapses and vanishes without a trace. It is a zero-rest-mass entity. All moving objects (*e.g.*, helium atoms) produce similar particle waves, whose frequency

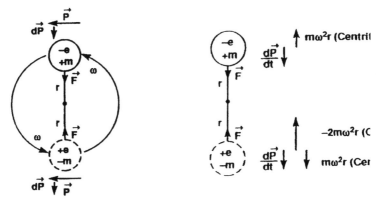

Figure 2. The basic P—H "zeron" excitation. A "particle" P (+m, −e) is raised out of the vacuum state, leaving behind a "hole" H (−m, +e). Without rotation the zeron P—H pair would
quickly de-excite, but when sent into an orbit where the centrifugal force is balanced by the
electrostatic attraction, the excitation is stabilized. The left figure shows the laboratory frame
momentum vectors, and the right figure shows the forces in the rotating frame. The motion of
H in one direction corresponds to a mass motion through H in the other direction. The zeron
rotates at the de Broglie frequency $\omega = mc^2/\hbar$, and its angular momentum is formally equal
to zero. The "antizeron" is $\bar{P}(+\bar{m},+e) - \bar{H}(-\bar{m},-e)$.

depends only on the object's mass and velocity, and not on internal frequencies
within the object.

In addition to the massless particle *waves*, the one truly massless *particle*
that we know about is the spin 1 photon, which is created out of "pure energy"
in (*e.g.*) atomic transitions or particle-antiparticle annihilation processes, and
which gives up its energy and angular momentum and vanishes when it is absorbed. Like the electron, the photon is accompanied by a wave—the electromagnetic wave—as it moves along. The analogies between the "photon-photon
wave" system and the "electron-electron wave" system are very close, and they
both share the enigma known as "particle-wave duality," wherein the "particle-
wave" system behaves like a wave when it is moving along, and like a particle
when it is stopped and detected. In trying to determine the nature of massless
particles and massless waves, we can employ two guidelines obtained from the
above studies of the electron: (*1*) the *spectroscopy* of the massless particle or
wave—its frequency, energy, angular momentum, and internal charge states—
gives valuable information about the structure of the physical system involved;
(*2*) the spectroscopic requirements may lead us to a *new type of mass* for the
system. In the case of the electron, we demonstrated that its spectroscopic requirements can be satisfied within the confines of our known theories of mechanics and electrodynamics by invoking a new kind of non-interacting "mechanical" mass that, together with the small magnetic self-energy, constitutes
the rest mass of the electron. In the case of the massless particle waves and
photons, we will now show that their spectroscopic requirements can be met by
invoking "particle-hole" excitations of the vacuum state, where the "holes"
function mechanically as "negative-mass" states, and appear electromagneti-

cally as electrically charged objects (in the same manner as semiconductor hole states). We denote these particle-hole (P-H) pairs as "zerons," since the P and H contributions to the total energy, angular momentum, and charge of the pair cancel out, and since the P-H excitations, which require rotational motion to maintain their stability, vanish without a trace in the de-excitation process. Electron-positron pairs can be created out of pure energy, but are positive-energy particles that require twice the energy of a single electron to create, and which release this energy when they annihilate.

If we think of empty space as being a total void, then it is not easy to envision the process in which a "particle" state is created—lifted out of the void—and leaves behind a matching vacancy denoted as a "hole." What is the hole in? But if so-called empty space in fact has some kind of residual mass, or if it represents a sea of zero-point excitations, as evidenced for example in the Casimir effect, then the hole represents a "gap" in this spatial continuum. As discussed above, recent astrophysical studies indicate that the universe appears to contain large amounts of attractive "dark matter" and repulsive "dark energy" spread out over vast regions of space. The nature of these invisible mass states is completely unknown, but their possible existence suggests that our ideas about "empty space" may need to be updated. In this context, the concept of "holes" in the vacuum becomes more plausible. By using particles and holes to account for massless photons and massless matter-waves, we may be able to examine the nature of the mass of the vacuum state on a microscopic as well as a galactic scale of distances.

The systematics of vacuum-state particle-hole excitations has been given elsewhere [34], and we summarize the pertinent results here. The basic P-H excitation, the rotating *zeron*, is displayed schematically in Fig. 2. The left figure shows the P and H masses and charges, together with the momentum vectors in the laboratory frame of reference. The right figure shows the equilibrium force vectors in the rotating frame.

The de Broglie wave equation is customarily used in the form $\lambda = h/m\mathrm{v}$, where m and v are the mass and velocity of the particle. Inserting $V\mathrm{v} = c^2$ and $\lambda/2\pi = V/\omega$, where V is the wave velocity, we obtain $\hbar\omega = mc^2$. Thus the frequency ω of the de Broglie electron wave depends only on the relativistic mass of the electron. One essential task in deducing a set of basis states for the de Broglie wave is to reproduce the frequency ω. Let us see how this result is accomplished by using the zeron P-H pair of Fig. 2. The zeron excitation process conceptually occurs when a "particle" P of mass $+m$ and charge $-e$ (which we will assume to be equal to the electron charge) is removed from its position in a negatively-charged "matter lattice" in the vacuum state, leaving behind a "hole" H that appears operationally as having mass $-m$ and charge $+e$. (Antizerons are similarly created when an "antiparticle" $\overline{P}$ of mass $+\overline{m}$ and charge $+e$ is removed from a positively-charged "antimatter lattice.") If not set into rotation, the P-H pair will recombine under the action of the electrostatic attraction. In the recombination process, the particle P is attracted by the apparent positive

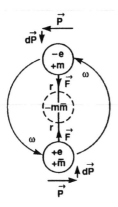

Figure 3. The zeron-antizeron quartet model of the photon. The hole states H and H̄ coalesce at the center, and the particle states P and P̄ revolve around them in a stable spin 1 orbit. This model accounts for the standard spectroscopic properties of the photon [34]. The electromagnetic wave that accompanies the photon is formed from zeron and antizeron pair excitations.

charge of the hole H and moves toward it. The negatively charged P does not act directly on H (since nothing is actually there), but instead acts on the negative charges that border the hole H, forcing the charge nearest to P into the hole, which cause H to move toward P. Hence the $+e$ state H moves under the action of an external *electrostatic* force as a positive-mass state would move, in the direction of the force (but with its momentum vector in the opposite direction); whereas H moves under the action of an external *mechanical* force as a negative-mass state (in the direction opposite to the force) [35]. In order to prevent de-excitation, the P-H pair must rotate, so that its outward centrifugal force can counterbalance the inward electrostatic force. The force equation for P is $m\omega^2 r = e^2/4r^2$, where $2r$ is the charge separation distance and ω is the angular velocity. We assume an angular momentum of $\hbar/2$ for P, which gives the angular momentum equation $\hbar/2 = mr^2\omega$. Eliminating m from these two equations and using the de Broglie wave equation $\hbar\omega = m_e c^2$ yields $e^2/2r = m_e c^2$: the electrostatic energy of the rotating pair is equal to the total energy of the electron. We also have $m/m_e = 2/\alpha^2$, where $\alpha \cong 1/137$, which shows that the rotating P mass m is much larger than the mass of the electron. These same equations apply to the hole state H of Fig. 2, where we must keep in mind that the motion of H to the right is in reality a streaming of particles to the left, which produces an outward centrifugal force. The rotating zeron is a stable structure that carries the frequency ω of the de Broglie electron wave. Since there are three excitation parameters—m, r, ω—and two constraining dynamical equations—centrifugal force and angular momentum, zerons can be constructed for any frequency ω.

Zeron rotation is dominated by the large mass values of P and H. Their mass energies and angular momenta essentially cancel out. Thus they can be easily excited and de-excited, as is required for the production of electron waves. The choice of $\hbar/2$ for the angular momentum of P seems arbitrary in the case of electron waves, but is more relevant in the case of the massless photon. The irreducible photon basis state is a rotating zeron-antizeron quartet, as shown in Fig. 3. This $Z - \bar{Z}$ quartet has particle-antiparticle symmetry, and it accounts for the spin angular momentum $\hbar$, transverse crossed electric and magnetic fields, and circular polarization of the photon. It revolves at the Einstein frequency $\omega = E/\hbar$. The overlapping H and $\bar{H}$ hole states, which are held together electrostatically, play no role in the dynamics, but are important for the overall energy balance. The electromagnetic wave that accompanies the photon is closely analogous to the electron wave that accompanies the electron, and is logically constructed of single zeron and antizeron excitations [34].

The zeron and antizeron particle-hole excitations of the vacuum state serve as basis states that can be used to account for the main properties of massless photons and massless electromagnetic and particle waves. One of the key problems here is to reproduce the large spin angular momentum $J = \hbar$ of the photon in terms of rotating masses, keeping in mind that the net rest mass of the photon is zero. This clearly requires the pairwise use of positive-mass and negative-mass (hole) states. Our main goal in the present paper is to see what this tells us about the vacuum state, rather than to focus on the particle properties of the photon. In order to create the zeron we require a spatial manifold of negatively charged positive-mass matter, and in order to create the antizeron we require another manifold of positively charged positive-mass antimatter. If we superimpose these two manifolds, we can envision that the negative potential energy of their electrostatic attraction offsets their positive-energy masses, thus keeping the overall spatial mass density at a reasonable value. This is the rather complex spatial continuum that we require in order to reproduce massless photons with large ($1\hbar$) spin values within the context of our present spectroscopic theories.

A dynamical basis for the de Broglie superluminal phase velocity

In order to reproduce the large spin value of the otherwise "point-like" electron, we introduced a new type of "mechanical mass" that is required to be non-interacting and very rigid (Fig. 1). Then, in order to reproduce the large spin value of the massless photon, we introduced P-H "particle-hole" pairs, denoted as "zerons" and "antizerons," that have matching positive and "negative" masses (Figs. 2-3). Since the photon travels at the luminal velocity c in a vacuum, the sum of the P and H positive and negative masses in a P-H pair is required to stay finite at this velocity. When we use zerons to act as basis states for de Broglie electron waves, we encounter the additional problem that the zerons move at superluminal velocities, $V > c$. The de Broglie velocity relationship is $vV = c^2$, where v is the (group) velocity of the electron (v $< c$) and V is the (phase) velocity of the electron wave. This velocity relationship can be derived by requiring relativistic invariance between the electron and its wave [36]. It can also be derived by requiring relativistic energy and momentum conservation in the excitation of zerons by moving electrons. This calculation has been published elsewhere [37], and we summarize the main results.

The initial state is an electron of mass m moving at velocity **v**. The final state is a zeron with a relativistic mass n (but no rest mass) moving at velocity

V at angle ϕ with respect to **v**, plus the electron with mass m' moving at velocity **v'** at angle θ. The solution to this problem in the perturbative limit where $n \ll m$ is straightforward, but is not to be found in any of the textbooks on special relativity. If $n < 10^{-5}\, m$, then [37]

$$\mathbf{v \cdot V} = c^2.$$

This result includes the de Broglie velocity relationship $vV = c^2$, but it goes beyond that. It shows that the forward velocity V_f of the scattered zeron with respect to the incident velocity **v** of the electron is c^2/v for all values of the zeron scattering angle ϕ. Hence the de Broglie phase wave is accurately planar (and not spherical). Also, we can combine the zeron model shown in Fig. 2, which reproduces the rotational frequency $\omega = mc^2/\hbar$ of the de Broglie phase wave, with the dynamically calculated phase velocity $V = c^2/v$, to directly obtain the de Broglie wavelength $\lambda = 2\pi V/\omega = h/mv$.

Mechanical masses and constituent quarks

An examination of the observed spin and magnetic moment of the electron, together with its point-like charge, led to the phenomenological conclusion that the electron is in the form of a relativistically spinning sphere of "mechanical" matter with a point charge e located on the equator of the RSS, as shown in Fig. 1. This raises the question of the nature of the masses of the other 200 or so elementary particles and resonances [9]. Are they also formed from mechanical matter? These particle states can be reproduced by a subset of "quark" states, which passes the mass question on from the particles themselves to the quarks of which they are composed. What are quarks made of?

The Standard Model of quantum chromodynamics (QCD) is widely accepted as the correct way to deal with the quark systematics, and it does an excellent job of correlating the principle quantum numbers of the observed particle states. However, it does not *per se* give us direct information about quarks masses. Kurt Gottfried and Victor Weisskopf (1984) summarized this situation as follows: [38]

> Unfortunately, QCD has nothing whatsoever to say about the quark mass spectrum, nor, for that matter does any other existing theory.

And Richard Feynman (1985) echoed their sentiments: [39]

> Throughout this entire story there remains one especially unsatisfactory feature: the observed masses of the particles, m. We use these numbers in all our theories, but we do not understand them—what they are or where they come from. I believe that from a fundamental point of view, this is a very interesting and serious problem.

The difficulty with quark masses is that the quarks cannot be removed from the particles and examined individually. We can only infer their properties from those of the particles within which they reside. The two basic types of quarks that have been considered are *constituent* quarks, wherein the masses of the quarks add up to form the mass of the particle, and *current* quarks, wherein the "gluon" currents that bind the quarks together contribute most of the mass at low energies and a substantial fraction of the mass at higher energies. The Standard Model features current quarks. However, if we look directly at the mass values of the metastable (longer-lived) particles, they have the linear relationships that we logically expect to find with constituent quarks [40, 41]. Also, the interval spacings between related low-mass particles reveal the existence of a basic constituent-quark excitation quantum X. We illustrate these statements with the minimal constituent-quark model shown in Table I. The lowest-mass strongly-interacting particle is the π meson, which is spinless and is reproduced as a quark-antiquark pair. The $\pi^{\pm}$ mass of 139.57 MeV matches the mass $m_e(2/\alpha) = 140.05$ MeV, and it suggests $m_e/\alpha = 70$ MeV as a basic constituent-quark mass quantum. We assume that this is the spinless mechanical mass $M(70)$. If we further assume that M is in the form of a Compton-sized uniform sphere of matter, and if we set it into RSS rotation, its mass becomes half again as large, as shown above, and it has a calculated spin of $\frac{1}{2}\hbar$. This gives $M_S(105)$ as the basic spin $\frac{1}{2}$ constituent-quark mass quantum. With these two basis states, and with the assumption of zero binding energy, we can accurately reproduce the absolute masses of the stablest mesons and baryons, and also the masses of the leptons, as is demonstrated in Table I. Furthermore, we can account for the combinations of constituent quarks that occur in the *low-mass* resonances in terms of the cross-over excitation $X(420)$, and we can reproduce the *high-mass* thresholds in terms of characteristic mass triplings of lower-mass quarks.

The quarks Q, S, C, and B correspond to quark states in the Standard Model. The spinless quark M and its spinning counterpart M_S do not form part of the SM systematics, but they generate the SM quarks, and they accurately reproduce the masses of the lowest (and therefore most basic) lepton and meson states. Two points to note in Table I are the overall accuracy of the mass values and the comprehensiveness of the results. All of the principle stable particles and particle-channel threshold resonances are accurately reproduced in Table I with no adjustable parameters. By way of contrast, the Standard Model requires, according to one count, 19 adjustable parameters in order to obtain its fits to the data, although it should be noted that the SM accounts for many more properties of the elementary particles than just their masses and quantum numbers.

In addition to reproducing the masses of particles in terms of spinless $M(70)$ and spinning $M_S(105)$ mass units, Table I also shows that the lowest-mass narrow-width resonances and particle states are spaced by mass intervals which are multiples of an excitation quantum $X(420)$. This quantum has an interesting phenomenological explanation. The mass transformation $3M(210) \Leftrightarrow 2M_S(210)$ is isoergic, which suggests a mechanism for transforming between spinless and spinning mass quanta. However, if we apply this to the generation of a $\mu\bar{\mu}$ (211) pair (see Table I), the $\mu\bar{\mu}$ pair is particle-antiparticle symmetric, whereas a combination of three M and $\bar{M}$ quanta is not. The lowest symmetric transformation quantum is $X = 6M = 4M_S = 420$ MeV, as

Table I. The minimal constituent-quark mechanical mass model. The basic CQ masses are the spinless quantum $M = 70$ MeV and its RSS spin 1/2 counterpart $M_S = 105$ MeV, which occur in particle and antiparticle forms, and are used here with zero binding energy. The excitation quantum $X \equiv M\bar{M}M \cdot \bar{M}M\bar{M} = 2M_S\bar{M}_S = 420$ MeV is the lowest-mass symmetric cross-over spin excitation, and it dominates low-mass particle production. The higher-mass threshold resonances feature mass triplings of the S quark. The particle mass values shown here are from Ref. [9].

Spin 0 low-mass mesons

$M(70)$	$M\bar{M}$	$M\bar{M} \cdot X$	$M\bar{M} \cdot XX$	$M \cdot X$
	$M\bar{M}(140)$	$4M\bar{M}(560)$	$7M\bar{M}(980)$	$7M(490)$
	$\pi^{\pm}(140)$	$\eta(547)$	$\eta'(958)$	$K(494)$

Spin ½ lepton and nucleon thresholds

$M_S(105)$	$M_S\bar{M}_S$	$M_S\bar{M}_S \cdot 4X$	$M_S\bar{M}_S \cdot 8X$
	$M_S\bar{M}_S(210)$	$9M_S\bar{M}_S(1890)$	$17M_S\bar{M}_S(3570)$
	$\mu\bar{\mu}(211)$	$p\bar{p}(1877)$	$\tau\bar{\tau}(3554)$

Spin ½ quark masses

$M_S(105)$	$3M_S(315)$	$M_S \cdot X(525) \equiv S$	$3S(1575)$	$3C(4725)$	$9B(42525)$
	$Q \equiv U,D(315)$	$S(525)$	$C(1575)$	$B(4725)$	$WZ(42525)$

Spin 1 meson threshold resonances

$M_S\bar{M}_S \cdot X$	$M_S\bar{M}_S \cdot XX \equiv S\bar{S}$	$3S\bar{S}$	$9S\bar{S}$	$9 \times 9S\bar{S}$
$3M_S\bar{M}_S(630)$	$S\bar{S}(1050)$	$C\bar{C}(3150)$	$B\bar{B}(9450)$	$9B\bar{B}(85050)$
$Q\bar{Q}(630)$	$\phi(1020)$	$J/\psi(3097)$	$\Upsilon(9460)$	$\frac{1}{2}(W+Z)(85806)$

Spin 0 mixed meson resonances

$Q\bar{C}(1890)$	$Q\bar{B}(5040)$	$S\bar{C}(2100)$	$S\bar{B}(5250)$	$C\bar{B}(6300)$
$D(1869)$	$B(5279)$	$D_S(1969)$	$B_S(5369)$	$B_C(6400)$

Baryon octet resonances

$QQQ(945)$	$QQS(1155)$	$QSS(1365)$	$SSS(1575)$
$p(938)$	$\Lambda(1116), \Sigma(1193)$	$\Xi(1321)$	$\Omega(1672)$

Charmed and bottom baryon resonances

$QQC(2255)$	$QSC(2415)$	$QQB(5355)$
$\Lambda_C(2285), \Sigma_C(2455)$	$\Xi_C(2466)$	$\Lambda_B(5624)$

shown at the top of Table I. Multiples of this mass spacing appear in the spinless low-mass mesons (X and $2X$), in the lepton and nucleon pair production thresholds ($4X$ and $8X$), and in the strange S quark ($S = M_S \cdot X$). Other quark masses are $Q = 3M_S$, $C = 3S$, $B = 3C$, and $WZ = 9B$, where WZ is a hypothetical quark which gives rise to the observed W and Z spin 1 resonances. (The actual existence of the WZ quark would be manifested as (e.g.) a $\bar{Q} \cdot WZ$ resonance at half the W and Z mass values.)

The U and D quarks of the Standard Model correspond to the Q quark in Table I. The mass splittings of different charge states in the same resonance (*e.g.*, D(1869)) can be attributed to slightly different U and D mass values. We note that the $\pi = M\overline{M}$ mass splitting is almost exactly equal to 9 electron masses, as shown at the bottom of Table I.

The mass values shown in Table I include all of the basic lepton, meson, and baryon threshold particles—roughly 22 states in all. The fact that one common set of masses accurately works for all of these various types of particles shows that they bear a deep relationship to one another, which from the constituent-quark point-of-view is suggestive of a common set of mechanical masses.

In jumping from the electron mass m_e to the 70 MeV mechanical mass M, we invoked a scaling in $\alpha \cong 1/137$. It is worthwhile to mention that the *lifetimes* of the metastable particles form groups that correlate with their quark structures [42], and these lifetime groups are spaced by powers of α, over a range of lifetimes that extends from the long lifetime of the neutron to the shorter lifetimes over a span of 11 powers of α, or 23 orders of magnitude [43].

The mass of the neutrino

In the mass analyses of the present paper, we demonstrated that the spin angular momentum of the electron can be reproduced by a Compton-sized relativistically spinning sphere of non-interacting "mechanical" matter. We also reproduced the spin angular momentum of the massless photon by the rotation of P-H "particle-hole" excitations of the vacuum state. These results suggest that a particle such as the neutrino, which carries a spin of ½, should logically have a mass that is responsible for its spin. If we start with the RSS electron model shown in Fig. 1, and then remove its charge, we are left with a spinning sphere of matter that has a spin of ½, that does not interact appreciably with other objects, and that carries the lepton quantum number. These are the salient properties of the neutrino. Hence the studies on the electron imply that spin ½ neutrinos have masses. At the time that these ideas were first set forth [14], the prevailing view was that neutrinos are chirally invariant particles which have zero rest mass and travel at the velocity c. Thus the recent confirmation of a finite mass for the neutrino [6] serves to reinforce the present notion that where there is spin angular momentum, there is also mass.

References

[1] A. Pais, *'Subtle is the Lord...'* (Oxford University Press, New York, 1982), p. 159.

[2] For a more complete discussion of the electron, and for additional references, see M. H. Mac Gregor, *The Enigmatic Electron* (Kluwer Academic, Dordrecht, 1992).

[3] See for example B. T. Feld, *Models of Elementary Particles* (Blaisdell, Waltham, 1969), pp. 362-4, where a lower limit of 5 GeV/c² for the quark mass is quoted.

[4] See P. J. E. Peebles, "Making Sense of Modern Cosmology," *Scientific American*, Jan. (2003), pp. 54-55.

[5] K. A. Milton, *The Casimir Effect: Physical Manifestations of Zero-Point Energy* (World Scientific, River Edge, N.J., 2001).

[6] See "Direct Measurement of the Sun's Total Neutrino Output Confirms Flavour Metamorphosis," *Physics Today*, July (2002), pp. 13-15.

[7] J. P. Ostriker and P. J. Steinhardt, "The Quintessential Universe," *Scientific American*, Jan. (2003), pp. 46-53.

[8] A. O. Barut, in *The Electron: New Theory and Experiment*, D. Hestenes and A. Weingartshofer, (eds) (Kluwer Academic, Dordrecht, 1991), p. 108.

[9] These values are taken from "Review of Particle Properties," *Phys. Rev.* **D66**, Part I, July 1 (2002).

[10] A.G. W. Cameron, in *The Crab Nebula*, R. D. Davies and F. G. Smith (eds) (Reidel, Dordrecht, 1971), p. 327.

[11] J. P. Ostriker, *Scientific American*, Jan. (1971), p. 48.

[12] C. Møller, *The Theory of Relativity* (Clarendon Press, Oxford, 1952), p. 318, Eq. 42.

[13] R. P. Feynman, R. B. Leighton, and M. Sands, *The Feynman Lectures in Physics, Vol. II* (Addison-Wesley, Reading, 1964), p. 34-3.

[14] M. H. Mac Gregor, *Lett. Nuovo Cimento* **4**, 211 (1970).

[15] P. Ehrenfest, *Phys. Z.* **10**, 918 (1909).

[16] E. Dewan and M. Beran, *Am. J. Phys.* **27**, 517 (1959); E. M. Dewan, *Am. J. Phys.* **31**, 383 (1963).

[17] M. H. Mac Gregor, *Lett. Nuovo Cimento* **30**, 427 (1981).

[18] M. H. Mac Gregor, *Lett. Nuovo Cimento* **43**, 49 (1985). Also see ref. [2], Ch. 12.

[19] M. H. Mac Gregor, *The Nature of the Elementary Particle* (Springer-Verlag, Berlin, 1978), pp. 79-81. Also see ref. [2], Ch. 7.

[20] See ref. [2], p. 103.

[21] See ref. [2], Ch. 7.

[22] See ref. [2], p. 124.

[23] See ref. [2], p. 72.

[24] See ref. [2], p. 122.

[25] See ref. [2], Ch. 16.

[26] See ref. [2], Ch. 8.

[27] See ref. [2], p. 73.

[28] R. P. Feynman, in *The Quantum Theory of Fields, Proceedings of the Twelfth Conference on Physics at the University of Brussels, October, 1961*, R. Stoops (ed) (Interscience, New York), pp. 75-76.

[29] Ref. [2], pp. 64-66.

[30] See *Physics Today*, Feb. 2003, p. 44. Also see J. D. Jackson, *Classical Electrodynamics* (Wiley, New York, 1962), p. 593.

[31] A. O. Barut, ref. [8], p. 109.

[32] Mott helical channelling calculations are discussed in detail in ref. [2], Ch. 16.

[33] Mott scattering experiments are described in ref. [2], Ch. 17.

[34] M. H. Mac Gregor, *Found. Phys. Lett.* **8**, 135-160 (1995); also in *The Present Status of the Quantum Theory of Light*, S. Jeffers, S. Roy, J-P. Vigier and G. Hunter (eds) (Kluwer Academic, Dordrecht, 1997), pp. 17-35.

[35] This result seems to be original with the present author (see first ref. in [34], pp. 140-141).

[36] C. Møller, *The Theory of Relativity* (Oxford Univ. Press, London, 1955) pp. 6-7, 51-52, 56-58.

[37] M. H. Mac Gregor, *Lett. Nuovo Cimento* **44**, 697 (1985); also in *Causality and Locality in Modern Physics*, G. Hunter, S. Jeffers and J-P. Vigier (eds) (Kluwer Academic, Dordrecht, 1998), pp. 359-364.

[38] K. Gottfried and V. F. Weisskopf, *Concepts of Particle Physics, Volume I* (Clarendon Press, Oxford, 1984), p. 100.

[39] R. P. Feynman, *QED, The Strange Theory of Light and Matter* (Princeton University Press, Princeton, 1985), p. 152.

[40] M. H. Mac Gregor, *Phys. Rev.* **D9**, 1259-1329 and **D10**, 850-883 (1974); also ref. [19].

[41] M. H. Mac Gregor, *Nuovo Cimento* **103A**, 983-1052 (1990).

[42] Ref. [41], Fig. 6.

[43] Ref. [41], Fig. 5.

The Electron in a (3+3)-Dimensional Space-Time

Paolo Lanciani[1] and Roberto Mignani[1–4]
[1]Dipartimento di Fisica "E. Amaldi"
Università degli Studi di Roma "Roma Tre"
Via della Vasca Navale, 84 – 00146 Roma, Italy
[2]I.N.F.N., Sezione di Roma III
[3]INDAM – GNFM
[4]e-mail: mignani@fis.uniroma3.it

We review the basic features of a model of the electron in a 6-dimensional space-time with two extra time dimensions. The electron is assumed to be a massless particle in this space, and it acquires mass when considered in the usual 4-D space-time. Its spin and magnetic moment are obtained by an effect of polarization, which gives rise—in the three-dimensional temporal space—to a helicoidal motion along the usual time axis with radius of the order of the Compton wavelength of the electron. In this model, chirality is nothing but the direction of rotation of the electron in the "time" space. It is also shown that a connection can be established between the 6-D space-time and a recently developed Kaluza-Klein-like formalism with energy as fifth dimension.

1. Introduction: time and the electron

1.1 How many times?

What is time? Many definitions exist, all very different one from the other, from psychological to physical. For example, Rovelli [1] has analyzed the distinct roles that the notion of time plays in different scientific theories and found ten versions of the concept of time, from "natural time" to "no time," all used in natural sciences. From a physical point of view and in the framework of special relativity, time is the fourth dimension of the Minkowski space-time, namely the component of the position four-vector $dx^\mu = (dt, \mathbf{dr})$ ($c = \hbar = 1$). This representation is so well established experimentally that a necessary requirement to be satisfied by any physically acceptable theory is its Lorentz covariance in a space-time with 4 dimensions, 3 spacelike and 1 timelike.

As is well known, there is no experimental evidence that time has more than one dimension in our Universe. This is obvious, since any matter with a different time direction would remain observable only for a very short time before disappearing almost immediately. For example, had particles been created during the Big Bang with more than one initial common time direction, they would have been immediately separated, and now only those with the same time would constitute our Universe. (The hypothesis by Stueckelberg [2] and

Feynman [3] that antimatter corresponds to particles travelling backwards in time can explain why the Universe does not seem to contain the same amounts of matter and antimatter.)

However, if space-time is a physical concept and not only a mathematical representation—as suggested indeed by Einstein's Special Relativity (SR)—the question arises: why do three space coordinates and only one time coordinate exist? The idea of a multidimensional time was considered in the second half of the past century, mainly as a possible tool to explain quantum features in a semi-classical framework [4–6]. In particular, the problem of a possible symmetry in the number of dimensions in space and time was noticed in the early seventies in the framework of the generalization of SR to faster-than-light speeds [7]. Indeed, the superluminal Lorentz transformations in four dimensions (introduced by Recami and one of the present authors (RM) [7]) change the sign of the space-time interval, thus mapping spacelike dimensions into timelike ones, and *vice versa*. This leads in a natural way to the hypothesis of an equal number of time and space dimensions. Following Demers [8] (who assumed a 3+3-dimensional space-time $M(3,3)$ and applied it to some aspects of trichromatism[*]), Mignani and Recami [9] introduced a time vector $\vec{t}$ (besides the space vector $\vec{r}$) in order to explain the appearance of the imaginary quantities entering into the 4-D generalized Lorentz transformations. A space-time point in the space $M(3,3)$ is therefore represented by a 6-vector $X_\mu = (\vec{r} \quad \vec{t})$ $(\mu = 1,...,6)[†]$. In agreement with Demers, it was assumed that only the modulus

$$t = \left(t_1^2 + t_2^2 + t_3^2\right)^{1/2} \tag{1.1}$$

is physically observable for bradyons (slower-than-light particles), whereas only the modulus of $\vec{r}$ is a meaningful quantity for tachyons (faster-than-light particles).

Since that time, much work on the "three-time" formalism has been done, with or without connection to SR [10–23].

For instance, by defining a 6-D vector electromagnetic potential $\Phi_\mu = (\vec{A} \quad \vec{B})$, one gets a 6×6 electromagnetic tensor given by [10,11]:

$$E_{\mu\nu} = \frac{\partial \Phi_\mu}{\partial X_\nu} - \frac{\partial \Phi_\nu}{\partial X_\mu}. \tag{1.2}$$

This made it possible to build up a formulation of electromagnetism in (3+3) dimensions valid for both bradyons and tachyons (thus confirming the hypothesis [7] that electrically charged tachyons behave as standard magnetic monopoles).

Averaging on time directions was also proposed [12] in order to overcome the non-linearity of the transformation equations for four-vector components.

---

[*] This kind of application relies on the well known fact that transformations in a complex 3-space are related to the unitary group *SU(3)*.

[†] Throughout the paper, an arrow denotes a three vector. Moreover, the indices 1,2,3 label the space components, whereas 4,5,6 refer to the time (with the index 4 labeling the usual time axis).

Other authors [12–14] have considered theories with a three-dimensional time on a different (often criticized) basis. In particular, Cole formulated a complete mathematical theory for a 6-D space-time, which includes the usual four-dimensional theory as a special case [17–23]. In his theory, the motion of a particle at a given space-time point $X \in M(3,3)$ in any frame is specified by the unit vector $\bar{\alpha}$ along the projection of its path in the time subspace and by its velocity $\bar{v} = d\bar{r}/dt$, where dt is measured along this path. A transformation between two inertial frames is expressed by the 6×6 matrix

$$A = \begin{pmatrix} \underline{\underline{D}} & \underline{\underline{P}} \\ \underline{\underline{Q}} & \underline{\underline{R}} \end{pmatrix}, \tag{1.3}$$

where $\underline{\underline{D}}, \underline{\underline{P}}, \underline{\underline{Q}}, \underline{\underline{R}}$ are constant 3×3 matrices. Specific transformations have been given by Cole and Buchanan [21] for the simplest non-trivial cases.

The link with the standard 4-D theory is obtained if all time vectors in a given frame are parallel to a given fixed vector $\bar{\alpha}_0$.

The basic features of any reasonable 6-D theory have been summarized by Cole and Starr [22]. It must:

a. Include the standard (3+1) theory as a special case;
b. Explain why only one time dimension is observed;
c. Make new testable predictions.

It is possible to show that the most recent version of the theory does fulfill point (i). However, the total lack of experimental evidence of a time direction other than the macroscopic, everyday dimension has limited interest in the three-time theory, which therefore has not been further explored.

Now recall that space-times with more than 4 dimensions have often been proposed, mainly in attempts to unify fundamental interactions. The first pioneering proposal is due to Kaluza [24] and Klein [25]; they introduced a five-dimensional space-time in order to unify gravitation and electromagnetism in a single geometrical structure. Their scheme, in which the coefficient of the fifth coordinate is constant, was later generalized by Jordan [26] and Thiry [27], who considered this coefficient to be a general function of the space-time coordinates. The Kaluza-Klein (KK) formalism was later extended to higher dimensions, also in the hope of achieving unification of all interactions, including nuclear (weak and strong) forces. Modern generalizations [28] of the KK scheme require a *minimum* of 11 dimensions in order to accommodate the Standard Model of electroweak and strong interactions; note also that 11 is also the *maximum* number of dimensions required by supergravity theories [29]. For a recent exhaustive review of Kaluza-Klein theories we refer the reader to an excellent monograph by Wesson [30]. In these models, the extra dimensions are always spacelike, and are not observable, because they are wrapped up in a very small region (*compactified dimensions*). There have also been other proposals where the fifth (*uncompactified*) dimension is linked with mass [31-33] or energy [34,35].

1.2   What is the electron?

The electron (discovered by J.J. Thomson in 1897) is the first and best-known elementary particle, and plays a fundamental role in many observed physical phenomena. Notwithstanding, it still remains an enigmatic object [36]. Its intrinsic features (like mass, charge, spin and magnetic moment) have been measured with a very high level of accuracy. Only the electron's size is still unknown, since many different values, from 10^{-11} to less than 10^{-16} cm, can be attributed to its radius, depending on the different phenomena under consideration. All the electron properties related to its interactions are well described in the framework of the Glashow-Weinberg-Salam electroweak theory. However, the values of the mass and of electric charge of the electron considered in Quantum Electrodynamics (QED) are not intrinsic, but result from its interaction with the vacuum. As a matter of fact, there exists presently no model of the electron able to fully account for its behaviour and features in a convincing way.

The above facts are well known and generally accepted as a consequence of the impossibility of describing the electron either classically or quantum-mechanically. This leads to the quite contradictory view that the electron is regarded as an electrically charged, point-like particle, but endowed with an unknown internal structure responsible for its spin and magnetic moment.

This situation, already summarized by Fermi in 1932 [37], was restated by Barut [38] in 1990 in the following terms: "If a spinning particle is not quite a point particle, nor a solid three-dimensional top, what can it be? What is the structure which can appear under probing with electromagnetic fields as a point charge, yet as far as spin and wave properties are concerned exhibits a size of the order of the Compton wavelength?"

Although most physicists believe that quantum theory explains everything that can be explained concerning the electron, many attempts have been and are being made to establish a model of the electron able to link together its static and dynamical properties.

Several models of the electron have been proposed [36,39–48]. They can be divided into roughly three classes, in which the electron is regarded as:

   i. A strictly point-like particle;
  ii. An actual extended particle;
 iii. An extended-like particle in which the position of the point-like charge is distinct from the particle center-of-mass.

Among the latter type of theoretical approaches (generalizations of the theory of relativistic spinning particles [43]), we may cite the following:

- Classical extended electron: Theories that use the Lorentz-Dirac equation [44] to derive information on the electron's experimental behaviour from parameters of its internal structure [45,46].
- Classical mechanical models (*e.g.*, the electron as a rigid rotating sphere with a point-like equatorial electric charge [36]).

- The electron as a point particle buffeted around, so that its actual position is spread out over a region of space vastly larger than the intrinsic size of the electron itself [36]. This motion corresponds to the *Zitterbewegung* of QED, explains the Lamb shift in atomic nuclei and is the basis of the most recent studies on electron models. In particular, if the trajectory of the electrically charged particle is a helix around the center of mass that moves like a relativistic (spinless point) particle, an oscillatory motion gives rise to the spin [42,47,48].

1.3   Does the electron carry three clocks?

Is there any connection between the two seemingly unrelated problems (namely, the number of time dimensions and a consistent model of the electron) we discussed above? The answer might be yes. Indeed, recently one of the present authors (PL) proposed a model of the electron (and of the other charged leptons) in a (3+3)-D space-time [49,50]. In this framework, the electron is assumed to be a massless point-particle subjected to an attractive field towards the (3+1) space-time. This attractive force is caused by the vacuum polarization arising when the charge of the electron is removed from the standard space-time. The electron trajectory in the "time" space (as seen in the electron rest frame) turns out to be a helix, with our usual time axis, and radius equal to the Compton wavelength. The electron's motion along the time axis (occurring at the speed of light c) creates an "arrow" of time.

Obviously, since only four different "time motions" are allowed (corresponding to the two time directions and the two directions of rotation), a four-component wave function is required in order to describe the motion of the electron in the usual 4-D space-time.

This model exhibits some interesting features, among which:

a.   It recovers the usual values for the spin and magnetic moment of the electron [49];

b.   Quantization of the electron mass follows from the existence of a discontinuity in the attractive potential [50];

c.   The time-energy uncertainty principle is recovered indirectly.

In this paper, we shall review this "three-time" model of the electron, and we will also show possible connections with the five-dimensional formalism (with energy as extra dimension) [34,35].

The content of the paper is as follows. The model of the electron in a 6-D space-time is reviewed in Section 2. Its foundations are given in Subsec.2.1. In Subsec.2.2 it is shown that the electron mass is derived as an integration constant from the Klein-Gordon equation for a massless particle in (3+3) space-time. The usual values of the spin and magnetic moment of the electron are obtained in Subsec.2.3 from the hypothesis of an attractive force caused by vacuum polarization. We show in Section 3 that the "time" spin of the electron can be identified with chirality. In Section 4 we establish a connection between the

6-D space-time and the Kaluza-Klein-like formalism with energy as the fifth dimension. Section 5 concludes the paper.

2. The electron in a 6-D space-time

2.1 Foundations of the model

The model of the electron as a particle moving in a space with three space and three time dimensions has its foundations in the following considerations [49]:

a. The existence of the spin (and therefore of the magnetic moment) of the electron can be associated with a two-dimensional motion. This motion cannot be interpreted as the spinning of the electron in ordinary space, and must therefore be thought of as motion in an internal space or in two dimensions additional to the standard (3+1) space. In the latter case, these two extra dimensions cannot be spacelike (due to the r^{-2} behaviour of the Coulomb law). The simplest hypothesis is therefore to assume *two extra time dimensions*.

b. The modern view of the problem of the electron mass (pioneered by Wheeler and Feynman [51]) assumes that it is not of electromagnetic origin (as stated by the Abraham-Lorentz-Poincaré classical model of the electron), but entirely mechanical [36]. In this case, however, the rest mass of the electron must follow from its equation of state, and not be added by hand as an external parameter. This contradiction can be overcome by assuming that the electron is a *massless* particle, which acquires mass by means of a suitable mechanism (an assumption made in the past by many authors [48,52,53]).

c. Conservation of electric charge holds strictly in the usual (3+1) space-time. As a consequence, if for any reason a charged particle leaves the standard space-time, the resulting vacuum polarization acts strongly on the particle itself in the form of *an attractive force* (which attempts to restore the original condition by bringing the particle back to its starting space-time point).

The fundamental assumptions of the model of the electron based on a 3-dimensional time are therefore [49,50]:

i. The electron (as well as any other charged lepton) is a massless particle moving in a (3+3) dimensional space-time as introduced by Demers [8], Mignani and Recami [9] and developed by Cole [12,17–22];

ii. It is subjected to an attractive force (due to vacuum polarization) toward the standard space-time if it moves in the extra-dimensional time plane. If a particle with charge $-e$ is allowed to leave our time direction toward the "time" space of the 6-D space-time, then a "hole" with charge $+e$ arises at the point of the 4-D space-time left by the particle. This gives rise to a potential $-e/\tau$ (with τ being the time distance between $-e$ and $+e$), which attracts the former charge toward the latter one.

2.2   6-D Klein-Gordon equation and the electron mass

We now discuss some implications of the above assumptions, in particular as far as the properties of the electron are concerned [49].

It follows from i) that the electron is described by a massless 6-D Klein-Gordon equation:

$$\left(p_\mu - e\Phi_\mu\right)\left(p^\mu - e\Phi^\mu\right)\varphi = 0 \tag{2.1}$$

($\mu = 1,\ldots,6$) where $p_\mu = -i\partial/\partial x^\mu \equiv \left(\vec{p}, \vec{E}\right)$ and $\Phi_\mu \equiv \left(\vec{A}, \vec{B}\right)$ are, respectively, the 6-momentum and the e.m. 6-potential.

If the spatial part of the 6-potential vanishes ($\vec{A} = 0$), equation (2.1) can be rewritten as:

$$\left(E^2 + e^2 B^2 - 2e\vec{B}\cdot\vec{E} + ie\nabla\cdot\vec{B} - p^2\right)\varphi = 0 \tag{2.2}$$

By taking (imaginary*) cylindrical time coordinates $(t, \tau, \theta)^\dagger$ and an Euclidean metric, and assuming only that $B_\tau \neq 0$, Eq. (2.2) becomes:

$$\left[\frac{1}{\tau}\frac{\partial}{\partial\tau}\left(\tau\frac{\partial}{\partial\tau}\right) + \frac{1}{\tau^2}\frac{\partial^2}{\partial\vartheta^2} + \frac{\partial^2}{\partial t^2} - e^2 B_\tau^2 + 2eB\frac{\partial}{\partial\tau} + \frac{1}{\tau}\frac{\partial}{\partial\tau}\left(\tau eB_\tau\right) + p^2\right]\varphi = 0. \tag{2.3}$$

Inserting in Eq. (2.3)

$$\varphi = f(\tau)g(t,\vec{r})e^{in\vartheta} \quad n = 0,\mp 1,\pm 2,\ldots \tag{2.4}$$

and separating variables, one gets the usual 4-dimensional Klein-Gordon equation for g and the following "time" radial equation for f:

$$f'' + \left(\frac{1}{\tau} + 2eB_\tau\right)f' + \left[e^2 B_\tau^2 + \frac{1}{\tau}\frac{d}{d\tau}\left(\tau eB_\tau\right) - \frac{n^2}{\tau^2} - m^2\right]f = 0. \tag{2.5}$$

Here, m is an integration constant that corresponds to the rest mass of the particle in the 4-D Klein-Gordon equation satisfied by g. Therefore, *the electron mass arises in this framework as an integration constant[‡]*, which reflects, in the standard (3+1) space-time, the influence of the electron motion in the two extra dimensions.

2.3   Spin and magnetic moment of the electron

According to assumption ii), we set $B_\tau = -Ze/\tau$ (where Z is a parameter characterizing the strength of the polarization attractive force). Then, Eq. (2.5) becomes:

$$\frac{d^2 f}{dz^2} + \frac{1 - 2Z\alpha}{z}\frac{df}{dz} - \left(1 + \frac{n^2 - Z^2\alpha^2}{z^2}\right)f = 0 \tag{2.6}$$

with $z = m\tau, \alpha = e^2$. The general solution of this equation is:

$$f = z^{Z\alpha}\left(AK_n(z) + BI_n(z)\right) \tag{2.7}$$

---

* We stress that the choice of imaginary time coordinates is mandatory in this framework, in order to obtain physically acceptable (*i.e.*, finite) solutions of Eq. (2.3) [49].

† In this coordinate system, t represents the usual time coordinate, whereas τ,θ parametrize position in the extra-dimensional plane.

‡ An analogous result is obtained in the model by Barut and Udal [48].

where A, B are constants and $K_n(z)$, $I_n(z)$ are the Bessel functions with imaginary arguments. The solution of Eq. (2.5) must vanish for $\tau = \infty$ and remain finite for $\tau = 0$; therefore, a physically acceptable solution is:

$$f = Az^{Z\alpha} K_n(z) \tag{2.8}$$

The MacDonald function $K_n(z)$ for $n \neq 0$ (which is required to yield a rotating motion) behaves as e^{-z} for $z \to \infty$ and as z^{-n} for $z \to 0$. Therefore, $f_n(0)$ is finite provided that $Z\alpha \geq n$. The lowest-order solution showing a dependence on ϑ and finite for $\tau = 0$ is obtained for $n = \pm 1$, and yields the lowest value 1 for $Z\alpha$.

Finally, the complete solution Ψ is obtained by taking a plane wave for the 4-D Klein-Gordon equation. In the rest frame of the particle, it reduces to:

$$\Psi = Ae^{-i(mt \pm \vartheta)} m\tau K_1(m\tau) \tag{2.9}$$

Generalizing the usual 4-D expression for the current density to the 6-dimensional case, we have

$$j_\mu \equiv \frac{1}{2}\left(\Psi^* p_\mu \Psi - \Psi p_\mu \Psi^*\right) - \frac{e}{c}\Phi_\mu \Psi \Psi^* \tag{2.10}$$

In the above hypotheses on the 6-potential, only the time-vector part of the current survives, and reads (for a particle at rest, described by Eq.(2.9)):

$$j_t = im\Psi^* \Psi \; ; \; j_\vartheta = \pm\frac{1}{m\tau} j_t \; ; \; j_\tau = \pm i j_\vartheta . \tag{2.11}$$

In ordinary units, if $R_c = \hbar/mc$ is the Compton radius of the electron, it is

$$j_\vartheta = \pm\frac{R_c}{c\tau} j_t . \tag{2.12}$$

Since, classically, $j_t = -ice$ is the charge density of the electron, we get:

$$j_\vartheta^{el.} = \pm ice \frac{R_c}{c\tau} . \tag{2.13}$$

The magnetic moment due to the rotation of the particle in the "time" space is given, by analogy with its definition for the usual space magnetic moment, by:

$$\mu = \frac{1}{2c}\left|\vec{j}^{el.} \times \vec{t}\right| = \frac{1}{2c}\left(\pm ice \frac{R_c}{c\tau}\right) ic\tau = \pm\frac{\hbar}{2}\frac{e}{mc} \tag{2.14}$$

(in the reference frame moving with velocity c along the t-axis). Expression (2.14) is the usual one for the magnetic moment of the electron when neglecting QED effects.

It is also possible to define a time angular momentum, similar to the usual angular momentum in space, given by:

$$s = \left|\vec{j} \times \vec{t}\right| . \tag{2.15}$$

In the reference frame moving with velocity c along the t-axis, the only component of $\vec{j}$ to be taken into account is $j_\vartheta = \pm imc(R_c/c\tau)$. However, the massless particle considered here can be seen as a charged photon, whose rotating motion is equivalent to the motion caused by a gravitational potential. Ac-

cording to General Relativity [54], the kinetic energy of this particle is $mc^2/2$ [*]
This implies that the "time" momentum of the electron is reduced by a factor
$\frac{1}{2}$. Therefore the time angular momentum (2.15) is

$$s = \pm \frac{1}{2} imc \frac{R_c}{c\tau} ic\tau = \pm \frac{\hbar}{2} \qquad (2.16)$$

and can obviously be interpreted as the spin of the particle.

Still on the basis of general relativistic arguments, the classical radius of
the circular orbit of a massless particle, moving at the speed of light in the
"time" space, and subject to a potential $-Ze^2/c\tau$, is given by [54]:

$$c\tau = (2Z\alpha)\frac{\hbar}{mc}. \qquad (2.17)$$

If $Z\alpha = \frac{1}{2}$, $c\tau = R_c$, the spin and the magnetic moment of the electron have the
expected values (2.16) and (2.14), respectively. Moreover, if
$KMm/c\tau = Ze^2/c\tau$ is a gravitational field equivalent to the given potential, the
gravitational radius is:

$$c\tau_g = \frac{2KM}{c^2} = \frac{2Ze^2}{mc^2} = (2Z\alpha)R_c \qquad (2.18)$$

which, for $Z\alpha = \frac{1}{2}$, coincides with R_c.

The electron moves in the time space along a helix, its axis the usual time
axis and radius equal to the Compton length.

On the other hand, the averaged value of $c\tau$, with respect to the wave
function (2.9), is given by:

$$\langle c\tau \rangle = R_c \frac{\int_0^\infty K_1^2(z)z^4 dz}{\int_0^\infty K_1^2(z)z^3 dz} \cong 1.2R_c. \qquad (2.19)$$

The slight difference between $\langle c\tau \rangle$ as calculated from Eq.(2.9) and R_c is
probably due to the fact that the assumed form of the potential, although accurate enough for τ sufficiently large, is inadequate for $\tau \to 0$. In fact, it must
be noted that the rest mass of the electron obtained as an integration constant is
not quantized, and any value for m is acceptable. However, it has been shown
that the quantization of the electron mass can be recovered by using for the
time radial potential an expression obtained when, in addition to the vacuum
polarization described by a potential of the form $-Z\alpha/\tau$, a virtual pair of particles is created between the rotating electron and its center of rotation [50].

3.   Time spin of the electron and chirality

According to relativistic quantum theory, in the standard 4-D space-time, the
average spin of the electron $2\bar{s}$ is represented by the space components of an
antisymmetric   tensor S_{ik} or   by   its   associated   axial   vector

---

[*] The reduction of the particle energy only occurs in the direction of the rotating motion in the 6-D
space, whereas it is still mc^2 in the usual 4-D space-time. This situation is similar to the one in which
relativistic effects appear only in the direction where v is comparable to c, and not in the directions in
which $v \ll c$ [49].

$a^i = -(1/2m)e^{ijkl}S_{jk}p_l$, where p_l is the energy-momentum of the particle. The space part of a^i corresponds to $2\overline{s}$ in the rest frame (but not in other reference frames).

Since the results of Subsection 2.3 were obtained from the Klein-Gordon equation in 6 dimensions, the electron spin, even though it has the same numerical value as in the Dirac theory, has a completely different physical origin. It is now a *time* angular momentum and therefore it is conserved *independently from* the usual space angular momentum. This reflects the fact that the Klein-Gordon equation describes a scalar field with no spin. It would be possible in principle to consider a 6-D Dirac equation for a spinor field [55]. However, in this case, two spins—spatial and temporal [56]—would arise (in addition to the time angular momentum caused by the rotation of the massless charge around the standard 4-D time axis). This is in contrast with the well-established experimental evidence that only one internal angular momentum exists (in the 4-D space-time).

We now consider the 6-D angular momentum tensor $M_{\mu\nu}$ (and the associated dual 6-vector $m_\nu = \varepsilon_{\nu\alpha\beta}M^{\alpha\beta}$) with $\nu,\alpha,\beta = 4,5,6$. In the rest frame of the particle (in the standard 4-D space-time), the only non-null component of $M_{\mu\nu}$ (after averaging during rotation around the standard time axis x_4) is M_{56}. The corresponding (time) component of m_ν, m_4, can be regarded as the time component of a 4-vector (in the standard 4-D space-time) with null space components. This is contrary to the usual assumption that the spin is a space 3-vector in the particle rest frame, and that the time component is different from zero only in reference frames in which the particle is moving. Obviously, under a Lorentz transformation, M_{56} does not transform as the time component of a 4-vector. However, this is not really different from the standard relativistic representation of the spin as a 4-vector whose space components, too, do not transform as the average tridimensional spin.

Let s_0 denote twice the value of the spin of a charged lepton of mass m in its rest frame. It can be considered the time component of a 4-vector, which, in a generic frame, is given by:

$$s^\alpha = \frac{s_0}{m}p^\alpha \qquad (3.1)$$

where p^α is the 4-momentum of the particle. There is a space component of s^α parallel to the momentum, which, for ultrarelativistic speeds, is almost equal to the time component. This situation is analogous to the case of the neutrino, whose spin is always parallel to momentum. Therefore, in a 4-dimensional representation, the time component of the spin must be equal to the space component in order to have $s^\mu s_\mu = 0$.

Definition (3.1) can be used when calculating the decay rate of the muon. As is well known, this rate is proportional to $\sum|M|^2$, where t and t' are the spins of the two neutrinos which are not observed and the matrix element M is given by:

$$M = \left[\bar{u}_{v_\mu} \gamma^\mu (1 - \gamma_5) u_\mu \right] \left[\bar{u}_e \gamma_\mu (1 - \gamma_5) v_{v_e} \right].$$ (3.2)

The evaluation of the squared invariant matrix element by standard methods yields

$$\sum_{t,t'} |M|^2 = 64 (p' - m_\mu s')^\alpha k_\alpha (p - m_e s)^\beta k'_\beta$$ (3.3)

where p', s', p, s, k, t and k', t' denote momentum and spin of the muon, of the electron and of the two neutrinos, respectively. Replacing s' and s with their expressions (3.1), one obtains:

$$\sum_{t,t'} |M|^2 = 64 (1 - s'_0)(1 - s_0) p'^\alpha k_\alpha p^\beta k'_\beta.$$ (3.4)

Summing over the electron spin orientations and averaging over the spin orientations of the muon yields a decay probability proportional to

$$\frac{1}{2} \sum_{s,s',t,t'} |M|^2 = \frac{1}{2} (2.2 + 0 + 0 + 0) 64 p'^\alpha k_\alpha p^\beta k'_\beta.$$ (3.5)

In an analogous way, it is possible to evaluate the squared matrix element for charged pion decay, thus getting:

$$\sum_t |M|^2 = 4 m_\mu^2 (1 - s') p^\alpha k_\alpha.$$ (3.6)

Expressions (3.4) and (3.6) show explicitly that the introduction of the term $(1 - \gamma_5)$ in the matrix element (3.2) (and the similar term for pion decay) causes only particles with spin equal to -1 to interact. Therefore, if this interpretation is valid, one can conclude that *chirality is nothing but the inverse of rotation of the charged lepton in the "time" space.*

This result is a sensible prediction of the 6-D model of the electron, susceptible to experimental verification, and is presently under investigation.

4.   5-D space-time and the 3-D time model electron

We now wish to show that it is possible to connect the (3+3) space-time model of the electron with the 5-dimensional theories [32-35], in particular with the Kaluza-Klein-like scheme [34,35] with energy as an extra dimension.

Recall that this formalism (*Deformed Relativity in Five Dimensions*, DR5) is based on a five-dimensional Riemann space—with energy ε as the fifth dimension and a metric whose coefficients depend in general on ε—and essentially seeks to provide a (local) metric description of fundamental interactions.

In order to establish this connection, we first notice that, when $Z\alpha = \frac{1}{2}$, Eq.(2.17) yields for the electron rest mass (in ordinary units)

$$m = \frac{\hbar}{c^2 \tau}.$$ (4.1)

Therefore, the following relation holds between the energy ε, as measured in the standard 4-D space-time, and the radial time τ:

$$\varepsilon^2 = c^2 p^2 + m^2 c^4 = c^2 p^2 + \frac{\hbar^2}{\tau^2} \qquad (4.2)$$

or

$$\tau = \hbar \left(\varepsilon^2 - c^2 p^2 \right)^{-\frac{1}{2}} . \qquad (4.3)$$

The interval in the (3+3) space-time reads

$$ds^2 = c^2 dt^2 + c^2 d\tau^2 + c^2 \tau^2 d\vartheta^2 - dr^2 . \qquad (4.4)$$

Obviously the term in $d\theta$ can be taken equal to zero over a complete rotation, so that (4.4) reduces to:

$$ds^2 = c^2 dt^2 + c^2 d\tau^2 - dr^2 . \qquad (4.5)$$

It follows from Eq.(4.3)

$$\frac{d\tau}{d\varepsilon} = \frac{-\varepsilon}{\left(\varepsilon^2 - c^2 p^2 \right)^{\frac{3}{2}}} . \qquad (4.6)$$

When (4.6) is substituted in (4.5), the interval becomes:

$$ds^2 = c^2 dt^2 - dr^2 + c^2 \hbar^2 \frac{\varepsilon^2}{\left(\varepsilon^2 - c^2 p^2 \right)^3} d\varepsilon^2 . \qquad (4.7)$$

Eq.(4.7) has the form of the interval in a 5-dimensional space with energy ε as an extra dimension, the fifth metric coefficient depending on the fifth coordinate.

In particular, for a particle in its rest frame, Eq.(4.7) takes the form

$$ds^2 = c^2 dt^2 - dr^2 + c^2 \hbar^2 \frac{d\varepsilon^2}{\varepsilon^4} . \qquad (4.8)$$

This interval corresponds to a metric which is obtained, in the formalism of DR5, as a solution of the Einstein equations in vacuum (viz., Class VIII solutions of the 5-D Einstein equations with $r = -4$ [34,35]).

5.   Conclusions

The main advantages of the model of the electron in a (3+3) space-time we have described can be summarized as follows:

a.   The mass of the electron (which is assumed to be a massless particle in the 6-D space) is obtained from the 6-D Klein-Gordon equation as an integration constant;

b.   The usual expressions (and therefore the experimental values) of the spin and the magnetic moment are derived by assuming an attractive force (due to vacuum polarization in the usual space-time) along the radial time direction in the extra-dimensional plane. The direction of both of them is always orthogonal to this plane, and therefore parallel to the standard time direction. The origin of the electron spin is due to the helicoidal motion of the electron in the time three-space;

c. Mass quantization follows from an assumed discontinuity in the attractive potential. This discontinuity can be generated, *e.g.*, by the presence of a virtual pair of charged particles somewhere between the center of rotation of the electron and its actual position in the "time subspace."

In this model, chirality is nothing but the inverse of electron rotation in the "time" space. We have also shown that a connection can be established between the 6-D space-time and a recently developed Kaluza-Klein-like formalism with energy as the fifth dimension.

In conclusion, the model of the electron with two extra dimensions seems to provide a good representation of its known properties. However, many questions remain still unanswered, in particular concerning the applicability of the model to the other charged leptons. It is also an open problem how neutrinos fit in this scheme.

A possible test of the reliability of the model involves chirality. Work is presently in progress to establish the feasibility of a specific experiment. In the meantime, the model of the electron presented here would appear physically sensible, and suggests that the (3+3) space-time is perhaps something more than a mere mathematical hypothesis.

References

1. Rovelli C., *Nuovo Cimento B* **110**, 81 (1995).
2. Stueckelberg E.C.G., *Helv. Phys. Acta* **15**,23 (1942).
3. Feynman R.P., *Phys. Rev.* **74**,939 (1948).
4. Bunge M., *Br. J. Philos. Sci.* **9**, 39 (1959).
5. Dorling J., *Am. J. Phys.* **38**, 539 (1970).
6. Kalitzin N., *Multitemporal Theory of Relativity* (Bulgarian Academy of Sciences, Sofia, 1975).
7. Recami E. and Mignani R., *Riv. Nuovo Cimento* **4**, 209 (1974), and references quoted therein.
8. Demers P., *Can. J. Phys.* **53**, 1687 (1975).
9. Mignani R. and Recami E., *Lett. Nuovo Cimento* **16**, 449 (1976).
10. Dattoli G. and Mignani R., *Lett. Nuovo Cimento* **22**, 65 (1978).
11. Vysin V. *Lett. Nuovo Cimento* **22**, 76 (1978).
12. Cole E.A. B., *Nuovo Cimento A* **40**,.171 (1977)
13. Pappas P.T., *Nuovo Cimento B* **68**, 111 (1982)
14. Patty C.E., *Nuovo Cimento B* **70**, 65 (1982)
15. Pavsič M., *J.Phys. A* **14**, 3217 (1981)
16. Ziino G., *Lett. Nuovo Cimento* **31, 629** (1981)
17. Cole E.A.B., *Phys. Lett. A* **75**,29 (1979)
18. Cole E.A.B., *Nuovo Cimento A* **60**, 1 (1980)
19. Cole E.A.B., *Phys. Lett. A* **76**, 371 (1980)
20. Cole E.A.B., *J. Phys. A* **13**, 109 (1980)
21. Cole E.A.B. and Buchanan S.A:, *J. Phys. A* **15**, 255 (1982).
22. Cole E.A.B. and Starr I.M., *Nuovo Cimento B* **105**, 1091 (1990).

23. Cole E.A.B., *Nuovo Cimento B* **85**, 105 (1985).

24. Kaluza Th., *Preuss. Akad. Wiss. Phys. Math.* **K1**, 966 (1921).

25. Klein O., *Z. Phys.* **37**, 875 (1926).

26. Jordan P., *Z. Phys.* **157**, 112 (1959), and references quoted therein.

27. Thiry Y., *C.R. Acad. Sci.(Paris)* **226**, 216 (1948).

28. See e.g. *Modern Kaluza-Klein Theories*, T. Appelquist, A. Chodos and P.G.O. Freund eds. (Addison-Wesley, 1987), and references quoted therein..

29. See e.g. *Supergravity in Diverse Dimensions*, A. Salam and E. Sezgin eds. (North-Holland & World Scientific, 1989), and references quoted therein.

30. Overduin J.M. and Wesson P.S., *Phys. Rept.* **283**, 303 (1997); Wesson P.S., *Space- Time- Matter – –Modern Kaluza-Klein Theory* (World Scientific, Singapore, 1999).

31. Pavsič M., *Lett. Nuovo Cimento* **17**, 44 (1976).

32. Wesson P.S., *Astron. Astrophys.* **119**, 145 (1983).

33. Wesson P.S., *Gen. Rel. Grav.* **16**, 193 (1984).

34. Cardone F., Francaviglia M. and Mignani R., *Gen. Rel. Grav.* **30**, 1619 (1998); **31**, 1049 (1999).

35. Cardone F., Francaviglia M. and Mignani R., *Found. Phys. Lett.* **12**, 281 (1999).

36. Mac Gregor M.H., *The Enigmatic Electron* (Kluwer, Dordrecht, 1992).

37. Fermi E., *Rev. Mod. Phys.* **4**, 87 (1932).

38. Barut A.O., in *The Electron: New Theory and Experiment*, D. Hestenes and A. Weingartshofer eds. (Kluwer, Dordrecht, 1991).

39. See e.g. *The Theory of the Electron*, J. Keller and Z. Oziewicz eds. (UNAM, Mexico City, 1997); *Electron Theory and QED*, J. Dowling ed., NATO ASI Series B Physics, Vol. 358 (Plenum, New York, 1997).

40. Salesi G. and Recami E., *Phys. Lett. A* **190**, 137 (1994); **195**, 389 (1994).

41. Bunge M., *Nuovo Cim.* **1**, 977 (1955).

42. Recami E. and Salesi G., *Phys. Rev. A* **57**, 98 (1998).

43. Barut A.O., *Mod. Phys. Lett. A* **7**, 1381 (1992).

44. Dirac P.A.M., *Proc. Roy. Soc. A* **167**, 148 (1938).

45. Gutkowski D., Moles M. and Vigier J.P., *Nuovo Cim. B* **39**, 193 (1977).

46. Caldirola P., *Riv. Nuovo Cim.* **2**, 1 (1979), and references therein.

47. Barut A.O. and Zanghi N., *Phys. Rev. Lett.* **52**, 2009 (1984).

48. Barut A.O. and Udal N., *Phys. Rev. A* **40**, 5404 (1989).

49. Lanciani P., *Found. Phys.* **29**, 251 (1999).

50. Lanciani P., *Found. Phys. Lett.* **14**, 541 (2001).

51. Wheeler J.A and Feynman R. P., *Rev. Mod. Phys.* **17**, 157 (1945). *ibidem*, **21**, 425 (1949).

52. Hestenes D., *Found. Phys.* **15**, 63 (1985); **20**, 1213 (1990); **23**, 365 (1993).

53. Pavšič M., *Lett. N. Cim.* **17**, 44 (1976).

54. Levi Civita T., *The Absolute Differential Calculus* (Blackie, London, 1954), p.403.

55. Patty C.E. and Smalley L.L., *Phys. Rev. D* **32**, 892 (1985).

56. Boyling J.B. and Cole E.A.B., *Int. J. Theor. Phys.* **32**, 801 (1993).

Can Studying the Inertial and Gravitational Properties of the Electron Provide us with an Insight into its Nature?

Vesselin Petkov
Science College, Concordia University
1455 de Maisonneuve Boulevard West
Montreal, Quebec, Canada H3G 1M8
vpetkov@alcor.concordia.ca

It is shown that both the classical spherical electron and a point-like electron considered in the framework of quantum electrodynamics predict the correct form of the expressions for the inertial and gravitational forces acting on a non-inertial electron. Surprisingly, the two models turn out to be indistinguishable if formulated in terms of a radically new idea of the electron structure (4-atomism), which explains why the correct forms of the expressions for the inertial and gravitational forces can be derived from the two models. That idea also appears to shed some light on a number of quantum mechanical puzzles.

1. Introduction

Studies of the nature of the electron deal, as a rule, with its quantum mechanical features. This paper takes a different approach. In Section 2 the inertia and gravitation of the classical electron are discussed. The classical model regarding the electron as a small charged sphere leads to correct expressions for the inertial force acting on an accelerating electron and for the gravitational force acting on an electron at rest in a gravitational field. As these forces originate from the interactions of different elements of the charged sphere (representing the classical electron) it is concluded that their derivation from the classical model is an indication that the real electron is unlikely to be a point-like particle of a size smaller than 10^{-18} m as experiments probing its scattering properties appear to suggest [1]. In Section 3 the inertial and gravitational properties of the electron in terms of quantum electrodynamics (QED) are studied. It is shown that a point-like model of the electron in QED yields expressions for the inertial and gravitational forces in which the forces are proportional to the acceleration and mass of the electron. Section 4 analyzes arguments that indicate that the electron cannot be a point-like object and considers a model (4-atomism) which appears to resolve several puzzles: (i) why is an electron not localized when not measured, (ii) the stability problem of the classical electron, (iii) the meaning of the scattering experiments according to which the electron appears to be a particle localized in an area of dimensions smaller that $10^{-18}\,m$ (which contradicts the wave-like nature of the electron), and (iv) why do *both* the classical electron and a point-like electron in QED lead to correct forms of the expressions of the inertial and gravitational forces acting on a non-inertial electron (accelerating and supported in a gravitational field, respectively). It is also shown in the last section that the 4-atomistic model of the electron may shed some light on a number of difficult quantum mechanical problems.

2. Inertial and gravitational properties of the classical electron

In the classical model of the electron, its charge is uniformly distributed on a spherical shell. This model offers a mechanism responsible for the electron's inertia and mass. The repulsion of the charge elements of an electron in uniform motion cancels out exactly and there is no net force acting on the electron. If, however, the electron is accelerated with respect to an inertial reference frame I its electric field distorts, which causes the repulsion of its elements to become unbalanced. As a result the non-inertial (accelerated) electron experiences a self force $F^a{}_{self}$ that resists its acceleration: it is precisely this resistance that we call inertia (why the repulsion of the different parts of an accelerating electron becomes unbalanced is explained elsewhere [2]).

The calculation of the self force is easier in the non-inertial (accelerated) reference frame N^a in which the electron is at rest. At first, it appears that the field of the electron is not distorted in N^a since it is at rest in N^a, which would mean that no force is acting on the electron. If this were the case, there would be a problem: an inertial observer in I and a non-inertial observer in N^a would disagree on whether or not the electron is subjected to a force. As the existence of a force is an absolute fact, all observers should recognize it. This problem disappears when a corollary of general relativity—that the average velocity of light in non-inertial reference frames is anisotropic—is taken into account in the calculation of the electron field in N^a [3]. The average anisotropic velocity of electromagnetic disturbances (for short light) in N^a is [3]:

$$c^a = c\left(1 - \frac{\mathbf{a} \cdot \mathbf{r}}{2c^2}\right), \tag{1}$$

where $\mathbf{a}$ is the proper acceleration of N^a and $\mathbf{r}$ is a radius vector representing the path of a light signal between two points separated by the distance $r = |\mathbf{r}|$. The anisotropic velocity of light (1) causes the distortion of the electric field of the electron in N^a. As a result the balance in the mutual repulsion of all pairs of charged elements de_1 and de_2 of the spherical shell representing the classical electron is disturbed and the resulting net (self) force acting on the electron turns out to be precisely equal to the inertial force:

$$\mathbf{F}^i{}_{self} = -m^i \mathbf{a}. \tag{2}$$

Detailed calculations of (2) are given in [4]. The coefficient of proportionality $m^i = U/c^2$ in (2) represents the inertial mass of the electron, where

$$U = \frac{1}{8\pi\varepsilon_o} \iint \frac{de_1 de_2}{r}$$

is the energy of the electron field. Therefore the inertial mass of the classical electron is electromagnetic in origin since it is the mass that corresponds to the energy of the electron field. Equation (2) is an important result for three reasons: (i) it reveals that both inertia and mass of the classical electron have electromagnetic origin [5-8]; (ii) it demonstrates that inertia is a local phenomenon contrary to Mach's hypothesis that the local property of inertia has a non-local origin [9], and (iii) it constitutes a derivation of Newton's second law $\mathbf{F} = m\mathbf{a}$ [10]—a law that had been considered so fundamental that after Newton postulated it few attempts were made to derive it.

Consider now a classical electron which is supported in the Earth's gravitational filed. The average velocity of light in the non-inertial reference frame N^g in which the electron is at rest is also anisotropic [3]:

$$c^g = c\left(1 + \frac{\mathbf{g} \cdot \mathbf{r}}{2c^2}\right), \tag{3}$$

where $\mathbf{g}$ is the gravitational acceleration. The anisotropic propagation of light in N^g distorts the electron field, which in turn disturbs the balance in the mutual repulsion of its charged elements. The resulting unbalanced repulsion gives rise to a net (self) force [4]

$$\mathbf{F}^g_{self} = m^g \mathbf{g}, \tag{4}$$

where $m^g = U/c^2$ is again the electromagnetic mass of the electron, interpreted in this case as its passive gravitational mass.

The self-force (4) which acts upon the electron on account of its own distorted field is directed parallel to $\mathbf{g}$ and is traditionally called the gravitational force. It resists the deformation of the electron field caused by the fact that the electron is at rest in the Earth's gravitational field and is therefore prevented from falling, i.e. prevented from following a geodesic path. The only way for the electron to keep its field from getting distorted is to compensate the anisotropy in the propagation of light in N^g by falling with an acceleration $\mathbf{g}$; the electric field of a falling electron turns out to be the Coulomb field, and therefore the electron does not resist its fall [3, 4]. This sheds some light on the fact that in general relativity a falling particle, represented by a geodesic worldline, is moving non-resistantly. Since a non-resistant motion is motion by inertia, a particle falling in a gravitational field is moving by inertia.

As a Coulomb field is associated with a non-resistantly moving electron (represented by a geodesic worldline) it follows that $\mathbf{F}^g_{self}$ is, in fact, an *inertial* (not gravitational) force since it resists the deviation of the electron from its geodesic path [11]. That is, $\mathbf{F}^g_{self}$ resists the deviation of the electron from its motion by inertia. Therefore, the nature of the force acting upon a classical electron at rest in a gravitational field is inertial and is purely electromagnetic in origin as seen from (4), which means that the electron passive gravitational mass m^g in (4) is also purely electromagnetic in origin. It is clearly seen from here why the inertial and the passive gravitational masses of the classical electron are equal. As the self-force (4) is inertial in origin it follows that what is traditionally called passive gravitational mass is, in fact, inertial mass. This becomes evident from the fact that the two masses are the measure of resistance an electron offers when deviated from its geodesic path. In flat spacetime, when the worldline of an electron is not a straight line, the force that resists the deformation of the worldline is $\mathbf{F}^i_{self} = -m^i\mathbf{a}$ [14], whereas in curved spacetime the same force that resists the deviation of the electron from its geodesic path is $\mathbf{F}^g_{self} = m^g\mathbf{g}$, where m^i and m^g are the measures of resistance (inertia) in these cases. The two resistance forces are equal for $|\mathbf{a}| = |\mathbf{g}|$, and therefore $m^i = m^g$. This equivalence also follows from the fact that m^i and m^g are the *same* thing—the mass associated with the energy of the electron field.

The self forces (2) and (4) originate from the unbalanced repulsion of the charged elements of the classical electron. The equations (2) and (4) represent the correct expressions for the inertial force acting on an accelerating classical electron and for the (gravitational) force to which an electron supported in a gravitational field is subjected. This fact appears to suggest that inertia and the inertial and passive gravitational mass of the real electron may be also caused by unbalanced self interactions of its charge which would not be possible if the electron were point-like. Let us now examine the inertial and gravitational properties of the electron in the framework of QED and see whether we can gain some additional insight into its nature.

3.   Inertial and gravitational properties of the electron in quantum electrodynamics

We have seen that both the inertial and gravitational forces acting on the classical electron originate from the self interaction of its charge through its distorted field.

In QED the quantized electric field of a charge is represented by a cloud of virtual photons that are constantly being emitted and absorbed by the charge. It is believed that the attraction and repulsion electric forces between two charges interacting through exchange of virtual photons originate from the recoils the charges suffer when the virtual photons are emitted and absorbed.

A free charge is not subjected to any self force since the recoils from the emitted and absorbed virtual photons constituting its own undistorted electric field cancel out exactly. Therefore, in terms of QED a charge is moving non-resistantly (and is represented by a geodesic worldline) if the recoils from the emitted and absorbed virtual photons completely cancel out.

The field of a non-inertial electron, however, is distorted. A distorted field in QED manifests itself in the anisotropy in the average velocity of the virtual photons comprising the electron field, which leads to the general relativistic red/blue shifts of the frequencies of the virtual photons that are absorbed by the non-inertial electron; the recoils from the emitted virtual photons always cancel out since they are emitted with the *same* initial frequency and wavelength as seen by the electron. Hence, virtual photons coming from different directions before being absorbed by the electron have different frequencies (and wavelengths) and therefore different momenta. But since it is the momentum of a photon that determines the recoil felt by an electron when the virtual photon is absorbed, the balance in the recoils a non-inertial electron experiences will be disturbed, and a self force acting upon the electron will arise.

This means that in QED the interaction of a non-inertial electron with its own distorted field also gives rise to a self force that is electromagnetic in origin. It should be stressed that the mechanism that gives rise to this self force is the *accepted* mechanism responsible for the origin of attraction and repulsion forces in QED, which in the case of a non-inertial electron, described in the non-inertial reference frame in which it is at rest, should take into account the anisotropic propagation of light there and the resulting frequency shift of the incoming virtual photons. Therefore QED and the general relativistic shift of the frequency of the virtual photons absorbed by a non-inertial charge do lead to a self force which acts on the charge—an effect that has been overlooked so far.

It seems that the inertial and gravitational properties of at least two models of the electron—a point-like electron and an electron whose charge has a spherical distribution—can be studied in QED. However, a semiclassical calculation of the simpler case—a point-like electron—can be carried out relatively easily in order to determine what kind of self force is acting on a non-inertial electron in this case.

Consider an accelerating electron. In the accelerating frame N^a where the electron is at rest the frequencies of the virtual photons coming from a direction $\mathbf{n} = \mathbf{r}/r$ toward the electron (as seen by the electron) can be written in the vector form

$$f^a = f\left(1 - \frac{\mathbf{a} \cdot \mathbf{r}}{c^2}\right),$$

where f is the frequency measured at $\mathbf{r} = 0$. Here $r = |\mathbf{r}|$ is the (half) distance traveled by a virtual photon during its lifetime. An incoming virtual photon of frequency f^a has energy

$$\Delta E^a = \Delta E\left(1 - \frac{\mathbf{a} \cdot \mathbf{r}}{c^2}\right).$$

As virtual particles are off-mass-shell particles their momenta are not equal to E/c. For this reason the momentum of a virtual photon approaching the electron can be written as

$$\Delta p^a = \frac{\Delta E^a}{c}\beta = \frac{\Delta E \beta}{c}\left(1 - \frac{\mathbf{a} \cdot \mathbf{r}}{c^2}\right),$$

where β is a real number.

In order to calculate the self force caused by the recoils the accelerating electron experiences during the absorption of the incoming virtual photons, several explicit assumptions should be made. Like the energy of the electric field of the classical electron, which is determined at a given moment of time, the energy of the electron field in QED can be defined as the energy of all virtual photons also determined at a given moment. This energy can be alternatively defined as the energy of the virtual photons that are absorbed during some characteristic time Δt. The second definition is equivalent to the first since it appears natural to assume that during the characteristic time the electron renews its field; so, for every Δt the energy of the absorbed virtual photons will be equal to the energy of all virtual photons at a given moment of time. We will also assume that a virtual photon is absorbed for a time $\Delta \tau$.

The lifetimes of the virtual photons absorbed by the electron can be expressed in terms of the characteristic time Δt as $\alpha \Delta t$, where α is a real number. By the uncertainty principle the energy of a virtual photon of lifetime $\alpha \Delta t$ is proportional to $\hbar / \alpha \Delta t = \Delta E^a / \alpha$. The distance traveled by a virtual photon during its lifetime $\alpha \Delta t$ is $\alpha r = \alpha c \Delta t$.

Assume that the number of the virtual quanta coming from a direction $\mathbf{n}$ within the solid angle $d\Omega$ which are absorbed during the characteristic time Δt is X. The total momentum of all X virtual photons is

$$\sum_{i=1}^{X} \Delta p^a_{\alpha_i \beta_i} \mathbf{n} \, d\Omega = \sum_{i=1}^{X} \frac{\Delta E^a \beta_i}{c \alpha_i} \mathbf{n} \, d\Omega = \sum_{i=1}^{X} \frac{\Delta E \beta_i}{c \alpha_i} \left(1 - \frac{\mathbf{a} \cdot \mathbf{r}}{c^2} \alpha_i \right) \mathbf{n} \, d\Omega .$$

The recoils from all X give rise to a force:

$$d\mathbf{F}^i{}_{self}{}^a = \sum_{i=1}^{X} \frac{\Delta p^a_{\alpha_i \beta_i}}{\Delta \tau} \mathbf{n} \, d\Omega.$$

The *unbalanced* recoils of all virtual photons, coming from all directions toward the electron and absorbed during the time Δt, produce a self force acting on the electron:

$$\begin{aligned}
\mathbf{F}^i_{self} &= \int \sum_{i=1}^{X} \frac{\Delta p^a_{\alpha_i \beta_i}}{\Delta \tau} \mathbf{n} \, d\Omega = \int \sum_{i=1}^{X} \frac{\Delta E \beta_i}{c \Delta \tau \alpha_i} \left(1 - \frac{\mathbf{a} \cdot \mathbf{r}}{c^2} \alpha_i \right) \mathbf{n} \, d\Omega \\
&= \int \sum_{i=1}^{X} \frac{\Delta E \beta_i}{c \Delta \tau \alpha_i} \mathbf{n} \, d\Omega - \int \sum_{i=1}^{X} \frac{\Delta E \beta_i}{c^3 \Delta \tau} (\mathbf{a} \cdot \mathbf{r}) \mathbf{n} \, d\Omega
\end{aligned} \qquad (5)$$

Due to symmetry the first integral in (5) is zero and for the self force we can write (noticing that $\mathbf{r} = \mathbf{n} \, r$ and $r = c \Delta t$):

$$\mathbf{F}^i_{self} = -\sum_{i=1}^{X} \frac{\Delta E \Delta t \beta_i}{c^2 \Delta \tau} \int (\mathbf{a} \cdot \mathbf{n}) \mathbf{n} \, d\Omega . \qquad (6)$$

For the integral in (6), which is similar to the one evaluated in the Appendix of [4], we have:

$$\int (\mathbf{a} \cdot \mathbf{n}) \mathbf{n} \, d\Omega = \frac{4\pi}{3} \mathbf{a} .$$

When this result is substituted in (6) the self force acquires the form

$$\mathbf{F}^i_{self} = -\frac{1}{3}\frac{\Delta t}{\Delta \tau}\frac{U}{c^2}\,\mathbf{a}\,,$$

where

$$U = 4\pi \sum_{i=1}^{X} \Delta E \beta_i$$

is the energy of all virtual photons approaching the electron from all directions of the solid angle 4π and absorbed during the time Δt. As the energy U was defined as the energy of the electron field the electromagnetic contribution to the inertial mass of the electron can be written as $m^i = U/c^2$. Finally, for the self force acting on the accelerating electron we can write:

$$\mathbf{F}^i_{self} = -\frac{1}{3}\frac{\Delta t}{\Delta \tau}\,m^i\,\mathbf{a}\,. \tag{7}$$

The self force (7) has the form of the inertial force, since (i) it is proportional to the acceleration, (ii) the coefficient of proportionality has the dimension of mass, and (iii) it has the correct sign.

The *unbalanced* recoils of the virtual photons that are absorbed during the time Δt by an electron at rest in a gravitational field also give rise to the self force

$$\mathbf{F}^g_{self} = \frac{1}{3}\frac{\Delta t}{\Delta \tau}\,m^g\,\mathbf{g}\,, \tag{8}$$

which has the form of what is traditionally called the gravitational force. The mass m^g in (8) is the electromagnetic contribution to the passive gravitational mass of the electron.

The equations (7) and (8) have the form of Newton's second law, but due to the factor of $\Delta t/3\Delta \tau$, they are not exact equations for the inertial and gravitational forces. A possible reason is that the mass of the electron may not be entirely electromagnetic in origin. The electron does not participate in strong interactions and does not have a strong charge and strong field. Therefore its mass should not contain a contribution from the strong interactions. As the electron participates in weak and gravitational interactions (in addition to electromagnetic interactions) it appears to follow that its mass should contain weak and gravitational contributions. At this moment it seems more realistic to expect only a weak contribution to the electron mass. So far all attempts to quantize the gravitational interaction have failed and there is no experimental evidence that its carrier—the graviton—exists. That failure may mean either that we are not dealing with the quantization of gravity properly or that gravitational interactions are not as fundamental as the electromagnetic, strong and weak interactions.

The semiclassical calculations of the self force to which a non-inertial electron is subjected were carried out to demonstrate that the unbalanced recoils of the virtual photons that are absorbed by a non-inertial electron do give rise to the forces (7) and (8) which have the form of the inertial force in the case of an accelerating electron and the gravitational force in the case of an electron at rest in a gravitational field. This shows that we may expect inertia and mass to be explained in the framework of the Standard Model in terms of unbalanced recoils from virtual quanta. An obvious question that should be addressed is about the rest masses of the carriers of the weak interaction. As shown in [15] this question does not affect the conclusion that the mechanism of unbalanced recoils from virtual quanta, considered here, should give rise to inertia and mass of elementary particles if the Standard Model and general relativity are correct.

4.   On the nature of the electron

We started with the intention to see whether the study of the inertial and gravitational properties of the electron can help us gain some understanding of its nature. Instead, it appears we have obtained contradictory results. The classical model of the electron leads to correct expressions for the inertial and gravitational forces acting on an accelerating electron and on an electron at rest in a gravitational field, respectively. Those expressions cannot be obtained if the electron is point-like. In QED, however, it is a point-like electron whose semi-classical treatment yields the correct form of the expressions for the inertial and gravitational forces acting on it. It is precisely this apparent paradoxical result that may provide some insight into the nature of the electron.

At first, it might seem that a point-like model of the electron may be closer to what the real electron is due to (i) the experimental evidence (putting an upper limit of 10^{-18} m on the electron size) mentioned in the Introduction, and (ii) the stability problem of the classical electron according to which it cannot be stable since its charge will tend to explode. However, there exist strong arguments against a point-like electron.

As one of the most difficult problems of the classical electron is its stability, one may conclude that the basic assumption in the classical model of the electron—that there is interaction between the elements of its charge—may be wrong. The very existence of a radiation reaction force, however, is evidence that there is indeed interaction (repulsion) between the different parts of the electron charge. "*The radiation reaction is due to the force of the charge on itself*—or, more elaborately, the net force exerted by the fields generated by different parts of the charge distribution acting on one another" [16]. In the case of a *single* radiating electron the presence of a radiation reaction force implies interaction of different parts of the electron. Therefore, not only does the classical model of the electron yield the correct expressions for the inertial and gravitational forces acting on a non-inertial classical electron (accelerating and at rest in a gravitational field, respectively), but also reflects an important and puzzling feature of the real electron—the self interaction of its charge.

The strongest argument against a point-like electron, however, comes from quantum mechanics. Despite all studies devoted to the nature of the electron (see, for instance, [17-19]) no one knows what an electron looks like before being detected and some even deny the very correctness of such a question. One thing, however, is completely clear: the experimental upper limit on the size of the electron ($<10^{-18}$ m) cannot be interpreted to mean that the electron is a point-like particle (localized in such a region) without contradicting both quantum mechanics and the existing experimental evidence.

Let us consider an example which clearly demonstrates why according to quantum mechanics the real electron cannot be a localized (point-like) particle. The hydrogen atom does not possess a dipole moment when its electron is in an s-state. It is difficult to explain why so little attention has been paid to the fact that this is *only* possible if the electron is *not* localized somewhere "above" the nucleus, but somehow occupies the spherical region (for short, shell) around the nucleus where its wavefunction is different from zero. It should be stressed that this example leaves us with no choice about the interpretation of the non-zero probability of finding the electron in the spherical shell—the electron must *actually* occupy the *whole* shell; otherwise, if it were as small as the scattering experiments seem to suggest, the hydrogen atom (with its electron in s-state) would certainly have a dipole moment.

Therefore, the experimentally determined size of the electron tells us very little about what the electron itself is and needs further studies in order to understand the meaning of that size.

The example with the s-electron of the hydrogen atom is important in two respects. First, it shows that the argument that there should be a stability problem in the case of the classical electron applies equally to an electron in s-state when we take into account that an s-electron must actually exist in the whole spherical shell where its wavefunction is different from zero (strictly speaking, it applies to all electrons in an atom). In that state the electron is like the classical electron—a charged spherical shell which should tend to blow up due to the repulsion of its different parts. But it does not, which further deepens the mystery of the electron. It seems there is no stability problem for an s-electron and no one knows why. This implies that there should be no such problem for the classical electron either.

Second, the hydrogen atom example clearly demonstrates that perhaps the greatest mystery we have to solve in order to understand the nature of the electron is: how can an electron occupy the whole region where its wavefunction is different from zero when it is not measured, whereas the electron is always measured as a localized entity (a point-like particle)? As this mystery has persisted for decades, *any* idea that is sufficiently radical to have some chance of surviving both theoretical and experimental scrutiny should be studied thoroughly.

I will briefly discuss such a radical idea here, since it offers a possible simultaneous resolution of the mysteries mentioned above: (i) how can an electron be an extended object before measurement, but is always measured as a point-like particle, and (ii) why is there no stability problem (why does a charged spherical shell not blow up)? The idea was proposed by Anastassov [20] in the 1980s but, unfortunately, remained unnoticed and untested. Its essence is bringing the idea of atomism to its logical completion—discreteness not only in space but in time as well (4-atomism). In the 4-atomistic model an electron is represented not by its worldline (as deterministically described in special relativity) but by a set of four-dimensional point-like objects (for short 4D points) modeled by the energy-momentum tensor of dust—in this case a sum of delta functions. We can regard these objects as the points of the disintegrated worldline of the electron. Those 4D points are scattered all over the spacetime region in which the wavefunction of the electron is different from zero.

The 4-atomistic approach sheds light on the physical meaning of the Compton frequency of the electron—for one second the electron is represented by 10^{20} 4D points which means that the constituents of the electron appear and disappear with the Compton frequency. It follows from here that during a given period of time, when not measured, the electron is everywhere in the region where there exists a nonzero probability of finding it. However, the electron is always measured as a point-like particle since when the first 4D point appears in the detector it is trapped there; there is a jump in the boundary conditions of the electron wavefunction and all consequent 4D points of the electron start to appear and disappear in the detector as well.

What is promisingly original in the 4-atomism hypothesis is its radical approach toward the way we understand the structure of an object. The present understanding is that an object can have structure only in space. The 4-atomistic model of the quantum object suggests that an object can be indivisible in space (like an electron) but *structured in time*. The idea of 4-atomism not only offers a possible (and nice) resolution of the extended/point-like electron mystery, but also gives an idea of why there appears to be no stability problem. The charge of an *s*-electron is not *continuously* smeared out on a spherical shell (since in this case the charged spherical shell will tend to explode). Instead, for one second the spherical shell is formed by 10^{20} 4D points. In a three-dimensional language such a spherical shell can be described in the following way. The spherical shell is not given *entirely* at *any* moment of time—the *s*-electron looks like a spherical shell only when the instantaneous locations of the 4D points are averaged over a given period of time (say, 1 s). At a given instant there is just one 4D charged point at a given place on the shell, at the next instant it disappears there and re-appears at another location; for one second this repeats 10^{20} times. So, when the electron is not measured for every second it will be represented by 10^{20} 4D points appearing and disappearing on the spherical shell. Each charged 4D point feels the repulsion from other previously existing constituents of the electron, but cannot be repelled since it exists just one instant. Therefore, a spherical distribution of the electron charge may be stable.

The 4-atomistic hypothesis is fully compatible with the scattering experimental data—the dimensions of the constituents of the electron (its 4D points) can be much smaller than 10^{-18} m.

This hypothesis also provides a hint as to why the classical (spherical) model of the electron and the point-like electron studied in Section 3 *both* yield the correct form of the expressions for the inertial and gravitational forces acting on a non-inertial electron. If the spherical shell of the classical electron is not a rigid shell, but is formed by the 4D points of a 4-atomistic electron, then the same calculations will give the same expressions for the self forces (2) and (4) since there is no fundamental difference between a spherical 4-atomistic electron and a point-like electron in the case of QED—in both cases it is the individual 4D points of the electron that emit and absorb virtual quanta no matter in what effective shape those 4D points are arranged.

I will briefly mention two other problems on which the 4-atomistic hypothesis also sheds some light. The first is the issue of whether the electron wavefunction describes a single electron or an ensemble of electrons. As the electron, according to this hypothesis, is itself an ensemble of 4D points, the electron wavefunction does describe a *single* electron. The second problem is the understanding of the superpositional state in quantum mechanics. It is not clear how one should understand the statement that a half-spin particle in Bohm's version [21] of the Einstein-Podolsky-Rosen argument, for example, is in a superpositional state. If that particle *continuously* existed in time, it would be really impossible to think of the particle as *actually* being in a superpositional state of spin up and spin down. However, if it is a 4-atomistic particle, each of its constituents is in either in spin-up state or spin-down state and the particle itself is in an *actual* superpositional state when no experiment to detect its spin is carried out.

The 4-atomistic hypothesis shows that it is not unthinkable to view the electron as an entity (i) that occupies the whole region where its wavefunction is different from zero, but is always localized when measured, (ii) that has different shapes in different situations, and (iii) that is free of the stability problem. Whether or not that hypothesis will turn out to have anything to do with reality remains to be seen, but the very fact that it offers conceptual resolutions to several open questions and goes beyond quantum mechanics (which cannot be discussed in this paper) by predicting at least two new effects that can be tested makes it a valuable candidate for a thorough examination. All agree that radical ideas deserve careful study, especially in such desperate times when quantum physics is unable to say anything about the nature of the quantum objects it studies. In spite of all the successes, quantum mechanics remains fundamentally incomplete, since it does not describe the quantum objects themselves, but only their states.

Conclusions

The correct forms of the expressions for the inertial and gravitational forces acting on a non-inertial electron are obtained from two different models of the electron—the classical spherical electron and a point-like electron considered in the framework of QED. It turns out that both models yield correct predictions since there is no fundamental difference between them if they are expressed in terms of a radically new idea of the electron structure (4-atomism). That idea also sheds some light on a number of quantum mechanical problems.

References

1. D. Bender *et al.*, *Phys. Rev. D***30**, 515 (1984).

2. R. P. Feynman, R. B. Leighton and M. Sands, *The Feynman Lectures on Physics*, Vol. 2, (Addison-Wesley, New York, 1964), p. 28-5.

3. V. Petkov, "Propagation of light in non-inertial reference frames," *gr-qc/9909081*.

4. V. Petkov, *Ph. D. Thesis*, (Concordia University, Montreal, 1997); see also "Acceleration-dependent self-interaction effects as a possible mechanism of inertia," *physics/9909019*.

5. M. Abraham, *The Classical Theory of Electricity and Magnetism*, 2nd ed. (Blackie, London, 1950).

6. H. A. Lorentz, *Theory of Electrons*, 2nd ed. (Dover, New York, 1952).

7. E. Fermi, *Nuovo Cimento* **22** (1921) 176; *Physik Z.* **23** (1922) 340.

8. F. Rohrlich, *Am. J. Phys.* **28** (1960) 639-643; *Classical Charged Particles,* (Addison-Wesley, New York, 1990).

9. E. Mach, *Science of Mechanics*, 9th ed., (Open Court, London, 1933). Around 1883 Mach argued that inertia was caused by all the matter in the Universe (no matter how distant it may be) thus assuming that inertia had a non-local cause.

10. The self-force $F'_{self} = -m'a$ is traditionally called the inertial force. According to Newton's third law the external force F that accelerates the electron and the self force F'_{self} have equal magnitudes and opposite directions: $F = -F'_{self}$. Therefore $F = m'a$, which means that Newton's second law can be derived on the basis of Maxwell's electrodynamics and Newton's third law.

11. It should be stressed that in general relativity the force acting on a particle deviated from its geodesic path due to its being at rest in a gravitational field is non-gravitational in origin. As Rindler put it "ironically, instead of explaining inertial forces as gravitational... in the spirit of Mach, Einstein explained gravitational forces as inertial" [12]. This is the reason why "there is no such thing as the force of gravity" in general relativity [13].

12. W. Rindler, *Essential Relativity*, 2nd ed. (Springer-Verlag, New York, 1977), p. 244.

13. J. L. Synge, *Relativity: the general theory*, (Nord-Holand, Amsterdam, 1960), p. 109.

14. The interaction of the charge of an uniformly moving electron with its own Coulomb (undistorted) field produces no net force acting on the electron as a whole; that is why an electron moving with constant velocity (represented by a straight worldline in flat spacetime) offers no resistance to its uniform motion.

15. V. Petkov, *Relativity and the Nature of Spacetime*, (Springer, Berlin, Heidelberg, New York, 2005), Ch. 10.

16. D. J. Griffiths, *Introduction to Electrodynamics*, 2nd ed. (Prentice Hall, London, 1989), p. 439.

17. M. H. Mac Gregor, *The Enigmatic Electron*, (Kluwer, Dordrecht, 1992).

18. D. Hestenes, and A. Weingartshofer (eds), *The Electron: New Theory and Experiment*, (Kluwer, Dordrecht, 1991).

19. M. Springford (ed.), *Electron: A Centenary Volume*, (Cambridge University Press, Cambridge, 1997).

20. A. H. Anastassov, *The Theory of Relativity and the Quantum Action (4-Atomism),* Doctoral Thesis, (Sofia University, Sofia, 1984), unpublished; *Annuaire de l'Université Sofia "St. Kliment Ohridski," Faculté de Physique,* **81** (1993) 135-163; *The Theory of Relativity and the Quantum Action (4-Atomism),* (Nautilus, Sofia, 2003), in Bulgarian.

21. D. Bohm, *Quantum Theory*, (Prentice-Hall, New York, 1952), Ch. 22, Sec. 16.

The Electron as a Periodic Waveform of Spin ½ Symmetry

Horace R. Drew
125 Charles Street
Putney 2112 New South Wales, Australia
email Horace.Drew@csiro.au

Authors write many things, and the common people cling to them through arguments made without experiment. –Roger Bacon, *On Experimental Science*, 1268.

I. Most current theories in physics or astronomy are not unique

For the purposes of this review, I would like to explain the general concepts which underlie my recent work in theoretical physics or astronomy, without going into technical details that may be found in the original papers (1-5).

Today in 2005, both modern physics and astronomy seem to be continuing in a state of crisis. Quantum mechanics and relativity remain as far as ever from unification; theories of light and matter such as QED remain unpalatable with their false infinities and *ad hoc* schemes of renormalization (6); high-energy particle physics has become increasingly expensive and unproductive, with hundreds of different unstable particles discovered, but only a highly parameterized QCD theory to explain them (7,8). Similarly, astronomy has become dominated by a cosmological concern to prove the Big Bang theory and to find an elusive "dark matter" (9,10).

More importantly perhaps, the possible *uniqueness* of currently accepted physical or astronomical theories seems highly doubtful. In physics, students are taught the Born-Bohr interpretation of quantum mechanics, as well as the Lorentz-Einstein view of special and general relativity, as if those were unique breakthroughs of a historical stature.

Yet any assertions of uniqueness are demonstrably not true. For example in quantum theory, many early workers including Schrödinger, deBroglie and Bohm disagreed vigorously with the proposal by Born and Bohr that the electron could be only a point-particle whose location in space or time might be found just probabilistically by squaring an abstract "wave function." More recently, Simulik and Krivsky (11) have shown that the electron may be treated as an electromagnetic scalar wave in a medium, with full accuracy of experimental prediction just as for the Dirac equation. In relativity theory, the Lorentz-Einstein view has had some success, yet has led also to many "paradoxes" between experiment and theory (*e.g.,* the twin paradox for time, or the spinning-disk paradox of Ehrenfest). Furthermore, it makes a first-order prediction concerning Thomas precession (*e.g.,* retrograde motion of any rotating object due to length contraction) that appears experimentally not to be true, a discrepancy noted by Phipps (12) and Galeczki (13).

Next in astronomy, students are taught the uniqueness of the Big Bang theory; and are even taught to regard "dark matter" as a real thing, just awaiting formal discovery. Yet a series of leading astronomers including F. Hoyle, G. Burbidge, M. Burbidge, J. Narlikar and H.C. Arp (14) argue that the Universe may not really be expanding at all, if the mass m and time-counting rate $f = 1/t$ of particles were lower billions of years ago than today, as a kind of natural "evolution." By a similar (and still quite heretical) view, quasars need not lie at huge Hubble distances from Earth, if they were to be composed of newly created matter, which might have a lower intrinsic m or f as for ancient matter in the past. Finally, the "dark matter" hypothesis to explain abnormally-large frequency shifts of spectral lines, across single galaxies or galaxies in clusters, becomes unnecessary if mass m and time-counting rates $f = 1/t$ are not absolute, but may vary slightly over broad astronomical scales, depending on the underlying energy of a zero-point vacuum.

II.   We should listen to experience as well as theory concerning quantum phenomena

Having suggested a lack of uniqueness for many accepted theories of physics or astronomy, let me now present a revised conceptual understanding of those two fields, based on my long experience in x-ray diffraction as well as molecular biology; on my correspondence with workers in many fields of science; as well as much study at home. When studying these ideas, we should remain open to the advice of Roger Bacon from 700 years ago and let experience be our guide. In other words, when people of 13th-century England saw a rainbow and wished to attribute it to mystical powers, Bacon would say, "no, it is not mystical at all: look at a waterfall with the Sun shining on it, or the range of colors created by light passing through a crystal, and you will see a similar thing."

In modern times, such practical advice is still useful. For example, we can see that many natural objects are "discretely coloured" as either red, blue, yellow or green, *etc.* From those simple visual observations, an uneducated man could deduce the general nature of quantum phenomena, even without knowing probabilistic quantum mechanics!

Thus from much direct experience, let me suggest gently to the reader that the photon cannot be a point-particle; nor can the electron be a point-particle. Rather, both the photon and the electron must possess some internal, highly-periodic structure through both space and time, in order to diffract as observed. Even outside of the laboratory we can watch light diffract as it passes by a sharp edge, or through a meshed fly-screen. Indeed, most natural fluid sub-

stances will break spontaneously into some periodic structure of their own accord, and will never remain random under all circumstances. From experience
we see that water will break into regular waveforms while running down a hill;
similarly, clouds will break into regular waveforms as they as pushed by air-
flows in the sky.

When x-rays pass through a crystal, some of them are deflected or "diffract" at various precise angles and intensities to one side. What actually seems
to happen, on a microscopic scale, is that the original x-ray waveform breaks
up into smaller undetectable fragments of precise phase, which are scattered
through space by electrons within crystalline atoms; then those many dispersed
fragments may join together to form a new detectable photon, probabilistically
in proportion to the *square* of their concentration in space: since at a fundamental level they seem to be composed of *dimeric* filaments.

Indeed, the very nature of "inertia" on a macroscopic scale may be derived from "loss of phase coherence" on a microscopic scale, although this
fundamental and important idea has not yet been widely recognized. For example, whenever an electron diffracts in-phase, or follows a closed path of integral (deBroglie) phase in atoms, no inertia seems to be observed. David Bohm
has included those strange, inertia-less phenomena in his alternative version of
quantum mechanics, and argues that particles moving under the influence of an
in-phase "guide wave" need not show any inertia whatsoever (which is just another way of saying the same thing). Bohm's views have been strongly supported by modern quantum experts such as J.S. Bell or S. Goldstein.

But the same process of photon or electron-creation in diffraction would
be treated very differently by Born or Bohr: they would formulate their model
in terms of some "point-photon" without any internal periodicity, which relo-
calizes instantly to a new spatial location that can be determined only by squaring an abstract "wavefunction." Next in order to convince a point-photon to
diffract, they have to invent an *ad hoc* principle known as "wave-particle duality," where the photon can sometimes be a wave with well-defined intensity
and phase, and sometimes just a dimensionless point.

Perhaps the Born-Bohr model seemed plausible in 1930, when it was first
applied to very simple particles such as an electron or photon; but today in
2005, how can we possibly explain the diffraction of large, multi-atom molecules by such an approach? For example, when a porphyrin ring from biology
shows diffraction phenomena in the gas phase, where might the information
come from to reconstruct accurately that multi-atom structure, if the entire porphyrin supposedly (according to Born and Bohr) "collapses" first to a dimensionless point; then moves instantly elsewhere without inertia; and finally is
"recreated" with perfect reliability in some new location? Surely in this limit of
diffraction by *large chemical structures*, the Born-Bohr model may be considered implausible or even absurd.

Alternatively, by the Bohm model for diffraction, a hypothetical (de-
Broglie) "guide wave" would be the only part of that multi-atom porphyrin to

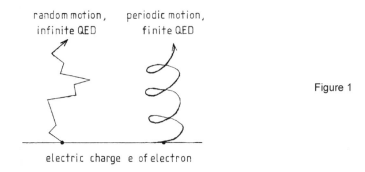

random motion,   periodic motion,
infinite QED   finite QED

Figure 1

electric charge e of electron

experience interference. The major part of that complex chemical structure would remain intact and undisturbed; and would just "follow" the in-phase guide-wave through one "slit" or another in an inertia-less fashion. Hence our multi-atom porphyrin would never need to "collapse" to a dimensionless point, nor be "recreated" in some new location.

Finally, the Bohm-deBroglie "guide wave" could plausibly represent some weak, externally detectable aspect of a much more energetic internal periodicity: not seen for a particle at rest, but seen for a particle in motion. In other words, the slightly higher mass m of a moving particle (see below) could cause its very energetic internal periodicity, not detectable for a particle at rest, to "spill over" into the space around itself; thereby causing wave-like diffraction phenomena for otherwise material particles.

In summary, as noted above, most current theories in physics or astronomy are not unique! We can choose between Bohm's "guide-wave" or Born-Bohr's "point-wave duality" seemingly at will. To this author, the former theory seems more plausible in terms of physical concepts.

Going now beyond simple concerns for uniqueness, the consequences of adopting a Born-Bohr model for modern quantum electrodynamics or QED are shown in Figure 1. There we can see that a randomly moving point-electron (left-hand side) will not easily be able to generate any finite energies in its electrical interaction with light. For example, a hypothetical point-electron of radius $r = 0$ will repel itself by an infinite amount $e^2/2r$. It will also interact with an infinity of photons of all possible wavelengths, as it travels randomly through space; yet it will not know how much of its total energy mc^2 to commit to any electron-light interaction. Hence that number (precisely $1/137.03599$) has to be put into the theory finitely "by hand." Nor will the point-electron even be able to proceed easily from one location in space or time to another, since it has no continuous structure in either dimension.

Infinite QED (quantum electrodynamics) is based precisely on those random, unstructured principles, and hence requires many essential parameters to be added "by hand." Moreover, the point-model generates at least three different kinds of spurious infinite energy that have to be removed by a mathematical trick known as "renormalization" (dividing one near-infinite quantity by another). Dirac was very knowledgeable concerning such faults, and therefore never accepted infinite QED as any more than an abstract calculating tool.

By contrast, we can next examine the consequences of adopting a continuous and periodic waveform for the electron, as shown in Figure 1 (right-hand side), by a philosophical approach more akin to the Bohm-deBroglie model for diffraction. Now the electron self-repulsion $e^2/2r$ becomes finite, because its radius r is finite. Also, the finite electron will only be able to interact with a limited number of photons of discrete wavelength from the outlying vacuum, as it exchanges some finite but mobile energy inside of itself (equal to $1/137$ of mc^2), with similar self-energies in other particles nearby. Hence this number need not be added "by hand." Its passage from one location in space or time to another will similarly be defined by the continuous periodic structure itself.

Having considered the favourable interaction of a wave-like electron with light, it should be noted further that our periodic electron is dynamically stable in a Casimir sense (15), whereas the random point-like electron is not. In other words, all particles should experience an outward force that depends on their finite size, due to inertial-motional influences as well as electrical self-repulsions; which may be balanced by an inward pressure due to zero-point waves from the surrounding vacuum. In order for any particle to be stable dynamically, those two terms "outward and inward" should balance closely, which they do for a typical periodic electron.

For example, considering the periodic electron shown below, one finds that $E(out) = 137 \times (e^2/2r)$ while $E(in) = hf/2 = hc/4\pi r$. Since $137 \times e^2 = hc/2\pi$ by the formula for fine-structure constant, one can see that $E(in) = E(out)$ to a first-order approximation (5). But for any point-like electron of $r = 0$, both the outward electrical self-repulsion $e^2/2r$ and the inward vacuum pressure $hc/4\pi r$ go to infinity! Only by omitting all mention of Casimir stability, an experimental certainty, can the electron point-theorists rescue an otherwise impossible situation.

III.   A periodic waveform of spin ½ symmetry

Based on the previous discussion, two general concepts should now be clear. First, physicists in the early 20th century had to make a *choice*: whether they would follow a Born-Bohr model assuming "point particles," or else a de-Broglie-Bohm model involving "guide waves." Secondly, having chosen a Born-Bohr model, those physicists wished to *extend* their point-model to encompass a wider range of phenomena, and so arrived at a strange and complex theory that we now know as "infinite renormalized QED," which is highly successful in a numerical sense, but an undisputed failure in a conceptual sense.

But what if the original choice made in 1930 was wrong? Next we should present some well-defined, plausible, wave-like model for the electron, in order to support the general assertions given above. Different contemporary workers have preferred different wave-like models: for example the electromagnetic

4-D model with spin 1/2 symmetry

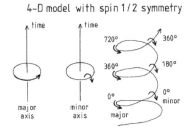

Figure 2

Self-electrical repulsions

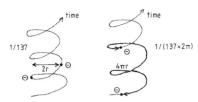

scalar wave of Simulik and Krivsky (11), or the three-dimensional spinning ring favoured by Parsons in 1915, and now back in favour. But as matter of personal choice, I prefer the model shown in Figure 2 (upper part), the drawing of which is intended to represent a doubly-spinning, continuous and periodic structure in four dimensions of both space and time (x, y, z, t).

However bizarre this four-dimensional wave-model might seem on first glance, there are several sound reasons for preferring a well-defined structure in four nearly-equivalent dimensions (x, y, z, t), rather than a well-defined structure in three spatial dimensions (x, y, z) which can advance through time t as a separate entity.

First, we know by experiment that the electron shows a periodic symmetry of "spin ½," which means that it must possess two separate axes of rotation. Thus it must advance about one axis (*e.g.*, the major) at twice the frequency that it advances about the other axis (*e.g.*, the minor): see Figure 2, upper left and right. That spin ½ symmetry is fundamental to building any sort of plausible model for an electron; just as by analogy, C2 space-group symmetry was essential to Watson and Crick for building a correct DNA model in 1953.

One could perhaps build some kind of three-dimensional model with the required spin ½ symmetry as a filamentous Moby's strip, yet other experimental aspects of electron behaviour suggest that the model should really be four-dimensional: for example $g = 2$ for magnetism, which suggests that the area of capture for external light may be twice as large through time as through space; or the anomalous part of magnetic moment as $1/(137 \times 2\pi)$, which represents a slight extra self-energy through time; or squaring of the wave-function to generate a new particle upon diffraction, as explained elsewhere in terms of a dimeric four-dimensional filament (3).

Next in the lower part of Figure 2, we can see that this particular model for the electron generates two principal forms of electrical self-repulsion.

Across the electron diameter through space, we see a well-defined self-repulsion $e^2/2r$ equal to 1/137 of mc^2, which is precisely the probability of any electron to exchange light with another particle at distance, so as to generate, for example a net energy of $1/(2 \times 137^2)$ for the $n = 1$ shell of hydrogen. Then along the electron path through time, we see another well-defined self-repulsion $e^2/4\pi r$ equal to $1/(137 \times 2\pi)$ of mc^2, which is the amount of electrical self-energy lost in the Lamb shift, when the electron experiences a full proton charge and loses some of its stable mass. That same $1/(137 \times 2\pi)$ is also the increment by which electron magnetic moment increases beyond its twice-classical value of $g = 2$: the slight extra self-energy thus gives a little extra magnetism.

These and other experimental quantities—Casimir stability, anomalous magnetic moments to high accuracy (16), the Lamb shift, and a possible topological nature for electron paths in atoms—have been calculated from a four-dimensional model (3, 5).

Finally, the slight extra energy of electricity ("magnetism") or gravity ("general relativity"), which is seen for two particles in relative motion, may follow naturally from the same four-dimensional model: if its minor plane shown in Figure 2 (upper) projects from "time into space" by an amount $\sin \theta = v/c$, as relative velocity v increases towards c; thereby providing an increased area in space for the two-way exchange of light or gravity waves.

Hence our major plane of spin could provide for electricity or gravity at rest, through exchange of light or gravity waves with a distant particle; while our minor plane of spin could provide for a slight extra exchange of light or gravity waves, when two particles are in motion relative to one another; thereby increasing the total energy from 1 to $(1 + v^2/c^2)$. That extra motional part would not, however, serve as the source for any further exchanges; and so total energy would remain constant at $(1 + v^2/c^2)$ without any possible "runaway" to infinity.

To conclude, it should be emphasized that I make no claims for the uniqueness of my own particular model over others of a suitable nature. We can only make models concerning Nature through a series of ever-improving approximations, and will never be sure of having the "right" answer.

I do intend to argue, however, that time t as it enters into modern physics through a series of infinitesimal time-slices in Hamiltonian theory, or through non-Euclidean geometry in relativity, has not been treated correctly. We imagine by our animal intelligence, having a brain that measures time t as a series of forward-moving events, that time "flows" forward in an absolute fashion. Yet I argue that the forward-flow of absolute time is just an illusion; and that on a subatomic scale particles may become structurally four-dimensional; while on an astronomical scale the flow-rate of time may become variable, depending upon the zero-point vacuum $E = hf/2$ within which any particle resides (e.g., through an effect on the Casimir equilibrium).

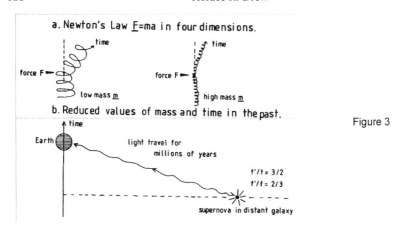

a. Newton's Law $\underline{F}$=ma in four dimensions.

b. Reduced values of mass and time in the past.

Figure 3

IV. Variable character of time on subatomic or astronomical scales?

One may address the notion of variable time rather simply, by considering a four-dimensional interpretation of Newton's Law $F = ma$ as shown in Figure 3, upper part. There we see how some lateral force F will induce greater acceleration in a low mass m than a high mass m'. Could the low mass m contain a lesser number of periodic turns, within a given volume of space and time, than the high mass m'? In other words, if we were to apply the same force F to a ball made of carbon *versus* a ball made of iron, would the greater acceleration a of the carbon ball be due to a lesser intrinsic rate of counting time as $m = f = 1/t$ on a subatomic scale, where $a = F/f(carbon)$ = high or $F/f(iron)$ = low?

This example, however trivial provides, a good introduction to what we will discuss next, which is the possibility of variable mass m or time t over broad astronomical scales. Many astronomers such as Arp and Hoyle (1, 14) have long argued that aberrant redshifts in astronomy, and even the Hubble redshift, might be due to altered mass m or time t at the distant source of such light, rather than due to Doppler shifts from receding motion as advocated by Big Bang enthusiasts. Only recently has that cosmological controversy come to a stage where it can be addressed through experimental data.

Thus a number of expert astronomers have recently made new measurements of supernovae lifetimes in distant galaxies, and have used those data to evaluate how cosmologically different such distant galaxies might be from modern Earth, in terms of dimensional parameters such as space-distance x or time-counting rate $f = 1/t$. The general result from their studies (17) may be expressed succinctly as $t'/t = f/f'$. Here t'/t tells how much longer it takes for some supernova in a distant galaxy to decay through its light-curve, the brightness of which reflects radioactive half-lives there, versus for a similar supernova near Earth. Similarly, f/f' tells how much more slowly the bulk of stars in that distant galaxy count time by a Hubble redshift there, versus for similar stars near Earth. Since the two experimental quantities t'/t and f/f' agree in many examples, one may conclude that we are really seeing a cosmological effect: but what might it mean?

One simple interpretation is shown in Figure 3 (lower part), where the supernova in that distant galaxy contains atoms which *count time more slowly* than on modern Earth, by a factor $t'/t = 3/2$ or $f'/f = 2/3$. Since it takes light at finite speed c a long time to reach Earth from the distant supernova—many millions or billions of years—perhaps the Universe may have evolved to a higher state of energy over such a long period, due to an increase in the energy of the underlying zero-point vacuum? That is a fundamental factor which would affect Casimir stabilities by an altered $E = hf/2$, leading to higher frequencies of periodic motion within particles on modern Earth today, versus particles in distant galaxies long ago.

The Big Bang cosmologists, however, interpret those same data in another way. They believe that the overall spatial structure of the Universe may be expanding uniformly over long periods of time, due to the initial momentum provided by some ancient, creational explosion; and so they argue that all *space-distances x are larger* on Earth today than on that distant galaxy in the faraway past. Hence light-wavelengths become longer or "stretched" like the rubber skin of a balloon, over millions of years of travel from that distant galaxy to Earth; and so we see those light-images on Earth today over greater lengths of time t and with lesser frequencies f, than when they were emitted at the source.

As an *ad hoc* addendum to the "stretched space" theory, we find that the apparent rate of stretching becomes greater when we look at later historical times near Earth. That is exactly the *opposite* of what one would expect from an ancient Big Bang, where the rate of stretching would decrease due to mutual gravitational attraction among distant galaxies, as time proceeds after an initial explosion. Hence recently we saw the proposal of another cosmological "discovery": the return of a repulsive cosmological constant or "quintessence" to explain why the Universe decides, in their view, to expand more rapidly as time goes on!

In response to "quintessence," Arp or Hoyle might say that the rate of energetic evolution of the Universe need be not linear with time. There is no reason why the zero-point vacuum should grow in energy-density with precise linearity over billions of years.

Yet another difficult aspect of the stretched-space theory is that the Hubble redshift of normal galaxies can no longer be treated as a Doppler shift, but rather must be treated as a uniform but very slow "stretching of space" which occurs all around us, even on commonplace scales of distance. As we proceed to shorter and shorter distances from Earth, an ambiguity arises: is any redshift due to a Doppler shift or due to stretched-space? A Doppler shift is mainly due to receding velocity as $(1 - v/c)$, and hence shows a much smaller time-dilation t'/t than for stretched-space.

As a final astronomical concern, one might expect that accurate measurements of proper motion—looking at angular motions of stars through the sky, as seen from Earth over many years—would give estimates of velocity that agree well with Doppler shifts as inferred from frequency shifts of spectral lines. Such is not the case however, since proper motions for stars on the outer edges of our Milky Way remain highly anomalous, when compared to apparent velocities as inferred from an interpretation of those frequency shifts as a Doppler effect (2, 18). Such anomalies lie at the heart of the "dark matter" problem, since if the zero-point vacuum can vary slightly across galaxies, to change masses m and rates of time-counting $f = 1/t$, no dark matter need exist (10).

V.   Variable character of time for particles in motion

To conclude this review, we will discuss the known variation in rates of counting time t for particles in rapid motion, and analyze those experimental results in terms of two distinct logical frameworks: perception versus dynamics (4).

The original derivation of special relativity in 1905 (ref. 19, Chapter 7a) was based on light-image distortions: when two observers in relative motion v exchange light-signals with one another at constant speed c, what might be the effects on length x, time t or mass m as inferred from viewing those light-signals at a distance? Lorentz and Einstein worked out that the perception of light-image distortions would co-vary reciprocally between any two moving observers, by numerical relations such as $t'/t = 1/\sqrt{(1 - v^2/c^2)}$ which are now familiar even to beginning physics students. The term "co-variant" therefore implies reciprocal and equivalent views by means of light-image distortions, without any specified changes to the intrinsic time-counting rates $f = 1/t$.

Those intrinsic rates of counting time could thus be regulated by a different physical mechanism, for example by Casimir stabilities within individual subatomic particles, as described above. Indeed with reference to the discussion above concerning Newton's law $F = ma$, one could say that carbon and iron balls count time intrinsically at different internal rates $f = 1/t$, even while at rest, and without any reciprocal light-images being exchanged between them. Each observer or ball would "see" the same external situation using different internal rates of counting time, while light signals need not be exchanged at all.

Furthermore, such intrinsic or internal rates of counting time should potentially change in a dynamic, energetic sense for any single particle moving through a vacuum; by a mechanism *similar but not identical* to the "co-variant" changes seen, when light signals are exchanged between two observers moving relative to one another. To be more precise, single particles when moving through a vacuum would seem to count intrinsic time t more slowly than when at rest; yet their intrinsic masses m will also increase in a proportional fashion; so that total energies (and atomic spectra) may remain "in-variant" or conserved, due to equal and opposite changes in both t and m.

It should be emphasized here that intrinsic distances x do not vary with motion at all; any changes to x are always perceptive. Indeed if it were otherwise, three intrinsic parameters t, m and x could never cancel exactly, so as to

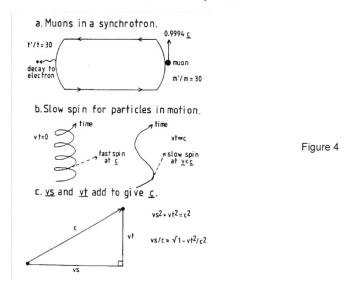

a. Muons in a synchrotron.

$t'/t = 30$

0.9994 c

muon

$m'/m = 30$

decay to electron

b. Slow spin for particles in motion.

$vt = 0$

time

fast spin at c

time

$vt \approx c$

slow spin at $v < c$

Figure 4

c. vs and vt add to give c.

c

vt

vs

$vs^2 + vt^2 = c^2$

$vs/c = \sqrt{1 - vt^2/c^2}$

produce the relativistic "invariance" which makes our Universe possible. We would see instead that electrical energies within atoms, or gravitational energies for planets about the Sun, would "run away" to infinity, since uncompensated increases in mass m would produce ever-higher energies of motion with each application of a supposed "relativistic formula"! But because changes of x are only perceptive (*i.e.*, expansion or shrinking of a light-image), they do not enter into dynamics, while reciprocal changes to t and m cancel.

As an actual example, let us consider what happens when muons are accelerated in a synchrotron to speeds of 0.9994c. Then it is observed that their mean-lifetimes increase by a factor of $t'/t = 1/\sqrt{(1 - v^2/c^2)} = 30$ or from roughly 1.5 to 50 microseconds: see Figure 4, upper part. This is clearly a dynamic or internal-intrinsic effect, since no reciprocal light-signals have been exchanged. Thus, the decaying muons are detected directly by emission of electrons to a counter. Nevertheless, the dynamic or "in-variant" time dilation seen here for fast-moving muons still follows the characteristic mathematical formula, derived by Einstein in 1905, for perceptive or "co-variant" time dilation by reciprocal light-signals.

Is it any wonder that those two kinds of physical phenomena have often been confused, if the same formula works sometimes in both cases? Yet in other cases of dynamic measurement, we see many failed predictions and paradox, as noted below.

Now if the speed of light c may be regarded as a limiting factor, in a dynamic sense just as in a perceptive, then we find easily for particle models which postulate some internal wave-like character, that internal periodic velocity v_s should decrease for any particle in translational motion v_t through a vacuum, to yield $t'/t = c/v_s = 1/\sqrt{(1 - v_t^2/c^2)}$ for the dynamic case as well as for the perceptive: see Figure 4, center and lower. In other words, some component of periodic motion vs will always lie perpendicular to the direction of translational motion v_t; and since speed c cannot be exceeded as the vector sum of vs and vt, we find that v_s must go to zero as v_t goes to c; likewise the periodic counting of internal-intrinsic time t comes to a stop.

A confusion between those two kinds of phenomena, perceptive *versus* dynamic, has led to many paradoxes and arguments concerning the nature of special relativity, all through the 20th century. For example in the well-known "twin paradox," one twin goes into space for many years at a speed close to c, then returns to Earth to find his partner aged beyond recognition. The non-reciprocal aging of those twins is clearly a dynamic process, where intrinsic rates of counting time are much less for the twin moving at close to speed c, rather than for the twin who remains home at speed $v = 0$.

Yet if at any time during that space voyage, the fast-moving twin had sent a television image back to Earth, in order to exchange news with his partner at home, the partner twin would have seen his fast-moving space friend as dilated to much slower time. Similarly, a television image sent from the near-stationary Earth twin, to his fast-moving friend in space, would be seen by his space friend as also dilated to much slower time.

Hence no true paradox really exists. One simply has to *specify* whether the *experimental circumstances* are dynamic-intrinsic or else perceptive-covariant. A spinning-disk paradox concerning length contraction x'/x may be resolved in a similar way; and the hypothetical Thomas precession need never be measured as a dynamic effect. Indeed, tests for it have always proven negative. (It should be mentioned here that Thomas precession is still invoked in many quantum mechanics textbooks, to resolve a factor-of-two error in the prediction of fine-structure spectra.)

Of course, according to a Born-Bohr model for diffraction, subatomic particles can be only "dimensionless points," and hence have no possible internal mechanism by which to count time intrinsically. And so we arrive at false infinities in QED, false infinities in astronomy, and paradoxes in special relativity. By contrast, according to a deBroglie-Bohm model for diffraction, subatomic particles and their "guide waves" would probably possess a periodic internal mechanism by which to count time intrinsically. And so we *could* arrive at finite versions of QED, finite versions of astronomy, and no paradoxes in special relativity, if the current generation of physicists would simply open their minds.

A clear distinction between dynamics and perception also clears up an important issue regarding possible *models for the electron*. By a perceptive, Lorentz-covariant view, only point-models would be allowed, especially due to a predicted length contraction x'/x for particles in rapid motion. But since we see now, that such concerns are associated only with light-image distortions and not with dynamics, there exists no real objection to proposing wave-like models of finite size, as has been done here and elsewhere (3,5,11).

Each individual particle will therefore count time in its own way, depending on Casimir relations between the outward self-energy of any particle (inertial and electrical), balanced against an inward pressure from the zero-point vacuum; which itself need not remain precisely constant over large scales of space or time. Hence "there may be as many times as there are inertial frames" (19), on Earth and perhaps across distant galaxies as well.

To conclude, when searching for an improved understanding of difficult subjects such as quantum theory, relativity or cosmology, we should listen to experience as well as theory, since most current theories in physics and astronomy are not unique. Also, we should try to understand all of these subjects together as a whole, so far as possible, in order to avoid the acceptance of theories in one field which contradict those in another.

References

1. Drew, H.R. (1997) On the nature of things as seen in the late 20th century. *Apeiron* **4**, 26-32. A survey of modern astronomy allowing for the possibility of variable time *t* or mass *m* over broad astronomical scales.

2. Drew, H.R. (1998) Zero proper motion in the outer parts of our galaxy. *Apeiron* **5**, 250-252. Some anomalous data from *Nature* are shown to fit a simple model, where stars along the outer edges of the Milky Way cannot be seen moving vertically due to its non-planar warp, because the actual motion of such stars lies close to zero, contrary to the "dark matter" hypothesis.

3. Drew, H.R. (1999) The electron as a four-dimensional helix of spin ½ symmetry. *Physics Essays* **12**, 649-661. A preliminary model for the electron is developed in terms of a periodic waveform or four-dimensional "helix" which remains continuous through both space and time, and contains only finite self-energies that explain electricity and magnetism.

4. Drew, H.R. (2000) If there is no Thomas precession, what then? Light-signal versus intrinsic relativity. *Apeiron* **7**, 217-224. Supposing the hypothetical Thomas precession as predicted by special relativity does not exist, what changes in our understanding would be required? A distinction between perceptive light-signal versus dynamic intrinsic phenomena is indicated.

5. Drew, H.R. (2002) A periodic structural model for the electron can calculate its outward properties to an accuracy of second or third order. *Apeiron* **9**, 25-66. A preliminary model for the electron is developed so as to explain Casimir stability, electron and muon anomalous magnetic moments, the Lamb shift, and a topological nature of electron paths in atoms.

6. Milonni, P.W. (1994) *The Quantum Vacuum: An Introduction to Quantum Electrodynamics*, Academic Press, San Diego. A thorough account of modern QED (quantum electrodynamics), emphasizing aspects of the fluctuating quantum vacuum: "Although the infinite self-energy calculated for an electron might be considered shameful, the extraction of finite numbers in fantastic agreement with experiment is one of the great triumphs of twentieth-century science."

7. Zerwas, P.M. and Kastrup, H.A. eds. (1993) *QCD: Twenty Years Later*, World Scientific, Singapore. A late 20th-century account of modern particle physics in terms of QCD (quantum chromodynamics) or quark-gluon theory, which was adopted from infinite renormalized QED: "although most of us are persuaded that it is the correct theory of hadronic phenomena, a really convincing proof still requires more work."

8. Pickering, A. (1984) *Constructing Quarks: A Sociological History of Particle Physics*, Edinburgh University Press, Scotland. A clear historical review of modern high-energy particle physics from an outside but intelligent perspective, which raised much controversy: "even when accepted fields of facts existed, a plurality of theories could be advanced for their explanation; none of those theories ever fitted the pertinent facts exactly."

9. Fukugita, M. (2003) The dark side. *Nature* **422**, 489-491. The latest news concerning a decade-long search for invisible dark matter, and speculations concerning its nature: "it seems impossible to discard the dark matter hypothesis without endangering our whole modern scheme of cosmology."

10. Irion, R. (2003) Do some galaxies lack shrouds of dark matter? *Science* **300**, 233. By looking at excited oxygen in planetary nebulae from elliptical galaxies, astronomers see very slow orbital mo-

tions which are consistent with "no dark matter at all," or little frequency change in going from one part of the galaxy to another.

11. Simulik, V.M. and Krivsky, I.Yu. (2002) Slightly generalized Maxwell classical electrodynamics can be applied to inner-atomic phenomena. *Annales de la Fondation Louis de Broglie* **27**, 303-328. By a generalization of Maxwell's equations, the electron may be modelled accurately in terms of a stationary electromagnetic-scalar wave within a medium.

12. Phipps, T.E. Jr. (1987) *Heretical Verities: Mathematical Themes in Physical Description*, Classic Non-fiction Library, Urbana, Illinois. A thorough and often humorous critique of 20th century physical theories, especially valuable for its discussion concerning Thomas precession and special relativity (Chapter 7.3).

13. Galeczki, G. (1996) Seventieth birthday of a non-effect: Thomas precession. *Apeiron* **3**, 120. A concise summary of Thomas precession and its chequered history.

14. Arp, H.C. (1998) *Seeing Red: Redshifts, Cosmology and Academic Science*, Apeiron, Montreal, Canada. An up-to-date summary of anomalies from experimental astronomy which do not fit into Big Bang theory, plus an exposition of alternative views.

15. Buks, E. and Roukes, M.L. (2002) Casimir force changes sign. *Nature* **419**, 119-120. A good discussion of the Casimir effect and whether it could be repulsive as well as attractive.

16. Cvitanovic, P. and Kinoshita, T. (1974) Sixth-order magnetic moment of the electron. *Physical Review* **D10**, 4007-4031. The calculation of anomalous magnetic moment for an electron to third-order accuracy in powers of $1/(137 \times \pi)$ by means of renormalized infinite QED; a good guide to carrying out finite calculations of similar kind.

17. Leibundgut, B. *et al.* (1996) Time dilation in the light curve of distant type Ia supernova SN1995K. *Astrophysical Journal* **466**, L21-L24. By studying the decay of light emission from a supernova in a distant galaxy as a function of time, and by comparing with references in our own galaxy, it was found that the decay of light proceeds more slowly in that distant galaxy as seen from Earth by a factor of $t'/t = 1.5$, which is just the reciprocal of its Hubble frequency $f'/f = 1/1.5$. Such a result could be interpreted in terms of a slower rate of counting time t' in that distant galaxy, as we see it billions of years ago using light which travels at finite speed c; or else in terms of an expanding universe where distances x' are greater on Earth today than in that distant galaxy long ago.

18. Smart, R.L. *et al.* (1998) Unexpected stellar velocity distribution in the warped Galactic disk. *Nature* **392**, 471-473. The proper motions of stars along the outer edges of the Milky Way do not show a large vertical component out of the plane of the disk, as might be expected for a substantial galactic warp, combined with very fast orbital speeds of 300 km/second due to "dark matter." I show elsewhere (*Apeiron* **5**, 250-252) that those outer-edge stars are hardly moving, and that the "dark matter" interpretation may be unwarranted.

19. Pais, A. (1982) *Subtle Is The Lord: The Science and Life of Albert Einstein*, Oxford University Press, England, Chapter 7a: "There are as many times as there are inertial frames."

A Geometrical Meaning of Electron Mass from Breakdown of Lorentz Invariance

Fabio Cardone[1,2], Alessio Marrani[3] and Roberto Mignani[2-5]
[1]Istituto per lo Studio del Materiali Nanostrutturati (ISMN-CNR)
Via dei Taurini, 18
I–00185 ROMA, Italy
[2]I.N.D.A.M. - G.N.F.M.
[3]Dipartimento di Fisica "E. Amaldi"
Università degli Studi "Roma Tre"
Via della Vasca Navale, 84
I – 00146 ROMA, Italy
[4]I.N.F.N. – Sezione di Roma III
[5]mignani@fis.uniroma3.it

We discuss the problem of the electron mass in the framework of Deformed Special Relativity (DSR), a generalization of Special Relativity based on a deformed Minkowski space (*i.e.,* a four-dimensional space-time with metric coefficients depending on energy). We show that, by this formalism, it is possible to derive the value of the electron mass from the space-time geometry via experimental knowledge of the parameter of local Lorentz invariance breakdown, and the Minkowskian threshold energy $E_{0,em}$ for the electromagnetic interaction. We put forward the suggestion that mass generation can be related, in DSR, to the possible dependence of mass on the metric background (*relativity of mass*).

1   Introduction

The problem of the mass spectrum of the known particles (leptons and hadrons) is still an open one from the theoretical side. As a matter of fact, the Standard Model of electromagnetic, weak and strong interactions is unable to say why a given particle has a given (experimental) mass. As to the carriers of the four fundamental forces, symmetry considerations would require they are all massless. However, it is well known that things are not so simple: weak quanta are massive. It is therefore necessary, in the framework of the Glashow-Weinberg-Salam model of electroweak interaction, to hypothesize the Goldstone mechanism, which gives weak bosons a mass by interaction with the (still unobserved!) Higgs boson.

Even the first, best known and most familiar particle, the electron, is still a mysterious object. In spite of the successes of the Dirac equation, which explains the spin value and the magnetic moment of the electron, the origin of its mass is far from being understood. The classical electron theory (with the works by Abraham, Lorentz and Poincaré) attempts to consider the mass of the electron as of purely electromagnetic origin, and is well known to be deficient in several respects. The basic flaw of this picture is due to the Ernshaw theorem, which states that it is impossible to have a stationary non-neutral charge distribution held together by purely electric forces. Moreover, a purely electromagnetic model of the electron implies the occurrence of divergent quantities. Such infinities can be dealt with by means of the renormalization procedure in Quantum Electrodynamics (QED). However, even in this framework, the value of the electron mass is not intrinsic, but only results from its interaction with the vacuum.

The modern view of the problem of electron mass was pioneered by Wheeler and Feynman[1], according to which it is not of electromagnetic origin but entirely mechanical[2]. In this paper, we show that the electron mass m_e can be obtained from arguments related to the breakdown of local Lorentz invariance, in the framework of a generalization of Special Relativity (*Deformed Special Relativity*, DSR), based on a "deformation" of Minkowski space (*i.e.,* with metric coefficients depending on energy). This assigns m_e a geometrical meaning, by expressing it in terms of the parameter δ of LLI breakdown.

The organization of the paper is as follows. In Sect. 2 we briefly introduce the concept of deformed Minkowski space, and give the explicit forms of the phenomenological energy-dependent metrics for the four fundamental interactions. The LLI breaking parameter δ_{int} for a given interaction is introduced in Sect.3. In Sect. 4 we assume the existence of a stable fundamental particle interacting gravitationally, electromagnetically and weakly, and show (by imposing some physical requirements) that its mass value (expressed in terms of $\delta_{e.m.}$ and $E_{0,grav}$ is just the electron mass. In Sect.5 we briefly introduce the concept of *mass relativity* in DSR. Sect. 6 concludes the paper.

2   Deformed Special Relativity in four dimensions (DSR)

2.1   Deformed Minkowski space-time

Deformed Special Relativity is a generalization of Special Relativity (SR) based on a "deformed" Minkowski space, assumed to be endowed with a metric whose coefficients depend on the energy of the process considered[3]. The deformation is intended essentially to provide a metric representation of the interaction governing the process considered (at least in the given energy range, and *locally, i.e.,* in a suitable space-time region)[3–6]. DSR applies in principle to *all* four interactions (electromagnetic, weak, strong and gravitational), at least as far as their non-local behaviour and non-potential part are concerned.

The generalized ("deformed") Minkowski space $\tilde{M}_4$ (DMS4) is defined as a space with the same local coordinates x of M_4 (the four-vectors of the usual Minkowski space), but with a metric given by the metric tensor*

---

* In the following, we employ the notation *"ESC on"* (*"ESC off"*) to mean that the Einstein sum convention on repeated indices is (is not) used.

$$\eta_{\mu\nu}(E) = diag\left(b_0^2(E), -b_1^2(E), -b_2^2(E), -b_3^2(E)\right) \overset{ESC\,off}{\equiv}$$

$$\equiv \delta_{\mu\nu}\left[\delta_{\mu 0}b_0^2(E) - \delta_{\mu 1}b_1^2(E) - \delta_{\mu 2}b_2^2(E) - \delta_{\mu 3}b_3^2(E)\right] \quad (2.1)$$

$\left(\forall E \in R_0^+\right)$, where the $\{b_\mu^{\,2}(E)\}$ are dimensionless, real, positive functions of the energy[3]. The generalized interval in $\tilde{M}_4$ is therefore given by $(x^\mu = (x^0, x^1, x^2, x^3) = (ct, x, y, z)$, with c being the usual light speed in vacuum) (ESC on)

$$ds^2 = b_0^2(E)c^2 dt^2 - \left(b_1^2(E)dx^2 + b_2^2(E)dy^2 + b_3^2(E)dz^2\right) =$$

$$= \eta_{\mu\nu}(E)dx^\mu dx^\nu \equiv dx * dx. \quad (2.2)$$

The last step in (2.2) defines the scalar product $*$ in the deformed Minkowski space $\tilde{M}_4^{\,*}$. It follows immediately that it can be regarded as a particular case of a Riemann space with null curvature.

We stress that, in this formalism, the energy E is to be understood as the energy of a physical process measured by the detectors *via* their electromagnetic interaction in the usual Minkowski space. Moreover, E is to be considered a dynamical variable, because it specifies the dynamical behaviour of the process under consideration, and, *via* the metric coefficients, it provides us with a dynamical map—in the energy range of interest—of the interaction ruling the given process. Let us recall that the use of momentum components as dynamical variables on the same footing as space-time variables can be traced back to Ingraham[9]. Dirac[10], Hoyle and Narlikar[11] and Canuto *et al.*[12] treated mass as a dynamical variable in the context of scale-invariant theories of gravity.

It was also shown that the DSR formalism is actually five-dimensional, in the sense that the deformed Minkowski space can be naturally embedded in a larger Riemannian manifold, with energy as fifth dimension[13]. Curved 5-d spaces have been considered by several authors[14]. In this regard, the DSR formalism is a kind of generalized (*non-compactified*) Kaluza-Klein theory, and resembles, in some aspects, the so-called "Space-Time-Mass" (STM) theory (in which the fifth dimension is the rest mass), proposed by Wesson[15] and studied in detail by a number of authors[16].

By putting $ds^2 = 0$, we get the *maximal causal velocity* in $\tilde{M}_4$ [3,21]

$$\vec{u}(E) \equiv \left(c\frac{b_0(E)}{b_1(E)}, c\frac{b_0(E)}{b_2(E)}, c\frac{b_0(E)}{b_3(E)}\right) \quad (2.3)$$

(*i.e.*, the analogue of light speed in SR) for the interaction represented by the deformed metric.

In DSR the relativistic energy for a particle of mass m subjected to a given interaction and moving along $\hat{x}_i$ has the form[3]:

---

* Notice that our formalism—in spite of the use of the word "deformation"—has nothing to do with the "deformation" of the Poincaré algebra introduced in the framework of quantum group theory (in particular the so-called κ-deformations)[7] In fact, the quantum group deformation is essentially a modification of the commutation relations of the Poincaré generators, whereas in the DSR framework the deformation concerns the metric structure of the space-time (although the Poincaré algebra is affected, too[8]).

$$E = m\, u^2{}_i(E)\tilde{\gamma}(E) = m\, c^2 \frac{b^2{}_0(E)}{b^2{}_i(E)} \tilde{\gamma}(E) \tag{2.4}$$

where u_i is the i-th component of the maximal velocity (2.3) of the interaction, and

$$\tilde{\gamma}(E) \equiv \left(1 - \tilde{\beta}_i^2\right)^{-\frac{1}{2}} = \left[1 - \left(\frac{v_i b_i(E)}{c b_0(E)}\right)^2\right]^{-\frac{1}{2}} ; \tag{2.5}$$

$$\tilde{\beta}_i \equiv \frac{v_i}{u_i}$$

In the non-relativistic (NR) limit of DSR, *i.e.*, at energies such that

$$v_i \cong u_i(E) \tag{2.6}$$

Eq.(2.4) yields the following NR expression for the energy corresponding to the interaction:

$$E_{NR} = mu^2{}_i(E) = mc^2 \frac{b^2{}_0(E)}{b^2{}_i(E)} \tag{2.7}$$

2.2 Energy-dependent phenomenological metrics for the four interactions

In terms of phenomenology, we recall that a local breakdown of Lorentz invariance may be envisaged for all four fundamental interactions (electromagnetic, weak, strong and gravitational), yielding *evidence for a departure of the space-time metric from Minkowskian* (at least in the energy range examined). The experimental data analyzed are for the following four physical processes: the lifetime of the (weakly decaying) $K^0{}_S$ meson[17]; the Bose-Einstein correlation in (strong) pion production[18]; the superluminal photon tunnelling[19]; the comparison of clock rates in the gravitational field of Earth[20]. A detailed derivation and discussion of the energy-dependent phenomenological metrics for all four interactions has been given [3-6]. Here, we limit ourselves to recalling their following basic features:

1. Both the electromagnetic and the weak metric show the same functional behaviour, namely

$$\eta_{\mu\nu}(E) = diag\left(1, -b^2(E), -b^2(E), -b^2(E)\right), \tag{2.8}$$

$$b^2(E) = \begin{cases} (E/E_0)^{1/3}, & 0 < E \le E_0 \\ 1, & E_0 < E \end{cases} =$$

$$= 1 + \theta(E_0 - E)\left[\left(\frac{E}{E_0}\right)^{1/3} - 1\right], E > 0 \tag{2.9}$$

(where $\theta(x)$ is the Heaviside theta function) the only difference between them being the threshold energy E_0, *i.e.*, the energy value at which the metric parameters are constant, *i.e.*, the metric becomes

Minkowskian ($\eta_{\mu\nu}(E \geq E_0) = g_{\mu\nu} = diag(1,-1,-1,-1)$); the fits to the experimental data yield

$$E_{0,e.m.} = (4.5 \pm 0.2) \text{ } \mu eV;$$
$$E_{0,weak} = (80.4 \pm 0.2) \text{ GeV;} \tag{2.10}$$

Notice that for either interaction the metric is isochronous, spatially isotropic and *"sub-Minkowskian,"* i.e., it approaches the Minkowskian limit from below (for $E < E_0$). Both metrics are therefore Minkowskian for $E > E_{0,weak} > 80$ GeV, and then our formalism is fully consistent with electroweak unification, which occurs at an energy scale ~100 GeV.

We recall that the phenomenological electromagnetic metric (2.8)-(2.10) was derived by analyzing the propagation of evanescent waves in undersized wave guides[17]. This accounts for the observed superluminal group speed in terms of a nonlocal behaviour of the wave guide, described by an effective deformation of space-time in its reduced part[5]. The weak metric was obtained by fitting the data on the mean lifetime of the K^0_S meson (experimentally known in a wide energy range ($30 \div 350$ GeV)[17]), thus accounting for its apparent departure from a purely Lorentzian behaviour[3,21].

2. For the strong interaction, the metric was derived[4] by analyzing the phenomenon of Bose-Einstein (BE) correlation for π-mesons produced in high-energy hadronic collisions[18]. In this approach the BE effect is explained as the decay of a "fireball" whose lifetime and spatial size are directly related to the metric coefficients $b^2_{\mu,strong}(E)$, avoiding the introduction of *ad hoc* parameters in the pion correlation function[4]. The strong metric reads

$$\eta_{strong}(E) = diag\left(b^2_{0,strong}(E), -b^2_{1,strong}(E), -b^2_{2,strong}(E), -b^2_{3,strong}(E)\right) \tag{2.11}$$

$$\left.\begin{array}{l} b^2_{1,strong}(E) = \left(\dfrac{\sqrt{2}}{5}\right)^2 \\[1.5em] b^2_{2,strong}(E) = \left(\dfrac{2}{5}\right)^2 \end{array}\right\} \forall E > 0,$$

$$b^2_{0,strong}(E) = b^2_{3,strong}(E) = \begin{cases} 1, & 0 < E \leq E_{0,strong} \\ \left(E / E_{0,strong}\right)^2, & E_{0,strong} < E \end{cases} = \tag{2.12}$$

$$= 1 + \theta(E - E_{0,strong})\left[\left(\frac{E}{E_{0,strong}}\right)^2 - 1\right], E > 0$$

with

$$E_{0,strong} = (367.5 \pm 0.4) \text{ GeV} \tag{2.13}$$

We stress that, in this case, contrary to the electromagnetic and weak cases, *a deformation of the time coordinate occurs*; moreover, *the three-space is anisotropic*, with two spatial parameters constant (but different in value) and the third one variable with energy like the time parameter.

3. The gravitational energy-dependent metric was obtained[6] by fitting the experimental data on the relative rates of clocks in the Earth's gravitational field[20]. Its explicit form is[*]:

$$\eta_{grav}(E) = diag\left(b_{0,grav}^2(E), -b_{1,grav}^2(E), -b_{2,grav}^2(E), -b_{3,grav}^2(E)\right) \qquad (2.14)$$

$$b_{0,grav}^2(E) = b_{3,grav}^2(E) = \begin{cases} 1, & 0 < E \leq E_{0,grav} \\ \dfrac{1}{4}\left(1 + \dfrac{E}{E_{0,grav}}\right)^2, & E_{0,grav} < E \end{cases} =$$

$$(2.15)$$

$$= 1 + \theta(E - E_{0,grav})\left[\dfrac{1}{4}\left(1 + \dfrac{E}{E_{0,grav}}\right)^2 - 1\right], E > 0$$

with

$$E_{0,grav} = (20.2 \pm 0.1) \ \mu eV. \qquad (2.16)$$

Intriguingly enough, this is approximately of the same order of magnitude of the thermal energy corresponding to the 2.7°K cosmic background radiation in the Universe[†].

Notice that the strong and the gravitational metrics are *over-Minkowskian* (namely, they approach the Minkowskian limit from above ($E_0 < E$), at least for their coefficients $b_0^2(E) = b_3^2(E)$).

3.   LLI breaking factor in DSR

The breakdown of standard local Lorentz invariance (LLI) is expressed by the LLI breaking factor parameter δ[23]. We recall that two different kinds of LLI violation parameters exist: the isotropic (essentially obtained by means of experiments based on the propagation of e.m. waves, *e.g.*, of the Michelson-Morley type), and the anisotropic ones (obtained by experiments of the Hughes-Drever type[23], which test the isotropy of the nuclear levels).

In the former case, the LLI violation parameter reads[23]

$$\delta = \left(\frac{u}{c}\right)^2 - 1, \qquad (3.1)$$

$$u = c + v$$

where c is, as usual, the speed of light *in vacuo*, v is the LLI breakdown speed (*e.g.*, the speed of the preferred frame) and u is the new speed of light (*i.e.*, the

---

[*] The coefficients $b_{1,grav}^2(E)$ and $b_{2,grav}^2(E)$ are presently undetermined at the phenomenological level.

[†] It is worth stressing that the energy-dependent gravitational metric (2.14)-(2.16) is to be regarded as a *local* representation of gravitation, because the experiments considered took place in a neighborhood of Earth, and therefore at a small scale with respect to the usual ranges of gravity (although a large one with respect to the human scale).

maximal causal speed in Deformed Special Relativity[3]. In the anisotropic case, there are different contributions δ^A to the anisotropy parameter from the different interactions. In the HD experiment, it is A = S, HF, ES, W, meaning strong, hyperfine, electrostatic and weak, respectively. These correspond to four parameters δ^S (due to the strong interaction), δ^{ES} (related to the nuclear electrostatic energy), δ^{HF} (coming from the hyperfine interaction between the nuclear spins and the applied external magnetic field) and δ^W (the weak interaction contribution).

All the above tests put upper limits on the value of $\delta^{(23)}$.

Moreover, at the end of the past century, a new electromagnetic experiment was proposed[24], designed to directly test LLI. It is based on the possibility of detecting a non-zero Lorentz force between the magnetic field **B** generated by a stationary current I circulating in a closed loop Γ, and a charge q, on the hypothesis that both q and Γ are at rest in the same inertial reference frame. The force is zero according to standard (relativistic) electrodynamics. The results obtained by this method in two experimental runs[25] admit as the most natural interpretation the fact *that local Lorentz invariance is in fact broken.*

The value of the (isotropic) LLI breaking factor determined by this electromagnetic experiment is[25]

$$\Delta \cong 4 \times 10^{-11} \tag{3.2}$$

and represents the present lowest limit to δ.

In order to establish a connection with the electron mass, we can define the LLI breakdown parameter for a given interaction, $\delta_{int.}$, as

$$\delta_{int} \equiv \frac{m_{in,int.} - m_{in,grav.}}{m_{in,int.}} = 1 - \frac{m_{in,grav.}}{m_{in,int.}} \tag{3.3}$$

where $m_{in,int.}$ is the inertial mass of the particle considered with respect to the given interaction*. In other words, we assume that the *local* deformation of space-time corresponding to the interaction, and described by the metric (2.1), gives rise to a *local violation* of the Principle of Equivalence for interactions different from gravitation. This departure, just expressed by the parameter δ_{int}, also constitutes a measure of the amount of LLI breakdown. In the framework of DSR, δ_{int} embodies the geometrical contribution to the inertial mass, thus discriminating between two different metric structures of space-time.

Of course, if the interaction considered is gravitational, the Principle of Equivalence holds strictly, *i.e.*,

$$m_{in,grav.} = m_g \tag{3.4}$$

where m_g is the gravitational mass of the physical object considered, *i.e.*, it is its "gravitational charge" (namely its coupling constant to the gravitational field).

Then, we can rewrite (3.3) as:

---

* Throughout the present work, "int." denotes a physically detectable fundamental interaction, which can be operationally defined by means of a phenomenological energy-dependent metric of deformed-minkowskian type.

$$\delta_{int.} = \frac{m_{in.,int.} - m_g}{m_{in.,int.}} = 1 - \frac{m_g}{m_{in.,int.}} \tag{3.5}$$

and therefore, when the particle is subjected only to gravitational interaction, it is

$$\delta_{grav.} = 0 \tag{3.6}$$

In the case of the gravitational metric (2.14)-(2.15), we have

$$\frac{b_{0,grav.}(E)}{b_{3,grav.}(E)} = 1, \forall E \in R_0^+ \tag{3.7}$$

Therefore, for $i = 3$, Eq.(2.4) yields, for the gravitational energy of a particle moving along the z-axis ($v_3 = v$):

$$E_{grav} = m_g c^2 \left[1 - \left(\frac{v}{c}\right)^2 \right]^{-\frac{1}{2}} = m_g c^2 \gamma, \tag{3.8}$$

with non-relativistic limit (*cfr.* Eq.(2.7))

$$E_{grav, NR} = m_g c^2 \tag{3.9}$$

namely, the gravitational energy takes its *standard, special-relativistic values.*

This means that the special characterization (corresponding to the choice $i = 3$) of Eqs.(2.4) and (2.7) within the framework of DSR relates the gravitational interaction with SR, which is—as well known—based on the electromagnetic interaction in its Minkowskian form.

4.   The electron as a fundamental particle and its "geometrical" mass

We now consider E the threshold energy of the gravitational interaction:

$$E = E_{0,grav} \tag{4.1}$$

where $E_{0,grav}$ is the limit value under which the metric $\eta_{\mu\nu, grav}$ *(E)* becomes Minkowskian (at least in its known components). Indeed, from Eqs. (2.14), (2.15) it follows that:

$$\eta_{grav}(E) = diag\left(1, -b_{1,grav}^2(E), -b_{2,grav}^2(E), -1\right)$$

$$\overset{ESC\,off}{=} \delta_{\mu\nu}\left[\delta_{\mu 0} - \delta_{\mu 1}b_{1,grav}^2(E) - \delta_{\mu 2}b_{2,grav}^2(E) - \delta_{\mu 3}\right], \forall E \in \left(0, E_{0,grav}\right) \tag{4.2}$$

Notice that at the energy $E = E_{0,grav}$ the electromagnetic metric (2.8),(2.9) is Minkowskian, too (because $E_{0,grav} > E_{0,e.m.}$).

On the basis of the previous considerations, it seems reasonable to assume that the physical object (particle) p with a rest energy (*i.e.*, gravitational mass) just equal to the threshold energy $E_{0,grav}$, namely

$$E_{0,grav} = m_{g,p} c^2, \tag{4.3}$$

must play a fundamental role for either e.m. and gravitational interaction. We can, *e.g.*, hypothesize that p corresponds to the lightest mass eigenstate which experiences both force fields (*i.e.*, from a quantum viewpoint, coupling to the respective interaction carriers, the photon and the graviton). As a consequence, p must be intrinsically stable, due to the impossibility of its decay into lighter mass eigenstates, even when the a particle is subject to weak interaction (*i.e.*, it couples to all gauge bosons of the Glashow-Weinberg-Salam group $SU(2) \times U(1)$, not only to its electromagnetic charge sector).

Since, as we have seen, for $E = E_{0,grav}$ the electromagnetic metric is Minkowskian, too, it is natural to assume, for p:

$$m_{in,p,e.m.} = m_{in,p}, \tag{4.4}$$

namely its inertial mass is that measured with respect to the electromagnetic metric.

Then, due to the Equivalence Principle (see eq. (3.4)), the mass of p is characterized by

$$p : \begin{cases} m_{in,p,grav} = m_{g,p} \\ m_{in,p,e.m.} = m_{in,p} \end{cases} \tag{4.5}$$

Therefore, for this fundamental particle the LLI breaking factor (3.3) of the e.m. interaction becomes:

$$\delta_{e.m.} \equiv \frac{m_{in,p} - m_{g,p}}{m_{in,p}} = 1 - \frac{m_{g,p}}{m_{in,p}} \tag{4.6}$$

$$\Leftrightarrow m_{g,p} = m_{in,p}\left(1 - \delta_{e.m.}\right)$$

Substituting (4.3) in (4.6) yields:

$$E_{0,grav} = m_{in,p}\left(1 - \delta_{e.m.}\right)c^2 \Leftrightarrow m_{in,p} = \frac{E_{0,grav}}{c^2}\frac{1}{1 - \delta_{e.m.}} \tag{4.7}$$

Eq.(4.7) allows us to evaluate the inertial mass of p from the knowledge of the electromagnetic LLI breaking parameter $\delta_{e.m.}$ and the threshold energy $E_{0,grav}$ of the gravitational metric.

Due to Eq.(3.1), we can relate the lowest limit to the LLI breaking factor of electromagnetic interaction, Eq.(3.3) (determined by the coil-charge experiment), with $\delta_{e.m.}$ as follows:

$$\delta = 1 - \delta_{e.m.} \cong 4 \times 10^{-11} \tag{4.8}$$

Then, inserting the value (2.16) for $E_{0,grav}$[*] and (4.8) in (4.7), we get

$$m_{in,p} = \frac{E_{0,grav}}{c^2}\frac{1}{1 - \delta_{e.m.}} \geq \frac{2 \times 10^{-5}}{4 \times 10^{-11}}\frac{eV}{c^2} = 0.5\frac{MeV}{c^2} = m_{in,e} \tag{4.9}$$

(with $m_{in,e}$ the electron mass) where the $\geq$ is due to the fact that in general the LLI breaking factor constitutes an *upper limit* (*i.e.,* it sets the scale *below which* a violation of LLI is expected). If experiment [25] *does indeed provide evidence* for a LLI breakdown (as seems the case, although further confirmation is needed), eq. (4.9) yields $m_{in,p} = m_{in,e}$. We find therefore the amazing result that *the fundamental particle p is nothing but the electron e (or its antiparticle*

---

[*] Recall that the value of $E_{0,grav}$ was determined by fitting the experimental data on the slowing down of clocks in the Earth gravitational field [20]. See ref.[6].

e^{+*}). The electron is indeed the lightest massive lepton (pointlike, non-composite particle) with electric charge, and therefore subjected to gravitational, electromagnetic and weak interactions, but unable to weakly decay due to its small mass. Consequently, e^- (e^+) shares all the properties we required for the particle p, whereby it plays a fundamental role for gravitational and electromagnetic interactions.

5.   Mass relativity in DSR

The considerations carried out in the previous Sections therefore relate the electron mass to the (local) breakdown of Lorentz invariance. Its mass would then be a *measure of the deviation of the metric from Minkowskian*. The minimum measured mass of a particle would be related to the minimum possible metric deviation compatible with its interactions.

This point can be reinforced by the following argument.

The maximum causal velocity $\tilde{u}$ defined by Eq.(2.3) can be interpreted, from a physical standpoint, as the speed of the quanta of the interaction locally (and phenomenologically) described in terms of a deformed Minkowski space. Since these quanta are associated with lightlike world-lines in $\tilde{M}_4$, they must be zero-mass particles (*with respect to the interaction considered*), by analogy with photons (*with respect to the e.m. interaction*) in the usual SR.

Let us clarify the latter statement. The carriers of a given interaction propagating with the speed $\tilde{u}$ typical of that interaction *are actually expected to be strictly massless only inside the space whose metric is determined by the interaction considered. A priori*, nothing prevents such "deformed photons" *from acquiring a non-vanishing mass in a deformed Minkowski space related to a different interaction*.

This might be the case of the massive bosons W^-, W^+ and Z^0, carriers of the weak interaction. They would therefore be massless in the space $\tilde{M}_4$ ($\eta_{weak}(E)$) related to the weak interaction, but would acquire a mass when considered in the standard Minkowski space M of SR (that, as already stressed, is strictly connected with the electromagnetic interaction governing the operation of the measuring devices). In this framework, therefore, it is not necessary to postulate a "symmetry breaking" mechanism (like the Goldstone mechanism in gauge theories) to allow particles to acquire mass. On the contrary, if measuring devices could be built based on interactions other than e.m., the photon might acquire a mass with respect to a non-electromagnetic background.

Mass itself would therefore assume a *relative nature*, related not only to the interaction concerned, but also to the metric background in which the energy of the physical system is measured. This can be seen if one considers that in general, for relativistic particles, mass is the invariant norm of 4-momentum,

---

[*] Of course, this last statement holds strictly only if the CPT theorem maintains its validity in the DSR framework, too. Although this problem has not yet been addressed in general on a formal basis, we can state that it holds true in the case we considered, since we assumed that the energy value is $E = E_{0,grav}$ corresponding to the Minkowskian form of both the electromagnetic and gravitational metrics.

and what is usually measured *is not* the value of an invariant, but the related energy.

6.   Conclusions

The formalism of DSR describes—among others things—in geometrical terms (*via* energy-dependent deformation of the Minkowski metric) the breakdown of Lorentz invariance at the local level (parametrized by the LLI breaking factor δ_{int}). We have shown that within DSR it is possible—on the basis of a simple and plausible assumption—to evaluate the inertial mass of the electron e (and therefore of its antiparticle, the positron $e^{\,1}$) by exploiting the expression of the relativistic energy in the deformed Minkowski space $\tilde{M}_4(E)_{E \in R_0^+}$, the explicit form of the phenomenological metric describing the gravitational interaction (in particular its threshold energy), and the LLI breaking parameter for the electromagnetic interaction $\delta_{e.m.}$.

Therefore, the inertial properties of one of the fundamental constituents of matter and the Universe find a "geometrical" interpretation in the context of DSR, when local violations of standard Lorentz invariance are admitted.

We have also put forward the idea of a *relativity of mass,* namely the possible dependence of the mass of a particle on the metric background where mass measurements are carried out. This could constitute a possible alternative mechanism of mass generation to those based on symmetry breakdown in Relativistic Quantum Theory.

References

1.   J.A Wheeler and R. P. Feynman: *Rev. Mod. Phys.* **17**, 157 (1945). *ibidem,* **21**, 425 (1949).

2.   M. H. Mac Gregor: *The Enigmatic Electron* (Kluwer, Dordrecht, 1992).

3.   F. Cardone and R. Mignani: *Energy and Geometry – An Introduction to Deformed Special Relativity* (World Scientific, Singapore, 2004), and references therein.

4.   F. Cardone and R. Mignani: *JETP* **83**, 435 [*Zh. Eksp. Teor. Fiz.***110**}, 793] (1996); F. Cardone, M. Gaspero, and R. Mignani: *Eur. Phys. J.* C **4**, 705 (1998).

5.   F. Cardone and R. Mignani: *Ann. Fond. L. de Broglie,* **23** , 173 (1998); F. Cardone, R. Mignani, and V.S. Olkhovski: *J. de Phys. I (France)* **7**, 1211 (1997); *Modern Phys. Lett.* B **14**, 109 (2000).

6.   F. Cardone and R. Mignani: *Int. J. Modern Phys.* A **14**, 3799 (1999).

7.   See *e.g.,* P. Kosiński and P. Maślanka, in *From Field Theory to Quantum Groups*, eds. B. Jancewicz and J. Sobczyk (World Scientific, Singapore, 1996), p. 11; J. Lukierski, in *Proc. of Alushta Conference on Recent Problems in QFT*, May 1996, ed. D. Shirkov, D.I. Kazakov, and A.A. Vladimirov (Dubna 1996), p.82; J. Lukierski, "κ-Deformations of relativistic symmetries: recent developments," in *Proc. of Quantum Group Symposium* (July 1996, Goslar), eds. H.-D. Doebner and V.K. Dobrev (Heron Press, Sofia, 1997), p. 173; and references therein.

8.   F. Cardone, A. Marrani and R. Mignani: *Found. Phys.* **34**, 617; 1155; 1407 (2004).

9.   R.L. Ingraham: *Nuovo Cim.* **9**, 87 (1952).

10.   P.A.M. Dirac: *Proc. R. Soc. (London)* **A333**, 403 (1973); *ibidem,* **A338**, 439 (1974).

11.   F. Hoyle and J.V. Narlikar: *Action at a Distance in Physics and Cosmology* (Freeman, N.Y., 1974).

12.   V. Canuto, P.J. Adams, S.-H. Hsieh and E. Tsiang: *Phys. Rev.* **D16**, 1643 (1977); V. Canuto, S.-H. Hsieh and P.J. Adams: *Phys. Rev. Lett.* **39**, 429 (1977).

13. F. Cardone, M. Francaviglia, and R. Mignani: *Gen. Rel. Grav.* **30**, 1619 (1998); *ibidem*, **31**, 1049 (1999); *Found. Phys. Lett.* **12**, 281; 347 (1999).

14. See V. G. Kadyshevsky, R. M. Mir-Kasimov and N. B. Skachkov: *Yad. Fiz.* **9**, 212 (1969); V. G. Kadyshevsky: *Nucl. Phys.* **B141**, 477 (1978); V. G. Kadyshevsky, M. D. Mateev, R. M. Mir-Kasimov and I. P. Volobuev: *Theor. Math. Phys.* **40**, 800 (1979) [*Teor. Mat. Fiz.* **40**, 363 (1979)]; V. G. Kadyshevsky and M. D. Mateev: *Phys. Lett.* B **106**, 139 (1981); *Nuovo Cim.* A **87**, 324 (1985); A. D. Donkov, R. M. Ibadov, V. G. Kadyshevsky, M. D. Mateev and M. V. Chizhov, *ibid.* **87**, 350. 373 (1985); and references therein.

15. P.S. Wesson: *Astron. Astrophys.* **119**, 145 (1983); *Gen. Rel. Grav.* **16**, 193 (1984).

16. See *e.g.,* J.M. Overduin and P.S. Wesson: *Phys. Rept.* **283**, 303 (1997). P.S. Wesson: *Space-Time-Matter – Modern Kaluza-Klein Theory* (World Scientific, Singapore, 1999), and references therein.

17. S.H. Aronson, G.J. Bock, H.-Y. Chang and E. Fishbach: *Phys. Rev. Lett.* **48**, 1306 (1982); *Phys. Rev. D* **28**, 495 (1983); N. Grossman *et al.*: *Phys. Rev. Lett.* **59**, 18 (1987).

18. For experimental as well as theoretical reviews, see *e.g.*, B. Lörstad: *Correlations and Multiparticle Production (CAMP),* eds. M. Pluenner, S. Raha and R. M. Weiner (World Scientific, Singapore, 1991); D.H. Boal, C.K. Gelbke and B. K. Jennings: *Rev. Mod. Phys.* **62**, 553 (1990); and references therein.

19. For reviews on both experimental and theoretical aspects of superluminal photon tunnelling, see *e.g.,* G. Nimtz and W. Heimann: *Progr. Quantum Electr.* **21**, 81 (1997); R.Y. Chiao and A.M. Steinberg: "Tunnelling Times and Superluminality," in *Progress in Optics,* E. Wolf ed., **37**, 346 (Elsevier Science, 1997); V.S. Olkhowsky and A. Agresti: in *Tunnelling and its Implications,* D. Mugnai, A. Ranfagni and L.S. Schulman eds. (World Scientific, Singapore, 1997), p.327.

20. C.O. Alley: "Relativity and Clocks," in *Proc. of the 33rd Annual Symposium on Frequency Control,* Elec. Ind. Ass., Washington, D.C. (1979).

21. F. Cardone, R. Mignani and R. M. Santilli: *J. Phys. G* **18**, L61, L141 (1992).

22. F. Cardone, A. Marrani and R. Mignani: *Found. Phys. Lett.* **16**, 163 (2003).

23. See C.M. Will: *Theory and Experiment in Gravitational Physics* (Cambridge Univ. Press, rev.ed.1993), and references therein.

24. U. Bartocci and M. Mamone Capria: *Am. J. Phys.* **59**, 1030 (1991); *Found. Phys.* **21**, 787 (1991).

25. F.Cardone and R. Mignani: *Physics Essays* **13**, 643 (2000); "On possible experimental evidence for a breakdown of local Lorentz invariance," in *"Gravitation, Electromagnetism and Cosmology: Toward a New Synthesis"* (Proc. Int. Conf. On Redshifts and Gravitation in a Relativistic Universe, Cesena, Italy, Sept. 17-20, 1999), ed. by K. Rudnicki (Apeiron, Montreal, 2001), p. 165; U. Bartocci, F.Cardone and R. Mignani: *Found. Phys. Lett.* **14**, 51 (2001).

On the Space-Vortex Structure of the Electron

Paramahamsa Tewari[*]

1. Introduction

It was Rene Descartes, the French Mathematician and Philosopher who, perhaps for the first time in a *scientific* sense, assigned a reality to the medium of space as a *property-less* fluid-entity, already known at that time as ether. According to Descartes, large cosmic ether vortices existed throughout the universe. One such vortex carried the planets around the sun, and countless smaller vortices aggregated into different sizes of universal matter, filling the whole of space. He explained gravity by the pressure and impact of ether on bodies; and framed the principles of the inertial tendencies of matter for straight line motion based on the property of the fluidity of a space-substratum filled with ether vortices. The transmission of the then known magnetic forces and the force of gravity between the earth and the planetary bodies found explanations in Cartesian philosophy with *physical contacts* between the interacting entities mediated by the intervening ether. The theory of Descartes at that time was the most convincing natural philosophy and was based on a single *dynamic ether* as the only reality of the universe. The theory remained in acceptance for almost a century after publication of Newton's *Principia*.

Newton's laws of motion took into account the principle of inertia for straight line motion as conceived by Descartes [1], and Galileo's experimental discoveries on freely falling bodies and their motion on inclined planes; but ether was not invoked to explain the properties of mass, inertia (which were introduced in Newton's laws of motion) and the force of gravity. Thus the medium of space, except for its utility as a continuous fluid-substratum for the transmission of light waves, was again made inert and inactive for transmission of forces; and this led to the reintroduction of the principle of "action at a distance." Based on this principle, R. G. Boscovich (1711-87) tried to explain all physical effects and, further, Coulomb and Ampère invoked it in explaining the mutual action of forces between charged bodies and electric currents. In contrast, Faraday's researches led him to the conclusion that electromagnetic induction cannot take place without the intervening medium (field). Faraday introduced the concept of continuously varying electric and magnetic fields, signifying that space is a continuous substratum and "action at a distance" is not the basic principle. He also suggested that an atom could be a structure of fields of forces—electric, magnetic, and gravitational, existing around its cen-

[*] Former Executive Director, Nuclear Power Corporation; Vinodini Nivas, Gotegali, Karwar-581317, Karnataka, India; email: ptewari1@sancharnet.in

tral point. On the existence of ether, Faraday's belief was that it may have its utility in other physical effects, in addition to providing a medium for transmission of light. Based on Faraday's concepts, Maxwell wrote equations using hydrodynamics to model ether, postulating that it was as an incompressible fluid. Helmholtz conceived the ether vortex filament as electric current, and W. Thomson believed [2] that 'the magnetic energy is the kinetic energy of a medium occupying the whole space, and that electric energy is the energy of strain of the same medium.' Atomic structure as a vortex motion was also proposed by Thomson and others, and after the electron's discovery (1897), Larmor concluded that the electron is a structure in the ether and that all matter consisted of electrons only.

Serious problems arose (1905) with the concepts of the vortex structure of atoms/electrons in an incompressible fluid. One problem was that of the dissipation of vortex motion, since the streamlines in a vortex may tend to dilate outward (W. Thomson). Another problem pertained to the difficulty of the transmission of an electromagnetic field in this fluid at the enormous speed of light, for which, if its properties are considered akin to matter, the elasticity should be near to that of steel! While these difficulties were yet to be overcome, Einstein's Theory of Relativity (1905), proposed around the same time, postulated the medium of space as an *empty extension,* which meant no point of space had a velocity-vector (or "velocity field"), thus making the very existence of ether superfluous. The space-vortex structure of the electron, based on this writer's works [3], and described in this paper, provides solutions to both the above problems. The high elasticity required for the fluid-ether, as pointed out above, is avoided by postulating it as a *nonmaterial* and incompressible fluid devoid of any known property of matter, such as mass, density, discreteness, viscosity, elasticity, or compressibility, *etc.* Further, if the properties of "mass" and "charge" of an electron must be derived from the *first principles* proposed by Descartes, Faraday, Maxwell, and Thompson, then a *massless* and *chargeless* fluid that, as a vortex, can form the structure of an electron, must be assumed. That the proof of this assumption—that the universal substratum of space with *nonmaterial** properties has real existence—is provided by deriving the basic properties of the electron (mass, charge, inertia, gravity, locality, *etc.*) from the space[†] vortex structure, and by explaining its behaviour in physical as well as quantitative terms as experimentally observed. The other problem, that of the outward dissipation of the vortex motion, is solved by introducing a *discontinuity* in the energy-distribution at the vortex center, as discussed later.

2. Postulates

1. The medium of space, throughout the universe, is an eternally existing, nonmaterial, continuous, isotropic fluid substratum.

---

* "Nonmaterial" signifies a massless, densityless, incompressible, non-viscous and continuous fluid.
† The absolute vacuum with non-material properties is termed as "Space."

2. The medium of space has a limiting flow speed equal to the speed of light relative to the absolute vacuum, and a limiting angular velocity, when in a state of circulating motion.

3. The medium of universal space is eternal and endowed with motion.

3.   Breakdown of fluid space

The creation of an electron requires a breakdown of the flow of the fluid medium of space (hereafter referred to as "space"). Fig. 1 shows an irrotational circular vortex of space with concentric streamlines. Consider an element of space of volume $dAdr$, as shown, on which a tangential velocity field u is acting. If this vortex pertains to a viscous fluid of density ρ, the mass of the element will be: $dm = \rho dAdr$. There will be a pressure differential on the two surfaces of the element as shown. The two equal and opposite forces acting on the element will be: (a) an inwardly directed, radial, net pressure force and (b) a centrifugal force, giving the relation:

Force   = net pressure force = centrifugal force = $dpdA$
    = $dm \times u^2/r = (\rho dA\, dr)u^2/r$, from which:

$$\frac{\text{Force}}{dm} = \frac{(dpdA)}{(\rho dAdr)} = \frac{u^2}{r} \qquad (1)$$

In an irrotational circular vortex, it can be shown that the velocity of a space-point at distance r from the vortex center is given by:

$$ur = \text{constant} \qquad (2)$$

When a vortex of *massless* space is considered, there is neither inward force (on the element) due to the pressure-differential, nor outward centrifugal force, because the property of mass is common to the origin of both these forces. On a circular streamline, and at each of its points, the velocity field u creates a radial outward acceleration field u^2/r that, acting simultaneously on diametrically opposite points, tends to create a *tearing action* to split open the *continuous* space. If the speed of the space-circulation reaches the limiting speed c, which is the speed of light in the absolute vacuum, and the velocity-field gradient around the center of the vortex becomes the postulated limiting angular rotation ω, the space breaks down, creating a spherical void (Fig. 2), which is defined as a field-less, energy-less and space-less volume of *nothingness* at the vortex center. The radius of the void created follows the relation, as determined by the ratio:

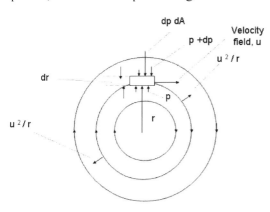

Fig. 1 Irrotational vortex

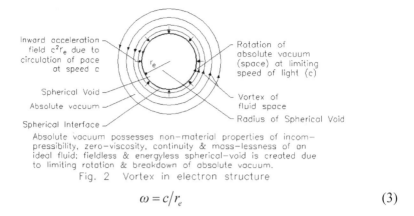

Inward acceleration
field c^2r_e due to
circulation of pace
at speed c

Rotation of
absolute vacuum
(space) at limiting
speed of light (c)

Spherical Void

Absolute vacuum

Spherical Interface

Vortex of
fluid space

Radius of Spherical Void

Absolute vacuum possesses non-material properties of incom-
pressibility, zero-viscosity, continuity & mass-lessness of an
ideal fluid; fieldless & energyless spherical-void is created due
to limiting rotation & breakdown of absolute vacuum.

Fig. 2   Vortex in electron structure

$$\omega = c/r_e \qquad (3)$$

4.   Stability of the void

Fig. 3 shows a diametrical cross section of the spherical void by the plane Y-Z. The circle C rotating around the Y-axis traces a sphere. The point P_z, at the intersection of C and the Z-axis, will have a tangential velocity c (down the paper) the velocity at which the flow of the fluid-space breaks down. The radius r_e of C, from (2), is determined by the ratio c/ω. Consider a point P at the circle C that has the Y-coordinate, $r_e\sin\theta$: it will have a tangential velocity $\omega r_e\sin\theta$ (down the paper at P) provided P too has the same angular velocity ω similar to P_z. The velocity gradient at P_z is c/r_e, which is also the velocity gradient at P, that is, $\omega r_e\sin\theta/r_e\sin\theta$, or ω.

Thus, though the tangential velocity of space varies from zero at P_y (located at the axis, Fig. 3) to the maximum value c at P_z in the diametrical plane, the velocity gradient for all the in-between points remains constant at ω (Postulate 2). Under these considerations the geometry of the void created at the vortex center due to the breakdown of the flow of space is concluded to be *spherical*. It is shown below that the void is dynamically stable. The creation of the void reverses the direction of the outward acceleration field[*] (Eq.1) that created the void; because the void (enclosed within a sphere, here referred as the *interface*) is an empty volume without any "circulating space" or "energy," it is now at zero potential relative to space surrounding it. Therefore, the acceleration field in Fig. 2 is shown inward. As described above, ω is the limiting *velocity gradient* c/r_e at the point P_z just prior to the creation of the void. At each point of the interface circle cut by a diametrical plane at right angles to the Y-Z plane (Fig. 3), the tangential velocity c produces maximum radial and inward acceleration, c^2/re.

The acceleration field at P is $(\omega r_e\sin\theta)^2/r_e\sin\theta$ along $r_e\sin\theta$. Although the interface is constituted of spinning fluid-space, due to the constancy of ω on each of its points, it rotates like a surface of a rigid spherical shell of negligible wall thickness. The stability of the void is due to the following two factors.

---

[*] The acceleration of fluid space at a point is termed "acceleration field."

Consider the circular section of the interface with the diametrical plane (Fig.2). The radial velocity gradient (ω) is c/r_e. If the void shrinks to a smaller radius, the value of ω increases proportionately; which is not possible according to Postulate 2; the void thus expands back to its original size. In the event the void tends to grow to a larger size, the *inward* acceleration field c^2/r_e opposes this increase and any increase in r_e decreases the velocity gradient ω to a lower value, which is no longer sufficient to sustain the void. The sphere of the void is thus reduced to its original size. The other factor is the property of the non-viscosity of space, which maintains the space-vortex eternally, except for its annihilation on meeting a similar vortex with an oppositely oriented velocity field (discussed later). Further, the energy-less-void being a region of zero potential, the inward acceleration field c^2/r_e on the interface prevents dilation of the streamlines, thereby, preventing dissipation of the space-circulation away from the interface. Thus, the void maintains its dynamic stability—its volume being regulated due to the constancy of ω and, consequently, the constancy of c and r_e, dictated by the absolute* properties of the medium of space.

5.   Fundamental particles of matter

If there is only one fundamental particle of matter, it is inconceivable that the universe has different kinds of "spaces" or many structures with varying basic properties. Hence, it is postulated that the most basic property of the universal medium of space is expressed by a single universal constant ω that limits its angular rotation and leads to the creation of a fundamental stable vortex. While the void of a *definite volume* is enclosed within the space-vortex, the vortex itself extends throughout the whole universal-space through its *velocity field*[†]. The space-vortex structure with a fixed volume of dynamically stable void at its center is defined as the fundamental particle of matter. The properties of "electric charge" and "mass" of the fundamental particle, and the "energy fields" associated with its structure are derived in the following pages.

6.   Generation of fields

The space in circulation at speed c within the volume of the spherical void prior to its creation is, qualitatively, the basic state of energy[‡]. At the instant of the creation of the void, this energy is pushed out from within the void, and distributed in continuous space as continuously varying gravity and electrostatic fields. The fields, so created, emanating from the interface of the fundamental particle, become integral with the whole of universal space. On account of the property of the non-viscosity of space, the void enclosed within the dynamically stable interface at the center of the vortex, and the above fields exist

---

* Properties of space, being non-material in nature, are defined to be absolute; unaffected by various conditions of temperature and pressure as applicable to material media.

† The motion of space leads to the generation of "the velocity field."

‡ The quantitative definition of energy is given later.

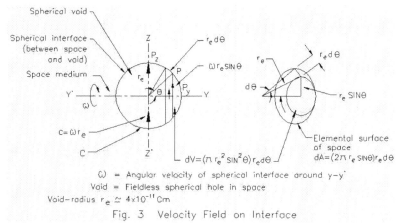

ω = Angular velocity of spherical interface around y–y'
Void = Fieldless spherical hole in space
Void–radius $r_e \simeq 4 \times 10^{-11}$ Cm

Fig. 3   Velocity Field on Interface

eternally without any loss of strength. The properties of the fundamental particle described above identify it as the electron itself.

7.   Unit electric charge

Electric charge is the effect of the space-circulation produced on the interface of a fundamental particle of matter. It is derived as follows. Refer to Fig. 3. Consider an elemental surface on the interface, which has an area: $dA = 2\pi r_e \sin\theta\, r_e d\theta$. The tangential velocity of space at each point of the elemental surface is $\omega r_e \sin\theta$. The electric charge on the elemental surface is defined from first principles as the *surface integral of the tangential velocity of space on each point of the surface*: $dq = 2\pi r_e \sin\theta r_e d\theta \omega r_e \sin\theta$. Substituting from (2), $\omega r_e = c$, in the above equation: $dq = 2\pi c r_e^2 \sin^2\theta d\theta$. Integrating for the total electric charge q_e, varying θ from 0 to π:

$$q_e = 2\pi c r_e^2 \sin^2\theta d\theta = \left(\pi/4\right)4\pi r_e^2 c \qquad (4)$$

The surface integral of the tangential space velocity on the interface is defined as the unit of electrical charge of the fundamental particle of matter. The dimensions of electric charge from (4) are: $q_e = L^3/T$. In CGSE system of units:

$$\mathrm{cm}^3/\mathrm{s} = CGSE - unit \qquad (5)$$

Substituting the experimentally determined value of the electric charge of an electron $(4.8 \times 10^{-10}$ CGSE) and the speed of light in absolute vacuum $(3 \times 10^{10}$ cm/s) in (4), and using the relationship (5), the radius of the interface enclosing the void is calculated as $r_e = 4 \times 10^{-11}$ cm. A comparison with the classical electron radius, which in modern textbooks is shown as 2.82×10^{-13} cm, reveals that r_e should be about 142 times smaller. However, the following quote supports the results obtained from (4). "There are several lengths that might aspire to be characteristic of the dimensions of the electron. If we proceed from modern theoretical electrodynamics, which has been established better than any other field theory, the conclusion seems to be that the electron has enormous dimensions, not 10^{-13} cm, as expected from classical physics, but 10^{-11}

cm (a hundred times greater!)." [4] This value of the electron radius (10^{-11}cm), and its closeness with the radius of the spherical void derived above from Eq.4, suggests that the "fundamental particle of matter" described above is the electron—already discovered by the close of the 19th century. An electron moving away from an observer (electron axis coinciding with the line of motion) is seen as a positron by another observer whom this electron is approaching. Fig. 4 shows, qualitatively, attractive and repulsive forces between these particles through interaction of their velocity fields, while quantitative relationships follow.

In (a) of Fig. 4, the velocity-field u between particles is increased due to the superposition of the fields. From (2), an increase in u results in a proportionate decrease of r, and hence the particles are brought closer by an attractive force between them. In (b) of Fig. 4, due to the decrease of the velocity field between the particles, r has to increase proportionately, and this causes a repulsive force between similar particles. Quantitative relationships are derived in a later section.

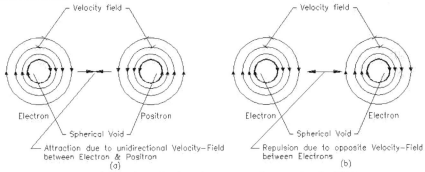

Fig. 4 Attractive & Repulsive Forces due to Velocity Fields

8.   Fundamental mass

The property of mass in the fundamental particle of matter (electron) arises due to the breakdown of space circulation at the center of the electron, and consequent creation of a dynamically stable spherical void associated with gravitational as well as electrostatic fields in space. The derivation of the mass of the electron from the vortex structure is as follows. (Refer to Fig. 3.) Consider an element of void volume, dV, within the spherical interface: $dV = (\pi r_e^2 \sin^2\theta) r_e d\theta = \pi r_e^3 \sin^2\theta d\theta$. The tangential velocity of space acting at the interface of this element is $\omega r_e \sin\theta$. The physical process of creation of mass, dm, of this element is due to volume dV of the fluid space being pushed out at the time of void creation at the speed $\omega r_e \sin\theta$ tangentially through the interface. The mass of the elemental void volume is defined from *first postulates* as $dm = dV(\omega r_e \sin\theta) = dV(c \sin\theta)$. Substituting the value of dV $dm = (\pi r_e^3 \sin^2\theta d\theta) \omega r_e \sin\theta = (4\pi/3) r_e^3 c$. Integrating for the total mass m_e, varying θ from 0 to π:

$$m_e = \left(4\pi/3\right) r_e^3 c \qquad (6)$$

Fundamental mass = Fundamental void volume $\times c$ $\qquad$ (7)

The volume-integral of space-circulation velocity within the void, at the instant of its creation, is the mass of the fundamental unit of matter (electron). A distinction between rest mass and relativistic mass is not made here, as explained. It was earlier shown that the void at the electron center is dynamically stable with radius r_e and space circulation c. This leads to the creation of only one size of stable void. Therefore all the particles of matter, nuclei and atoms will have their masses in exact multiples of electron mass (analyzed further below). The mass of the electron during motion relative to space will remain constant up to speed c because the fluid-space ahead of a moving electron can be displaced up to a maximum speed c only. Thus the volume of the void remains constant; therefore electron mass, which is proportional to the volume of the void (7), also remains constant. The relativistic increase in electron mass at speeds closer to light speed, as experimentally observed, is due to the reaction of the fluid space against the central interface in electron structure resulting from production of an additional acceleration field, discussed elsewhere [3]. The proportionality of mass to the limiting velocity field c and also to the volume of the central void (6) shows that *mass is not energy. "Mass is proportional to energy"* is a more accurate statement.

9.   Dimensions and the unit of mass

The dimensions of mass from Eq.6 are: $m_e = L^4/T$. Therefore, in the CGS system of units, the unit of mass is: cm^4/s. With the use of the experimentally determined mass of the electron, the computed mass of a molecule of water, and the known numbers of molecules in one cm^3 of water; a relationship between "cm^4/s" and "gram" is approximately determined below. From the charge equation (4), the electron radius is:

$$r_e = \left(q_e/\pi^2 c\right)^{\frac{1}{2}}. \qquad (8)$$

The electron charge is experimentally determined as 4.8×10^{-10} CGSE. Expressing CGSE as cm^3/s from (5), $q_e = 4.8 \times 10^{-10}$ cm^3/s, and substituting this value of electron charge and the value of c in (8), we obtain

$$r_e = \frac{\left(4.8 \times 10^{-10}\,cm^3/s\right)^{1/2}}{\left(\pi^2 3 \times 10^{10}\,cm/s\right)^{1/2}} = 4 \times 10^{-11}\,cm \qquad (9)$$

With the above radius of the interface (void), its volume is $V_e=(4\pi/3)(4\times10^{-11}$ $cm)^3 = 2.67 \times 10^{-31} cm^3$. The mass of the electron, experimentally determined, is 9.11×10^{-28}g. Although the concept of density in its structure is not applicable because of the central void, the ratio of the electron mass and the volume of its void will be indicative of the proportionality of the "quantity of mass" within a "unit volume" of void. From above, this ratio, m_e/V_e is 9.11×10^{-28}g/$2.67 \times 10^{-31} cm^3 = 3.42 \times 10^3$g/$cm^3$. One molecule of water is about

2.88×10^{-23}g. Since the mass of a water molecule has to be an exact multiple of the electron mass, the ratio, m_e/V_e, calculated above for the electron, will also be applicable to the water molecule. From this ratio, the void volume in the water molecule is $V_H = (2.88 \times 10^{-23}\text{g})/(3.42 \times 10^3\text{g/cm}^3) = 8.4 \times 10^{-27}\text{cm}^3$. One cm^3 of water has 3.34×10^{22} molecules, the void-volume in one cm^3 of water can be calculated as $(3.34 \times 10^{22})(8.4 \times 10^{-27}\text{cm}^3) = 2.8 \times 10^{-4}\text{cm}^3$. From the mass-equation (6), and mass and void-volume relationship (7), the equivalent mass of one cm^3 of water due to its void content is $(2.8 \times 10^{-4}\text{cm}^3)(3 \times 10^{10}\text{cm/s}) = 8.4 \times 10^6\text{cm}^4\text{/s}$. Since the mass of one cm^3 of water is one gram, from above, we have the relationship:

$$\text{gram} = 8.4 \times 10^6\,\text{cm}^4/\text{s} \qquad (10)$$

Alternatively, the above relationship can be found through a simpler method as follows. Substituting the values of electron radius r_e from (9) and the experimentally determined mass in mass equation (6), we have 9.11×10^{-28} $g = (4\pi/3)\,(4 \times 10^{-11}\,\text{cm})^3\,(3 \times 10^{10}\text{cm/s})$. From which:

$$\text{gram} = 8.8 \times 10^6\,\text{cm}^4/\text{s} \qquad (11)$$

The results obtained in (10) and (11) are close; from the average of both:

$$\text{gram} \approx 8.6 \times 10^6\,\text{cm}^4/\text{s} \qquad (12)$$

10. Energy in electron structure

Linear and accelerating motion of space are the basic states of energy. The circulation of space, forming the electron's interface and spreading throughout the universal space, is the structural energy of the electron; it is computed as follows. Refer to Fig.3. Consider, within the interface, an elemental "disc of void" of volume $dV = (\pi r_e^2 \sin^2\theta)r_e d\theta = \pi r_e^3 \sin^2\theta d\theta$, which is created due to the displacement of space through the interface at the tangential velocity, $\omega r_e \sin\theta$, or, $c\sin\theta$ (since $\omega r_e = c$), at the instant of the electron's creation. The mass of this disc element, as defined in (7) is:

$$dm = dV\left(c\sin\theta\right) = \left(\pi r_e^3 \sin^2\theta d\theta\right)c\sin\theta = \pi c r_e^3 \sin^3\theta d\theta \qquad (13)$$

The disc element has an area at the interface equal to $(2\pi r_e \sin\theta)r_e d\theta$, and has an inward radial acceleration field at each point on it such that $a_f = \omega^2 r_e^2 \sin^2\theta/r_e\sin\theta = c^2\sin\theta/r_e$. Consider the process opposite to void creation: the case of collapse of the interface to zero radius (as happens during annihilation, which is discussed later), when each point at the interface of the elemental disc will be displaced along the radius $r_e\sin\theta$ with the above inward acceleration field acting on it. The energy released due to collapse of the void-disc-element is defined as $dE = dm \cdot a_f$ (field displacement) $= (\pi c r_e^3 \sin^3\theta d\theta)(c^2\sin\theta/r_e)r_e\sin\theta = \pi c^3 r_e^3 \sin^5\theta d\theta$. Integrating, varying θ from 0 to π, to obtain the total energy released due to the collapse of the spherical void yields the creation energy

$$E = \left(4/5\right)\left(4\pi r_e^3 c/3\right)c^2 = \left(4/5\right)m_e c^2 \qquad (14)$$

which is obtained when the mass-equation (6), is used and $(4\pi r_e^3 c/3)$ is substituted for m_e. Here we see an equation discovered by Einstein (and others). However, the physical reason why the speed of light c appears in the mass-energy equation is now explained. It signifies the actual maximum possible space-circulation in the structure of fundamental matter, even when it is stationary relative to the medium of space.

11. Angular momentum of electron vortex

The intrinsic angular momentum of the spinning interface of the electron is found as follows. Refer to Fig. 3. Consider an element of void-volume $dV = \pi r_e^2 \sin^2\theta r_e d\theta$, which, at the interface, has the tangential velocity of space, $\omega r_e \sin\theta$. Its mass from (6) will be $dm = dV\omega r_e \sin\theta = (\pi r_e^3 \sin^2\theta d\theta)c\sin\theta = \pi c r_e^3 \sin^3\theta d\theta$ and angular momentum, $dL = dm(\omega r_e \sin\theta)r_e \sin\theta = (\pi c r_e^3 \sin^3\theta d\theta)c r_e \sin^2\theta = \pi c^2 r_e^4 \sin^5\theta d\theta$. Integrating, varying θ from 0 to π, to obtain the angular momentum for the whole interface, we obtain

$$L = \pi c^2 r_e^4 \sin^5\theta d\theta = (4/5)\left[(4\pi/3)r_e^3 c\right]cr_e = (4/5)m_e cr_e \qquad (15)$$

in which m_e has been substituted for the quantity within the bracket as per the mass-equation (6).
The intrinsic angular momentum of the electron is directly proportional to its mass, radius, and the speed of light.

12. Spin magnetic moment

Refer to Fig.3. Consider an infinitesimal ring-element of charge $dq = dA\omega r_e \sin\theta$. The Magnetic moment due to this charge element is defined as $d\mu = dq(\omega r_e \sin\theta)r_e \sin\theta = (2\pi r_e \sin\theta r_e d\theta)(\omega r_e \sin\theta)(\omega r_e \sin\theta)r_e \sin\theta = 2\pi c^2 r_e^3 \sin^4\theta d\theta$. Integrating, varying θ from 0 to π, to obtain total magnetic moment of the electron, we obtain

$$\mu = (2\pi c^2 r_e^3)(3\pi/8) = (3/4)(\pi/4)(4\pi r_e^2 c)cr_e = (3/4)q_e cr_e \qquad (16)$$

The magnetic moment of electron is directly proportional to its charge, radius, and speed of light.

13. Electrostatic field energy

An expression for the electrostatic field of the electron at a point in space is derived below from the vortex structure of the electron. Refer to Fig. 5. Consider a sphere of radius r, cut by a plane parallel to the X-Z plane containing a circle C of radius $p_1 y_1$. The radius r (op_1) passes through the interface of the electron at point p, and meets C at p_1. In the diametrical plane X-Z of the interface (void), the point z at the interface will have a tangential velocity of space ωr_e, that is c (down the paper); the tangential velocity of space at the point z_1 (in the plane X-Z) down the paper, from (2), will be cr_e/r. The velocity of space u_2, at p, tangential to the circle C_1, is $\omega r_e \sin\theta$, whereas, at p_1 tangential to the circle

C, the velocity of space from (2) is $u_1 = (\omega r_e \sin\theta) r_e \sin\theta / r \sin\theta = c r_e \sin\theta / r$. The inward acceleration field at p_1, along $p_1 y_1$ is:

$$a_f = \frac{u_1^2}{r\sin\theta} = \frac{(cr_e\sin\theta/r)^2}{r\sin\theta} = \frac{c^2 r_e^2 \sin\theta}{r^3} \tag{17}$$

The component of a_f along the radius op_1 from (17) is $a_r = a_f \sin\theta = c^2 r_e^2 \sin^2\theta / r^3$. The electric field E at p_1 along the radius op_1 is defined to have the following relationship with the radial space acceleration field a_r derived above:

$$\frac{dE}{dr} = a_r = \frac{c^2 r_e^2 \sin^2\theta}{r^3} \text{ from which } E = \frac{-c^2 r_e^2 \sin^2\theta}{2r^2} \tag{18}$$

which is an *inward* field created by the electron (also by a positron, if the same is considered) with the minimum value of r equal to r_e, because the void is *field-less*.

The magnitude of E at the interface, along the Y-axis, for $\theta = 0$, is zero; and in the transverse plane (E_{tr}) for $\theta = \pi/2$, at the point z_1 distant r from the origin is

$$E_{tr} = -c^2 r_e^2 / 2r^2 \tag{19}$$

The maximum value of E is at the interface in the transverse plane X-Z for $\theta = \pi/2$, and $r = r_e$

$$E_{max} = -c^2/2 \tag{20}$$

The electric potential ϕ at z_1 from (19) is given by $d\phi/dr = E_{tr}$, from which, $d\phi = E_{tr} dr = (c^2 r_e^2 / 2r^2) dr$, and $\phi = -c^2 r_e^2 / 2r$. In an irrotational vortex, from (2), $cr_e = ur$. Substituting this in the above equation, we have,

$$\varphi = \frac{-cr_e(ur)}{2r} = \frac{cr_e u}{2} \tag{21}$$

From (21) it is seen that in a space vortex, the velocity field u, is the most fundamental field in the universe, which creates the electrostatic potential. Attraction between an electron and a positron (Fig.4a) can be calculated by using Coulomb's equation for interaction between charges with the concept of the electric field derived above, and also explained through superposition of velocity fields as stated earlier. Coulomb's law, which was experimentally determined, can be derived from (19) as follows. Multiplying and dividing

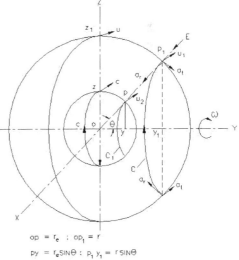

op = r_e : op_1 = r

py = r_e SINθ : $p_1 y_1$ = r SINθ

Fig. 5   Electric Field of Electron

the right-hand side of (19) by $(\pi/4)4\pi$ and rearranging terms: $E_{tr} = -c^2 r_e^2 (\pi/4)4\pi/2r^2 (\pi/4)4\pi = -2c[4\pi r_e^2 c\pi/4]/\pi 4\pi r^2$. Replacing the quantity in the bracket by q_e from the charge-equation (4), we have,

$$E_{tr} = \frac{-2/\pi\left(c/4\pi\right)q_e}{r^2} \qquad (22)$$

The above equation shows that the electric field, that is, "force per unit charge," is directly proportional to the charge, and inversely proportional to the square of the distance from the charge, in agreement with Coulomb's law, and for spherically symmetric charge distribution is

$$E = \frac{\left(1/4\pi\varepsilon_0\right)q_e}{r^2} \qquad (23)$$

14. Dielectric constant, permeability constant, Gauss' law

Using equations (20, 23), and charge equation (4), we derive the dielectric constant of the vacuum [3] as

$$\varepsilon_0 = \frac{\pi}{2c} \qquad (24)$$

The vacuum dielectric constant is inversely proportional to the speed of light. A check can be made for the above equation by substituting $\pi/2c$ in (23) in place of ε_0, yielding $E = 1/4\pi(\pi/2c)q_e/r^2 = (c/2\pi^2)q_e/r^2$.

Expressing q_e in CGSE and inserting the value of c, $E = [(3 \times 10^{10}\text{cm/s})/2 \times (3.14)^2]4.8 \times 10^{-10}$ CGSE/$r^2 = (0.73)$CGSE/r^2. Two CGSE unit charges, located 1 cm apart, require that the above computed coefficient, 0.73, should be 1; the difference is negligible.

From Maxwell's equation it follows that $c = 1/(\mu_0\varepsilon_0)^{1/2}$, where μ_0 is the permeability constant of the vacuum. (From this basic relationship it is possible to predict that light is an electromagnetic effect). When ε_0 is expressed in terms of c as derived in (24), the above equation becomes $c = 1/(\mu_0\pi/2c)^{1/2}$; from which we have:

$$\mu_0 = 2/\pi c . \qquad (25)$$

It is seen that like the dielectric constant, the *permeability constant of the vacuum is also inversely proportional to the speed of light.*

Using equation (18) for the electric field, charge equation (4), and relationship (24) for the dielectric constant, we derive Gauss' law [3] as $\Phi_E = (-2/3)q_e/\varepsilon_0$.

15. Electrostatic energy in electron vortex

The electrostatic energy U in the velocity field of the electron vortex is calculated [3] from the electric field(18), the dielectric constant (24), and mass equation (6), as

$$U = (\pi/10) m_e c^2 \tag{26}$$

In the integral to compute the above energy U, the lower limit of the radius from the electron center is the interface radius r_e of the electron, not zero, as is the case with a point-charge, which would lead to infinite energy in its electrostatic field. The electrostatic energy (26) is less than the total electron creation energy in space derived in the mass-energy equation (14). The difference (about $(1/2)m_e c^2$, given below) should appear as the electron's gravitational energy in space.

16. Gravitation

Gravitational effects arise from the very structure of the electron. As a result of the creation of the spherical void at the electron center due to the limiting speed of space-circulation, universal space is gravitationally energized (Fig.6) through the transmission of gravitational potential, a process starting from the interface of the electron and proceeding outwards at speed c, the limiting speed for transmission of fields/potentials in space. The energy used to create each electron is retained in space as gravitational/electrostatic potential, there being no reduction in the overall content of the universal energy due to the creation of electrons. The creation of electron voids requires energy (14) of magnitude $(4/5)m_e c^2$, out of which, from (27), $(\pi/10)m_e c^2$ is distributed in space as electrostatic energy, whereas, the remainder, about $(1/2)m_e c^2$, stays in space as gravitational potential. As shown in the figure, the gravitational field, g, of the electron is derived [3] as

$$g = \frac{(k/4\pi c) m_e}{r^2} \tag{27}$$

in which k is a "constant of proportionality" with dimensions $1/T^2$, so that the dimensions of g from (27) are: L/T^2. Since the electron is identified as the fundamental particle of matter, (27) is the equation of the gravity field applicable to all nuclei, atoms and matter in general. A gravitational constant for an atom of average atomic mass has been derived [3] from (27).

17. The annihilation of electrons and positrons—the fundamental nature of light

With the discovery of the positron (1932) a new phenomenon of the annihilation of electrons and positrons was observed. During this process, the spherical

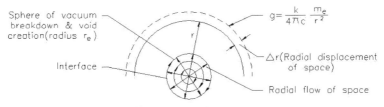

Fig. 6   Gravitation

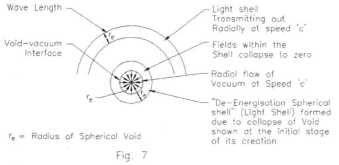

Wave Length

Void–vacuum
Interface

r_e

r_e = Radius of Spherical Void

Fig. 7

Light shell
Transmitting out
Radially at speed 'c'

Fields within the
Shell collapse to zero

Radial flow of
Vacuum at Speed 'c'

"De–Energisation Spherical
shell" (Light Shell) formed
due to collapse of Void
shown at the initial stage
of its creation

interfaces of the particles, under strong electrical attraction, are brought together and at a very close range, the particles super-impose on each other; thus stop-ping the oppositely directed space-circulations around their interfaces which leads to a collapse of their central voids. In this process mass vanishes and light is produced. It is evident that the *void interiors* within the interfaces of the electron and positron, being energy-less, cannot *emit* any kind of energy (such as photons). The energy (velocity and acceleration fields) in the vortex struc-ture of these particles pervades the whole of universal space both before anni-hilation; and following annihilation. Following the annihilation, the process in which the electromagnetic and gravitational potentials are reduced to zero, a single shell of light, seen as a pulse, initiates from the superimposed interfaces. (Fig.7).

When the interfaces of the particles superimpose, there is only one spheri-cal-void common to both particles; space flows radially at its maximum speed c into the void (Fig.7). The duration of collapse is $\Delta t = r_e/c$. During this period, a shell of radial width, Δtc, that is, $(r_e/c)c = r_e$, is formed, and transmitted out-ward at speed c relative to space. Within the wavelength, the space points un-dergo acceleration: $c/(r_e/c)$, which is c^2/r_e. (For light produced due to thermal radiation, acceleration of points within the wavelength is c^2/λ, where λ is the wavelength [3]) The transmission of the shell is a process that de-energizes the space medium, erasing for all the time the gravitational and electrostatic poten-tials that were created at the time of the creation of the now non-existent elec-tron and positron. *The spherical shell produced due to the dying of potentials, a process of de-energizing of the space substratum due to electron / positron an-nihilation, is the fundamental phenomenon known as light.*

The wavelength of the annihilation light (Fig.7) is equal to the electron radius. *The concept of frequency is not applicable to this light, with a single shell.* In the event several annihilations take place at a point one after another *without absolutely any time gap between the successive annihilations*, the fre-quency can be defined as the number of shells formed in unit time. Also, if the time for the formation of a *single* shell is Δt, then frequency f can be defined as: $f = 1/\Delta t$. This mathematical operation does not mean that the single-shell-light has the property of frequency in the conventional definition of frequency ($c = \lambda f$). The interrelationship between light and gravity and the derivation of the gravitational and Planck constants have been analyzed elsewhere [3].

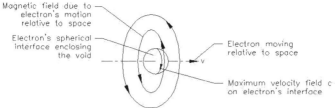

Fig. 8 Magnetic field of electron in motion relative to space

18. Magnetic fields

The electron has an axis of rotation at right angles to the diametrical plane of its space vortex (Figs. 2, 3). The pattern of the circular magnetic field distribution observed around a current carrying conductor, though of a representative nature, gives an indication that the natural motion of an electron in an electric current flowing in a conductor is along the axis of its vortex rotation, because the streamlines of the fluid-space in the electron vortex are concentric with the electron axis (Fig.2). Given the similarity between the velocity-field in the space vortex of an electron and the magnetic field produced in a conductor due to its motion relative to space, the fundamental nature of the magnetic field associated with a moving electron can be determined [3]. In Fig. 8 an electron is shown moving linearly at uniform velocity v relative to space. It is seen that the direction of the maximum velocity field c at the interface is opposite to the magnetic field produced due to the electron's motion. The analysis [3] shows that the magnetic field is an effect produced due to the reaction from the fluid space against the velocity field in the vortex on account of the electron's motion relative to space. It has also been shown that a point on a circle of radius r concentric with the axis (Fig.8) in the electron vortex will have magnetic field; $B = vr_e/r$; which shows that B falls inversely to r.

Given this relationship, the charge equation (4) and relationship (25), Ampere's law can be derived [3]. Due to the opposite direction of the magnetic field vector compared to the spin-direction in the electron vortex (Fig. 8), two electrons in parallel motion in the same direction will magnetically attract, while, at the closest range (about 10^{-10} cm) they will electrically repel.

19. Atomic Structure

The limitation on the creation of only one size of stable-void in the space vortex that produces stable fundamental mass and charge as basic units very much simplifies the theory of atomic structure with the electron as the fundamental particle of the atomic nucleus. It follows that all stable particles will possess mass in exact multiples of electron mass—there being no difference between rest-mass and relativistic mass. Further, no *stable* particle with mass less than electron mass can ever be found naturally or created through artificial means in laboratory. *Unstable* particles with masses different from the electron mass are presumed to be some intermediate stage in the formation of stable particles like

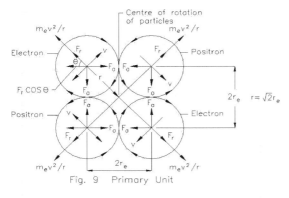

Fig. 9   Primary Unit

neutrons. Stable particles such as protons and alpha particles are enclosed in space-vortices that have the property of charge.

The *unstable* particles, with charge, will also be enclosed within space vortices of varying strengths for the duration of their lifetime. A neutral particle, like a neutron, does not a overall space vortex around it and hence, without an electric charge, it remains neutral. All stable particles, neutral or charged, will have spin-axes of rotation. The charge of a particle, from the charge-equation, will be the surface integral of the velocity field on its surface. An electron and a positron at closest possible range (about 10^{-10} cm) will undergo annihilation under electrical attraction, unless, the particles are translating *relative to space* and, thereby, producing a magnetic force of repulsion between them (Sec. 14).

Just as an electron is subjected to an "inward acceleration field" on its interface, all charged particles and nuclei, with space-circulation around them, will have an "inward acceleration field" tending to crush the particles. This inward force arises due to the existence of a void at the electron center, the vortex structure, and space-circulation around charged particles and the nuclei of atoms. Based on the above guiding principles, arising from the space-vortex structure of the electron, its observed properties and behaviour, the possible structures of nuclear particles are described below.

19.1 The primary unit

In Fig.9 an assembly of two electrons and two positrons is shown. The velocity fields between the particles are unidirectional, but in the region external to the assembly (not shown in the figure), will be in opposition. Therefore, this assembly (designated "primary unit") will show overall electrical neutrality. The particles repel diagonally (F_r) due to similar charges, whereas, there is attraction between the adjacent particles (F_a) due to dissimilar charges. In addition, if the particles are also spinning around the center of their assembly, there will be a radial force, $m_e v^2/r$, which will reinforce the diagonal electrostatic repulsive force F_r. If the component force, $F_r \cos\theta$, balances the attractive force F_a, the primary unit will be stable. Approximate computation [3] of the forces in the primary unit shows that *if the assembly rotates at speed c*, repulsive and attractive structural forces are nearly equal.

19.2 Neutrons

If a primary-unit is enclosed within a space vortex, it will be electrically charged and will be subjected to an inward acceleration field on the surface,

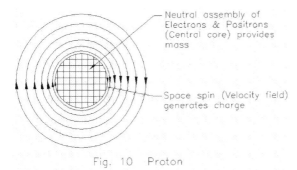

Neutral assembly of
Electrons & Positrons
(Central core) provides
mass

Space spin (Velocity field)
generates charge

Fig. 10   Proton

thus making it a stable building block of matter. A neutron core can be assembled with several such charged units, in a similar pattern as electrons and positrons assemble into a neutral primary-unit. For a spherical assembly of equal numbers of electrons and positrons with a total of n particles, the radius is $r = (n)^{1/3} r_e$. For a neutron, which should have 919 electrons and an equal number of positrons for overall neutrality with the superposition of their velocity fields, the radius is:

$$r_n = (1838)^{1/3} r_e \approx 12 r_e \qquad (28)$$

Calculations [3] show that electrical repulsive forces in this assembly are about two times less than the electrical attractive forces between the adjacent primary units. The neutron should therefore be a stable particle, but for the fact that it is known to have angular momentum; which signifies that it undergoes rotation.

It is found that a neutron rotating around its axis at speed c at the periphery (which will account for its maximum possible angular momentum), will not be stable; and therefore, its constituents (electron/positron) may be dislodged due to outward centrifugal force, and emitted outward. This explains beta-decay, and shows why a neutron has a short half-life of only about 15 minutes.

19.3   Protons and the hydrogen atom

The proton structure contains a neutron enclosed within a space-vortex (Fig. 10), which accounts for the charge of the proton and in addition, creates an inward acceleration field. In the proton structure, the inward acceleration field on its core (neutron's surface) makes the proton an ultra stable particle. Like the electron, the proton's maximum velocity field is confined within the diametrical plane at right angles to the axis of rotation. From (2), for an irrotational vortex, ur is constant. Therefore, the maximum tangential velocity (u_p) of space at the surface of the proton's core in the diametrical plane transverse to the axis of rotation is found from $u_p r_n = c r_e$, where c is the tangential velocity at the interface of electron of radius r_e. From this we obtain

$$u_p = c r_e / r_n = c r_e / 12 r_e = c/12 \qquad (29)$$

The electric charge of the proton due to u_p is computed from the relationship similar to the charge equation (4) as

$$q_p = (\pi/4) 4 \pi r_n^2 u_p = (\pi/4) 4 \pi (12 r_e)^2 (c/12) = 12 \pi^2 r_e^2 c \qquad (30)$$

which is 12 times the electron charge. A hydrogen atom (Fig.11), which has a proton and an electron, is neutral because of cancellation of the magnetic moments as shown below. The orbiting electron is located at a distance that reduces its velocity field to the same value as at the surface of the proton core

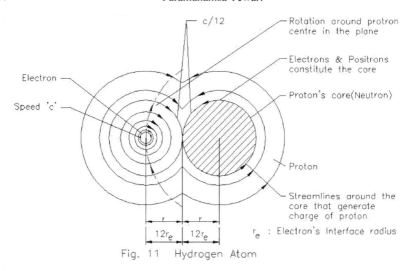

Fig. 11   Hydrogen Atom

$cr_e = (c/12)r$, where r is the distance of the electron center from the surface of the neutron; from this we have $r = 12r_e$, which is equal to r_n from (28). Thus, the radius of the electron orbit is $2r_n$. The magnetic moment of the orbital electron is due to its intrinsic spin (16) and its orbital velocity v_{orb}. The total of the magnetic moments is

$$\mu_e = \frac{(3/4)q_e cr_e + q_e V_{orb}(12r_e + 12r_e)}{2} = q_e r_e\left[\left(\frac{3c}{4}\right) + 12V_{orb}\right] \tag{31}$$

The intrinsic magnetic moment of the proton, from an expression similar to the electron (16) is $\mu_p = (3/4)[q_p(c/12)12r_e]$. Substituting, $q_p = 12q_e$, from (30), we have

$$\mu_p = (3/4)\left[12q_e(c/12)12r_e\right] = 9q_e cr_e \tag{32}$$

Equating the magnetic moment of the electron (31) to the magnetic moment of the proton (32) in order to achieve an electrically neutral hydrogen atom, we obtain $q_e r_e[(3c/4) + 12v_{orb}] = 9q_e cr_e$, which gives: $v_{orb} = 0.69c$. In the hydrogen atom, the radius of the electron orbit is $24r_e$, about 10^{-9} cm, and its orbital velocity is 69% of light speed. With this high rotational speed, the orbital electron completes one orbit in a time of $(2\pi)10^{-9}$cm/$(0.69)3 \times 10^{10}$ cm/s, that is, 3×10^{-19}s, providing an outer shield to the hydrogen atom with its spinning interface that can not be penetrated.

The binding force provided by the velocity fields of the oppositely spinning vortices of the orbital electron and proton maintain the assembly with no energy loss from the system since the vortices are formed in non-viscous space.

The Hydrogen nucleus (a neutron within a proton vortex) has an inward acceleration field of strength $(c/12)^2/12r_e$, or $(1/12)^3c^2/r_e$. This inward field, which is $(1/12)^3$ times less than the maximum possible field (c^2/r_e) on the electron interface, makes it a highly stable particle, as stated before. In a similar manner, two protons and two anti-protons (with opposite direction relative to the proton vortex), enclosed within an overall space-vortex, can assemble an alpha particle, a helium nucleus. When several alpha particles are assembled, with four in each unit (similar to the assembly of primary units in the neutron structure), and enclosed within an outer vortex, all nuclei of atomic mass higher than helium can be built. This process requires that nuclei should have equal numbers of neutrons and protons, which, however, is not the case. For example, the ratio of neutrons to protons in the Uranium nucleus is 1.586. This leads to the conclusion that, in addition to the alpha particles, neutrons are also *independently* present, as required by the atomic masses of the nuclei. The emission of alpha particles from radioactive nuclei provides solid proof of their existence within nuclei in an *independent* condition. The presence of electrons and positrons in nuclei is confirmed by beta particle radiation. For simplicity in the analysis of the stability of nuclear structure, we can assume that protons and neutrons exist independently in a dynamic assembly, and each proton exerts a repulsive force on the rest of the protons in the nucleus which is enclosed within an outer space-vortex [3]. The space-vortex enclosing the nucleus creates an inward field acting on the nucleus and it has a maximum value in the diametrical plane at right angles to the axis of rotation of the nucleus; given by u_n^2/r_n, where u_n is the tangential velocity of space at the nuclear surface in the diametrical plane, transverse to the axis of rotation, and r_n is the nuclear radius. Since from (2), u_n varies inversely as r_n, the *inward* acceleration field on the nucleus falls inversely as the cube of r_n. The *outward* electrical repulsive forces within the nucleus trying to disrupt its structure (due to the presence of protons) fall inversely as the square of r_n. Since the inward acceleration field falls faster, nuclei with more protons and a larger radius become radioactive. By equating the outward electrical force in the nucleus with the inward force it is concluded [3] that *stable* nuclei with protons more than 100 cannot exist in nature.

20.   Interaction of orbital electrons in an atom with a wave-pulse (shell) of light

With the nuclear structure described above, the nuclear radius of an average atom (120 times proton mass) is computed [3] as $r_n = 2.37 \times 10^{-9}$ cm. The maximum velocity field at the nuclear surface from (2) is $u_n = 5 \times 10^8$ cm/s. In the atomic vortex around the nucleus, this velocity field will fall off inversely with distance to $v = 1.2 \times 10^8$ cm/s at a radial distance of 10^{-8} cm, which is assumed to be the orbital radius of the outermost electron. The orbital electron in the space vortex will be subjected to an inward acceleration field $a_f = v^2$/orbital radius $= (1.2 \times 10^8 \text{cm/s})^2/10^{-8} \text{cm} = 1.44 \times 10^{24}$ cm/s^2. Suppose a light shell of wavelength λ, and an acceleration-field a_l, across the wavelength (directed towards the source) meet the orbiting electron at an instant when both the above acceleration fields are in line. Since the direction of a_l is opposite to that of a_f, the two acceleration fields will nullify and the electron will be released from the vortex if $a_l = a_f$. As stated earlier, $a_l = c^2/\lambda$. Substituting the values of the acceleration fields, we have $(3 \times 10^{10} \text{cm/s})^2 / \lambda = 1.44 \times 10^{24} \text{cm/s}^2$, from which $\lambda = 6.25 \times 10^{-4}$cm, corresponding to a frequency of 0.48×10^{14} cycles/s. (For metallic sodium, the threshold frequency for the photoelectric effect is about $5 \times 10^{14} \text{sec}^{-1}$). The orbital electron, moving with velocity v, will be released with the kinetic energy that it *already* possesses, $E = (1/2)m_e v^2 = (0.5 \times 10^{-28} \text{gm}) (1.2 \times 10^8 \text{cm/s})^2 = 7.2 \times 110^{-11}$ ergs. Experiments show that the kinetic energy of photoelectrons is about 8×10^{-11} ergs, very close to the above computed value! Considering the approximate nature of the assumption made as to the electron's orbital radius and computation of the nuclear radius for an atom of average mass, better results could not be expected. It is concluded that light (photons) does not impart energy to the photoelectron for its release. The kinetic energy of a released photoelectron is its own energy of motion in the space vortex of an atom. Light simply disturbs the stability of the forces under which an electron is stable in its orbit.

Conclusion

The medium of space in dynamic states creates matter and its associated fields. The properties of mass and charge, the gravitational, electromagnetic, and nuclear fields are produced from the most fundamental field, the velocity field, and unified in the electron structure. The property of inertia arises [3] due to the reaction from space on the central void in the electron's vortex structure. The velocity of light relative to the space-medium is a common factor in all the *basic* universal constants so far experimentally determined. The electron is the fundamental particle of matter.

Acknowledgements

The author thanks his friends and colleagues Toby Grotz and James Sheppard for their editing and suggestions during the preparation of this paper.

References

1. Alexandre Koyre & I. Bernard Cohen, *Isaac Newton's Principia*, page: 28
2. Sir Edmund Whittaker, *A History of the Theories of Ether and Electricity*, p 317
3. Paramahamsa Tewari, *Universal Principles of Space and Matter—a Call for Conceptual Reorientation*, (2002)
4. George Yankovsky, *Philosophical Problems of Elementary Particles Physics*, Progress Publishers, Moscow (1968).

Solving Nature's Mysteries: Structure of the Electron and Origin of Natural Laws

Milo Wolff
Technotran Press
1124 Third Street
Manhattan Beach, CA 90266
milo.wolff@quantummatter.com

Geoff Haselhurst
Space and Motion Productions
RMB 1153, Nornalup
W. Australia 6333
haselhurst@wave-structure-of-matter.org

The structure of the electron is investigated and found to be the origin of the natural laws. The natural laws have been measured for hundreds of years but no one knew how Nature creates them. The origins had been proposed earlier by Clifford and Schrödinger as a Wave Structure of Matter (WSM), to explain natural laws. Einstein also wrote: "Physical objects are not in space, but these objects are spherically extended. In this way the concept of 'empty space' loses its meaning." (*Ideas and Opinions*, Crown Paperbacks, 1954)

Using the WSM quantitative origins have been found based on the wave structure of the electron described here. It is shown that the quantum wave medium is the single entity underlying electron structure and the laws. Two Principles are found describing the wave medium, enabling calculation of properties of particles and the laws. The predictive power of the WSM is shown by deriving the previously unknown physical origin of electron spin. The WSM has important implications for research, industry, and humans' role in the universe.

Part I - Introduction

Einstein was once asked if he could understand the meaning of the enormous number of hadron particles being generated in giant accelerators. He replied, "I would rather know what an electron is." Answering his question is the purpose of this article. At the same time, the reader will gain, in hindsight, an understanding of the deep meaning of his reply; not only his disinterest in accelerators, but also because the forces of the electron extend to infinity, revealing relationships of the universe and the natural laws which govern it.

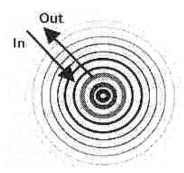

Figure 1. The Electron. The electron is composed of spherical waves which converge to the center and reverse to become outward waves. The two waves form a standing wave whose peaks and nodes are like the layers of an onion. The wave amplitude is a scalar number like a quantum wave, not an e-m vector. The center is the apparent location of the electron.

1.   Natural laws

Our knowledge of science is based on the natural laws that describe the behaviour of particles. The laws are the rules for calculating electricity, gravity, relativity, quantum mechanics, and conservation of energy and momentum. The origins have been unknown. Now the origin of the natural laws is found to be a quantitative result of a Wave Structure of Matter (WSM). The basic concept is very simple: The ancient Greek notion of a point particle, still in use today, is replaced with a *spherical wave structure*, which had already been predicted by Clifford[1] and Schrödinger[2] long ago. Figure 1 shows the structure of an electron. It is an inward wave that converges to a center, spherically rotates creating 'spin' then becomes an outward diverging wave. Together they form a standing wave. The endless wave combinations are like the eight note musical scale that becomes the grand symphonies of Wagner and Beethoven.

The rules of wave combination are of great importance to science because the rules and quantum spin determine the Atomic Table, that contains the varied forms of matter: metals, crystals, semi-conductors, and the molecules of life. The deep understanding of basic physics that is revealed opens a door to broad fields of applied technology such as integrated circuits, medicine, and commercial energy. It reveals a universe of real quantum wave structures in a *space medium* that we live in but seldom are aware of. This medium is the basis of matter and the Universe because its properties underlie the wave properties.

2.   Space, Human senses and survival

We don't easily see the space wave medium because our survival as an animal species depended mostly on our ability to fight with other animals seeking food, and to compete for mates to produce children, not closely related to the quantum space medium. Our sensory mechanisms evolved to directly aid our survival, not to be aware of quantum waves. In our self-focused human perspective few of us are even aware of the wave medium in which we exist. For survival, it doesn't matter what space is, or whether we can observe it—it exists unseen. This situation is much like the life of a fish, which cannot comprehend the existence of water because he is too deeply immersed in it. Like the fish, traditional scientists have tended to comprehend the universe in terms of their local experiences.

Our misperceptions are revealed by biological evolution, which teaches that the quantum wave universe is not as helpful to survival of our personal genes as recognizing apples we can eat and avoiding tigers who want to eat us.. Thus it was not necessary that nature equip us to observe quantum waves, although as will be seen below we do observe their presence and effects. Lacking direct personal experience of simple quantum waves, people chose to imagine that the electron is a discrete "particle," like a bullet. Laboratory evidence does not support this human-oriented idea. Accordingly, belief must change from discrete particles to true quantum wave structure.

Human perspective has another bias. We tend to see space as three rectangular dimensions, one of which is the vertical gravity vector of Earth, plus two other vectors perpendicular to it, shaped like the houses we live in. But in the cosmos, the shape of the enormous universe is *spherical* whose important dimensions are *inward* and *outward*, the direction of waves in space. In the vast expanse of the real universe, gravity occurs so rarely, that its direction is inconsequential in the larger scheme of things, despite its local importance to us. Unfortunately, we feel comfortable with rectangular coordinates and tend to ignore the spherical universe.

The proof of the WSM is that the physical structure of the electron, and the empirical natural laws can be obtained mathematically from two basic principles describing the wave space medium. In other words, all the experimental measurements of historical physics that described natural behaviour are now predicted by two fundamental principles. The laws and the principles agree with each other—each is the proof of the other.

Physics of the wave structure of matter is simple. In contrast, old discrete particle-structured physics required dozens of assumptions plus many more arbitrary constants to explain the operation of the laws. Many properties and laws, like electron spin, were puzzling with no understanding. The puzzles are now swept away. Particle-structured physics can be compared to the theory of epicycles of the planets around the Earth before Galileo found that the planets traveled around the Sun. Discrete particles satisfy our human prejudices but do not explain the measured facts.

3. Comfortable physics and physical reality

Many people, from the old Greek philosophers such as Democritus and Pythagorus, up to the colleagues of Albert Einstein, sought to understand the structure of the tiny atoms and molecules in our everyday world. Until recently, most answers have been speculation created in analogy to human scale objects around us; like baseballs and bullets, and grains of sand. Atoms were imagined to move like other familiar objects such as moons around planets and toy tops spinning on a table. These analogies made us feel comfortable. so we preferred to ignore strange new ideas. As Churchill said, "We often stumble onto the truth but most of us brush ourselves off and pretend it did not happen."

Serious thinkers, such as Einstein, Dirac, Schrödinger, and Ernst Mach, realized that the analogies were wrong. Instead, experimental measurements showed that the structure of matter was closely related to the properties of the apparently empty *space* around us, and that the elements of matter had to:

a. have a spherical symmetry.
b. be extended in space.
c. Possess a means of exchanging energy
d. Possess wave properties founded on wave equations.

Their thinking produced conclusions that in hindsight were prophetic. For example, Einstein rejected the discrete point particle and stated: "Matter must be spherical entities extended in space." Erwin Schrödinger[8] understood the requirements of particle structure when he wrote in 1937: "What we observe as material bodies and forces are nothing but shapes and variations in the structure of space. Particles are just *Schaumkommen* (appearances)." He believed that quantum waves were real, not probability distributions with a particle hidden inside. He saw that abolishing the discrete point particle would remove the paradoxes of 'wave-particle duality' and the 'collapse of the wave function'. They arrived at their valid conclusions by painstaking analysis and careful adherence to the rules of logic, and the philosophy of truth. But their thinking was ignored for sixty years; Truth is no match for belief. Machiavelli understood this human behaviour 500 years ago [1513]: "There is nothing more difficult to plan, more doubtful of success, more dangerous to manage than the creation of a new system. The innovator has the enmity of all who profit by the preservation of the old system and only lukewarm defenders by those who would gain by the new system."

The predictions of these pioneers, verified and described here, is that matter is a *wave structure embedded in space*. This result not only satisfies the experimental work, but surprisingly displays an immense but simple tapestry of the physical universe. Awe-inspiring connections between matter, ourselves, and the cosmos are found. The application of the electron wave structure reaches out, on the one hand, to unsuspected fields of cosmology such as the big bang, the redshift, and the structure of the universe. On the practical side, a new tool is provided that will enable us to deeply understand and improve industrial devices such as computers, micro circuits, and the efficient transmission of electric energy.

The authors, and others who have contributed recognize that wave structures describe physical reality for the first time; We are aware that this is a major and remarkable claim: The discovery of physical reality has been a holy grail of intellectual thought for thousands of years. Our hope is that readers will gain information and perspective so that they will confirm for themselves that this sensible theory deduces the laws of nature (reality) as observed.

This does not mean this is written only for the scientist and intellectual academic. Because the truth has deceived great minds, many people could assume that this subject is too difficult for them. This is not the case. The Wave Structure of Matter has an underlying simplicity that makes it easy to understand. But it is also an exciting mystery of how a century of science, was thwarted not by scientific complexity, but by the frailties of human emotions, economic ambition, and the power of politics.

4.   Finding laws, space, and the structure of the electron

Finding the structure of the electron was a key to finding the origin of the natural laws. Let us look at how the process of deduction has proceeded to find the origins from the electron.

The origin of natural laws. The business of physics is the abstract description and quantification of facts observed in nature. The rules we form for expression of the observed facts are the laws of nature. Since past laws were obtained by measurement of nature rather than derived from other knowledge, they are by definition *empirical* and "of unknown origin." Therefore if we seek to find the origins of laws we cannot use the existing laws themselves but must use other observed facts together with logic and established mathematics to find the origins. The old empirical laws are only a guide and not the source.

Accordingly the search for origins must probe deeper into nature than heretofore and we must be prepared to find new perspectives. The unexplained puzzles of nature are attractive sources of input data in the search .Finally, the *proof* of the origins of laws is a match between the observed empirical rules and the predictions of the new origins.

Circular reasoning. When seeking origins, it is important not to inadvertently use existing rules (laws) to deduce them. Such *circular reasoning* can occur if, for example, a mechanical model is assumed to be the structure of an electron. Such common errors are the use of toy tops, sheets and rings of charge, masses in orbit around each other, This is because the quantum laws of quantum particles can be extrapolated to large macro-objects but the inverse is not possible. Logically, finding the origins of existing laws (rules) requires forming <u>new</u> concepts that satisfy observed data.

The new perspective. The discovery of these origins creates a new perspective of the physical world: quantum mechanics and relativity are united, a single origin of forces is found, puzzles and paradoxes are explained and, most important, relationships between microphysics (electrons and other 'particles') and the Universe (cosmology) are seen to be a result of an all-pervading *space* filled with oscillating quantum waves of the matter in the universe. We exist in a sea of quantum waves.

Part II - The connections of the electron and the laws

This part is a discussion of the relationships between electrons, the natural laws and the Universe. Let us examine the meaning of these words. Our concept of

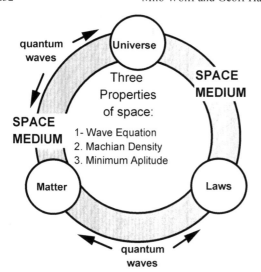

Figure 2. The Connected Universe. Laws particles, and the cosmos are interconnected by the quantum waves in the medium of space. Nature has created a reciprocity among the three: Matter creates the medium and the medium tells matter how it must behave. These interconnections are described by three Principles that define the properties of the space medium.

"Universe" is a collection of particles and their distribution. Thus without particles to populate a Universe, the Universe concept has no meaning. Accordingly, our concept of our universe depends on understanding the particles in it. Especially we need to understand the connections with the electron and proton, because the charge waves of those two particles extend throughout the Universe. The "natural laws" have no meaning without particles because laws require the presence of particles, upon which the laws operate. "Particles" are also meaningless without laws to identify them. We conclude that understanding of the connections between particles, laws, and the Universe is essential to understand the whole. Each requires the existence of the others. Therefore, we cannot understand cosmology unless we also understand the relationships within the trilogy shown in Figure 2.

5.   Measurement is a property of an *ensemble* of matter

A particle entirely alone in the universe cannot have dimensions of time, length or mass because these dimension are undefined without the existence of other matter. Dimensions can only be defined in *comparison* with other matter. For example, at least six separated particle-centers are necessary to crudely define length in a 3D space: four to establish coordinates and two being measured. Thus the measurement concept requires the existence of an *ensemble* of particles. The required ensemble must include all observable matter in our Universe, because there is no way to choose a special ensemble. Recalling that time, length and mass are the basic unit set that describes all scientific measurements we surprisingly concludes that the physical basic of all science depends on the Universe!

6.   Particle properties require perception-communication between particles

If there were no means for each particle to sense the presence of other matter in its universe, the required dimensional relationships above could not be established. How can a particle possess a property that is dependent on other particles, if there is no way for the particles to impart their presence to each other? Without communication, each particle would be alone and without meaning. Therefore continual two-way perceptive communication between each particle and other matter in its universe is needed to establish the laws of nature. Spherical quantum waves forming the particles are the means of communication. The laws are then established in terms of the dimensions (units) established by waves of the entire ensemble of matter.

7.   The measurement of time requires a cosmological clock

Using the reasoning above, but for the dimension of time, we conclude that time measurement requires the existence of cyclic events among the particles of the universe; a kind of clock. Those properties that involve time, notably velocity, mass, and frequency, cannot have a meaning if particles have no common scale of time. That is, the particles must have a way to compare their own cyclic events with other particles. Therefore, there must exist a standard cosmological clock. The proposal by DeBroglie was an oscillator (clock) contained in every electron. Evaluating his proposal in a wave medium, we see that electron quantum resonances (oscillators) provide the necessary property. Because of the near uniformity of space (the oscillator medium) the clock frequencies would be nearly alike throughout the universe. We note that a nearly uniform space medium is required in nature, otherwise different clocks in different places would produce time-chaos of the natural laws.

8.   We live in an inter-connected universe

The above discussions of the requirements of the laws of science make it clear that inter-connections must exist between matter, the laws, and the Universe. Independence of objects is not possible, for example, no planet, star, or galaxy, can exist without the rest. We know this because astronomical measurements show that the same laws apply in the farthest galaxy, as here on Earth. And we also know because 'time' has the same meaning from one moment to the next. These requirements are not the fantasy of a supreme law-maker who declares that the Standard Model applies everywhere (as is present believed by most of the physics community.) The only logical conclusion is that matter and laws are inter-connected throughout the universe by a physical mechanism—waves. As Lee Smolin writes (*Life of the Cosmos*, Phoenix books, 1998): "It can no longer be maintained that the properties of any one thing in the Universe are independent of the existence or non-existence of everything else."

In later sections it will be seen that the natural laws originate from the properties of the quantum waves of the electron and proton. It will also become clear that the 'Standard Model' point-particle of charge and mass substance without wave structure cannot satisfy the logic of science. The model is an historical relic.

9.   New physics and old physics

Study of the wave structure of the electron and other matter is a new adventure where you find the origin of the natural laws, and have a revealing window on science, cosmology, technology and ourselves. But first old mainstream attitudes must be discarded. For instance, a conventional quantum physicist expects that all quantum phenomena must derive from Schrödinger's Equation. No. It is the other way around; Schrödinger's Equation is derived from the quantum wave structure of matter.

Some concepts must be changed, for example, the meaning of charge and mass are not inherent properties of particles. Instead Nature has chosen, as Schrödinger deduced, that charge and mass are *properties of the wave structure*. And, as shown above, natural laws here on Earth depend on the matter of the rest of the Universe. Goals of research need to be changed knowing that the building blocks of the universe are the waves of the electron and positron. Accordingly, to be fruitful, physics must study the properties of the wave medium, not build accelerators.

Part III - A short history of the Wave Structure of Matter

10.   The pioneers

A wave structure of matter was proposed 130 years ago by the famous English geometer, William Clifford[1], who spoke before the Cambridge Philosophical Society in 1870, "All matter is simply undulations in the fabric of space." He developed this concept as three-dimensional dynamics that reduces to four-dimensional kinematics describing matter, electromagnetism and kinetic energy as curvature of a dynamic Riemannian space. His work was the progenitor of the WSM and General Relativity. In Clifford's thoughts, mass and charge substances do not exist but are properties of a wave structure in space. In short, space waves were real, while mass and charge points are mere appearances of the wave structure. His proposals and those of Schrödinger[2] were consistent with present day quantum theory, since quantum mathematics does not depend on a belief in particle or charge substance.

Ernst Mach[3] and Bishop Berkeley had proposed about 1890, that the law of inertia depended on all the matter of the universe. This is known as *Mach's Principle*. It was the first recognition that a natural law depends on cosmology. Albert Einstein was greatly influenced by it when he deduced the General Theory of Relativity (GTR). Now, Mach's Principle, in a more exact form, has become Principle II (below) of the Wave Structure of Matter.

Paul Dirac[4] was never satisfied with the discrete point particle because the infinity of the Coulomb force law had to be corrected by 'renormalization'. He wrote, "This is just not sensible mathematics. Sensible mathematics involves neglecting a quantity because it turns out to be small, not neglecting it because it is infinitely large and you do not want it! Of course the inference is that the basic equations are wrong and radical changes need to be made." Dirac seemed to foresee the WSM.

In 1945 Wheeler and Feynman[5] (W&F) sought the cause of the radiation from an accelerated charge. Their calculation assumed that the charge generated equal amplitudes of advanced (inward) and retarded (outward) spherical electromagnetic waves. The outward waves evoked a *response of the universe*; that is, the production of inward waves from absorbing charges elsewhere in the universe. The absorber waves began *before* arrival of the source waves. The calculated forces due to combined local and absorber waves agreed with Dirac's empirical formula and appeared to be the cause of energy transfer. Their remarkable result attracted much attention. However, W&F pointed out that the derivation had not been rigorous. Especially, there are no electromagnetic wave solutions in spherical coordinates. In hindsight, the success of the W&F calculation was in part due to suppression of the vector character of the electromagnetic waves so that in effect they were calculating *scalar* (quantum) waves. Below it will be shown that a scalar wave equation can rigorously produce two solutions; namely inward and outward spherical waves that have all the properties of positrons and electrons.

11.   The calculation by Wheeler and Feynman (W&F)

W&F wished to verify the empirical formula for the force of radiation used by Dirac[6]

$$\text{Force} = \left[\frac{d\mathbf{a}}{dt}\right]^2 \frac{e}{3c^2}$$

where e is the electron charge, c is the velocity of light and $\mathbf{a}$ is the acceleration. The mechanism of the force was unknown. They discussed this problem with Einstein, who suggested a proposal by Tetrode[7] that light (energy) transmission was *not* a one-way process, but two-way *communication* between a source molecule or atom and a receiver molecule utilizing inward and outward waves. This proposal was not popular since it appeared to violate the causality concept: *Actions should not appear before their causes,* since the inward waves appeared to be traveling backward in time.

Electromagnetic waves were assumed generated by the acceleration, using special solutions of the electric –wave equation:

$$\nabla^2 \mathbf{E} - \left(1/c^2\right)\partial^2 \mathbf{E}/\partial t^2 = 0$$

Half the difference between inward and outward waves was prescribed. Outward traveling spherical waves encountered absorber charges in the universe that produced spherical inward waves that returned to the initial charge—a *response of the Universe*. The inward waves from the universe were assumed to begin before the acceleration occurs.

Dirac's empirical formula for the force of acceleration was verified and found to be independent of the properties of the absorber provided that absorption was complete. Remarkably, no inward waves appeared to violate causality because the inward waves from the absorber were cancelled upon arrival by interference with waves from the source charge. The remaining wave fields gave the disturbance demanded by experience in agreement with the prescription of Dirac.

W&F described their inward waves: *Absorber charges at a large distance produce spherical waves toward the source. At the moment the source is accelerated, these waves just touch the source. Thus all the waves from the absorber charges form an array of approximately plane waves marching towards the source. The (Huygens) envelope of these plane waves is a spherical in-going (advanced) wave. The sphere collapses on the source, and then pours out again as a divergent outward wave.*

12. Applications of the wave electron

W&F's work has implications beyond an explanation of radiation forces because the transfer of energy and the motion of matter are the most fundamental processes of science. Further, the concept that spherical waves from all matter of the universe perform the roles of charged particles suggests that the whole universe of particles is involved, i.e., *every charged particle is a structural part of the universe and the whole universe contributes to each charged particle.* Their work pioneered the concept that every particle sends quantum waves outward, and receives an inward response from the universe. In hindsight, if they had used scalar quantum waves entirely, this chapter would have appeared 55 years ago.

Research on wave structure. After 1945, particle physicists mainly worked on WWII weapons-related contracts. Research on wave structure stopped until 1985 when Milo Wolff[8,9], using a scalar wave equation with spherical quantum wave solutions, found the Wave Structure of Matter described here. It successfully predicted the natural laws and the properties of the electron, and, as shown below derived a physical origin of spin that accords with quantum theory and the theoretical Dirac Equation .

The wave electron in electrodynamics: Akira Tonomura of the Hitachi Corp published in 1998 *The Quantum World Unveiled by Electron Waves* (World Scientific press), a beautifully illustrated book that discusses the quantization of flux at low temperatures in a closed loop of real electron waves. Quantization occurs because the waves of the circulating electrons must join in phase, otherwise they cancel each other.

Prof. Carver Mead, an engineer at Cal-Tech investigated electron waves in his 2000 book *Collective Electrodynamics*. He recognized that the electron is not a point particle but a wave structure, so that e-m approximations, especially in magnetism, do not work at quantum dimensions. He derived a vector potential to correct the flawed magnetic terms of Maxwell's Equations, using meas-

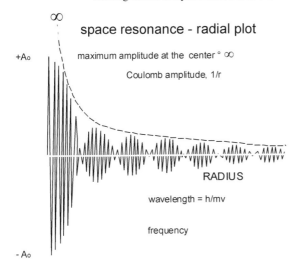

space resonance - radial plot

maximum amplitude at the center ° ∞

Coulomb amplitude, 1/r

RADIUS

wavelength = h/mv

frequency

+Ao

- Ao

∞

Figure 3. Electron composed of IN and OUT waves. This is a plot of the superimposed inward and outward waves of an electron. The envelope of the electron waves is observed in the lab as the charge potential. The potential of the Coulomb law and the wave electron are the same at large radius, but near the center only the wave model matches experiment. This difference causes the *Lamb Shift* of spectroscopy.

urements of electron waves in closed loops. His book, very popular in Silicon Valley, shows correct ways to solve the electromagnetics of transistor circuits. MIT awarded him a $500,000 prize.

Part IV - The Wave Structure of Matter

The wave-structured electron, Figures 1 & 3, is termed a *space resonance* (SR). Space, that supposed void of which we formerly knew little, is the medium of the waves and the leading player in this new physics of the universe. The properties of electrons, other matter, and the laws they obey are derived from properties of the medium, i.e. *space*. Thus space, described by three principles, underlies our knowledge of science.

13.   Principle I - A wave equation

This Principle describes how quantum waves are formed and travel in the space medium. The wave amplitudes are scalar numbers. If the medium is uniform, typical nearly everywhere, the equation allows only spherical waves. If observed in relative motion, Doppler modulation and elliptical waves appear. If the medium is locally dense, as in the central region of a proton, waves *circulate* like sound waves in a drum or a crystalline sphere.

Principle I: *Quantum matter waves exist in space and are solutions of a scalar wave equation:*

$$\nabla^2 \Phi - (1/c^2)\partial^2 \Phi / \partial t^2 = 0 \tag{1}$$

where Φ is a scalar amplitude, c is the velocity of light, and t is the time. Its solutions, Figure 1, are a *pair* of spherical in/out waves that form the structure of the electron or positron:

$$\text{Outward wave=}\Phi_{out} = \Phi_0 \exp(iwt - ikr)$$
$$\text{Inward wave= } \Phi_{in} = \Phi_0 \exp(iwt + ikr) \tag{2}$$

There are only *two* combinations of these two waves. They have opposite phase and spin rotation to form electrons and positrons:

$$\text{electron}=\Phi_{in} - \Phi_{out} + CW \text{ spin}$$
$$\text{positron}=\Phi_{out} - \Phi_{in} + CCW \text{ spin}$$

(3)

Thus matter is constituted of *binary* elements—like computer hardware.

Figure 3 is a plot of the wave amplitude as a function of radial distance from the wave center. The waves decrease in intensity with increasing radius, like the forces of charge and gravity. Note that the envelope of the amplitude matches the Coulomb Law but at the center reaches a maximum A_0 and does not go to infinity like the Coulomb Law. Experimental data matches this wave model but not the Coulomb Law. Although the variety of molecules and materials populating the universe is enormous, the building bricks are just two.

14.   Properties of electron and positrons

The two wave combinations contain all experimental electron-positron properties. Briefly: Charge polarity depends on whether there is a positive or negative amplitude of the in-wave at the center. If a resonance is superimposed upon an anti-resonance, they annihilate. The amplitude at the center is finite as observed. The properties of quantum mechanics (QM) and special relativity (SRT) are the result of the *Doppler* motion of one space resonance relative to another. The Doppler shifted waves contain QM and SRT for a moving particle; that is, the DeBroglie wavelength of QM and the relativistic mass and momentum changes, are exactly as experimentally measured. Details are in the Math Appendix.

15.   Energy transfer and the action-at-a-distance paradox

An important property not previously known is the mechanism of energy exchange. Experience tells us that communication or acquisition of knowledge of any kind occurs only with an energy transfer. Storage of information, whether in a computer disk or in our brain, always requires an energy transfer. Energy is required to move a needle, to magnetize a tape, to stimulate a neuron. There are no exceptions. This rule of nature is embedded in biology and our instruments. Finding the energy transfer mechanism between particles is essential to understanding the natural laws.

The first hint of the mechanism of cosmological energy transfer was Ernst Mach's observation[3] in 1883. He noticed that the inertia of a body depended on the presence of the visible stars. He asserted: "Every local inertial frame is determined by the composite matter of the universe" and jokingly, "When the subway jerks, it is the fixed stars that throw us down." His deduction arose from two different methods of measuring rotation. First, without looking at the sky one can measure the centrifugal force on a rotating mass m and use the inertia law $\mathbf{F} = m\mathbf{a} = mv^2/r$ to find circumferential speed v and position, as in a gyroscope. The second method is to compare the object's angular position with the fixed (distant) stars. Both methods give exactly the same result. The inertia law appears to depend on the fixed stars!

Mach's Principle was criticized because it appeared to predict instantaneous *action-at-a-distance* across empty space. How can information travel from here to the stars and back again in an instant? As Einstein observed (*Ideas and Opinions*, Crown paperbacks, 1954): "Forces acting directly and instantaneously at a distance, as introduced to represent the effects of gravity, are not in character with most of the processes familiar to us from everyday life."

Action-at-a-distance does not actually occur because Nature's energy exchange mechanism is now seen as the interaction of space resonances with the ever-present universal medium of *space*. Space is not empty because it is the quantum wave medium produced (See Principle II below) by waves from every particle in the universe as implied by Mach's Principle. The energy exchanges of inertia, charge, and other forces are mediated by the presence of the space medium. There is no need to travel across the universe.

16.   Principle II - Space Density Principle

This principle defines the quantum wave medium—space. It is fundamentally important because the properties of waves depend on properties of their medium. But, since the natural laws depend on the waves, we deduce that the natural laws in turn depend on the medium. Thus, the medium—space—is the wellspring of everything.

> **Principle II:** *At each point in space, waves from all particles in the universe combine their intensities to form the wave medium of space.*

$$\text{Space density} \propto mc^2 = hf \propto \sum_{1}^{N}\left[\Phi_n / r_n\right]^2 \tag{4}$$

In other words, at every point in space, the frequency f or the mass m of a particle depends on the sum of squares of all wave amplitudes Φ_n from the N particles inside the "Hubble universe." Amplitude decreases inversely with their range r_n squared. The "Hubble Universe" has a radius $R = c/H$, where H is the Hubble constant.

This principle is a quantitative version of Mach's Principle because the space medium is the inertial frame of the law $\mathbf{F} = m\mathbf{a}$. When mass or charge is accelerated, energy exchange takes place between its waves and the space medium. In hindsight, this is the mechanism of charge radiation, sought by Wheeler and Feynman[5] in 1945, using mixed e-m and quantum waves.

The number of particles, $N \sim 10^{80}$ in the Hubble universe, is large, thus the medium density is nearly constant everywhere and we observe a nearly constant speed of light. But close to a large astronomical body like the Sun, its large space density produces a measurable curvature of the paths of the inward and outward waves and thus of light and the motion of matter. We observe the curved paths as the effect of *gravity* described by Newton and also *space curvature* of Einstein's general relativity. Both due to Equation 4.

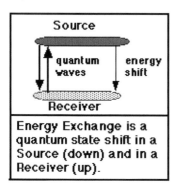

Figure 4. Energy exchange. The spherical IN and OUT waves of the source and receiver oscillate in two-way communication until a minimum amplitude condition is obtained. The decrease of energy (frequency) of the source will equal the increase of energy of the receiver. Thus energy is conserved. We observe 'e-m waves' as a large number of such quantum changes.

17. The appearance and origin of charge

Schrödinger and Clifford predicted that charge was due to wave structures in space. Charge 'appears' at wave-centers because the spherical waves of an electron resonance are very large at the center due to the $1/r^2$ dependence. Centers have a high density due to the large wave amplitude. The dense space at the central region is *non-linear*, which causes energy transfer or *coupling* between two resonances. We observe this process and call it 'charge.' But as Clifford and Schrödinger wrote, there is no charge substance involved. It is a property of the wave structure at the center.

The high-density wave centers appear to us as the location of point charges because force interactions occur there that we call 'electric'. The center wave-amplitude is *finite*, as shown in Figure 3, and as experimentally observed. It is not infinite as in the puzzling Coulomb law.

Producing particle motion. The in-waves of a particle, on arrival at the wave-center, produce the position and motion of the center that we observe as the 'particle.' If any matter nearby changes the medium density, this changes the in-wave speed and motion, and moves the particle location. We observe this motion and describe it as the result of electric forces. Motion (and acceleration) of matter also changes the apparent wavelengths (Doppler effect) and produces an energy exchange to the wave medium, similar to the W&F *response of the universe*. Inertial motion (F=ma) produces Mach's Principle because the space density is formed by all the matter of the Universe.

18. Equation of the Cosmos

Can this mechanism be tested? Yes. If a resonance's self-waves can dominate in its local space, then at some local radius, r_0 from the center, self-wave density must equal the total density of waves from the other N particles in the Universe. Evaluating this equality[8] yields

$$r_0^2 = R^2 / 3N \tag{5}$$

The best astronomical measurements, $R = 10^{26}$ meters, $N = 10^{80}$ particles, yield $r_0 = 6 \times 10^{-15}$ meters. To satisfy the test, r_0 should be near the classical radius,

e^2/mc^2 of an electron, which is 2.8×10^{-15} meters. It is. The test is satisfied verifying Mach's Principle.

This is called the *Equation of the Cosmos* a relation between the 'size' r_0 of the electron and the size R of the Hubble Universe. Astonishingly, it describes how all the N particles of the Hubble Universe create the space medium and the appearance of charge and mass of each electron as a property of the universal space. It also implies the inter-connectedness of all matter. Other properties of the cosmos can be derived from it.

19.   Principle III - Minimum Amplitude Principle (MAP)

This third principle can be obtained from Principle II, but because it is a very useful law of the universe, which simply describes how interactions take place and how wave structures will move, it will be written out separately:

Principle III: The total amplitude of particle waves always seeks a minimum at each location.

$$\sum_{1}^{N} \Phi_n = \text{a minimum}$$

This principle is the disciplinarian of the universe. That is, energy transfers take place and wave-centers move in order to minimize total wave amplitude. Amplitudes are additive, so if two *opposite* resonances move together, the motion will minimize total amplitude. This explains empirical rules such as, "Like charges repel and unlike charges attract," because those rules minimize total amplitude. The origins of other rules are also now understood. For example, MAP produces the *Pauli Exclusion Principle*, which prevents two identical resonances (two fermions) from occupying the same state. Two identical states are not allowed because total amplitude would be a maximum, not a minimum. The operation of MAP is seen in ordinary situations like the water of a lake, which levels itself, and in the flow of heat that always moves from a hot source to a cold sink, which are examples of the *increase of entropy principle.*

20.   Conservation of energy

The energy transfer mechanism occurs at the high density wave-centers, which permits coupling or changes of their wave frequency. When the waves of a potential source and a potential receiver pass through each other's centers, and an allowed transition exists between them, MAP minimizes the total of both amplitudes by choosing the transition. In the source, the frequency (energy) of the wave state shifts downward. In the receiver, there is an equal shift upward, as in Figure 4. Only wave states (oscillators) with equal frequencies 'tuned' to each other can couple and shift frequency. Accordingly, the frequency (energy) changes must be equal and opposite. This is exactly the content of the *Conservation of Energy law*, *n*ot too different from rules of tuning up an orchestra matched to the 'A' played by the first violin.

This mechanism also describes the proposal noted by Einstein in the writing of Tetrode[5,7]: "When I see a star 100 light-years away, the star knew that its

Huygens wave front

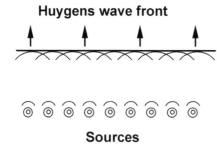

Figure 5. Plane wave formed by Huygens combination of wavelets. The wavelets from a line of sources combine, at a distance from the sources, to form a new wave front which repeats the geometry of the sources.

Sources

light would enter my eye, 100 year ago—before I was born!" Tetrode was predicting that energy exchange is a two-way symmetrical resonance exchange between source and receiver—not a one-way photon.

If you keep the traditional assumption that matter consists of points of mass and charge substance and that energy exchange is a one-way e-m photon traveling between particles, you are doomed to the paradoxes of: causality violation, wave-particle duality, Copenhagen errors, Heisenberg uncertainty, redshift, and others. Only the two-way exchange of the WSM matches observation.

Even though Einstein had originally proposed the 'photon' he never understood them. In 1954, he wrote to his friend Michael Besso expressing his frustration, "All these fifty years of conscious brooding have brought me no nearer to the answer to the question, 'What are light quanta? Nowadays every Tom, Dick and Harry thinks he knows it but he is mistaken." Einstein also came to realize that matter could not be described by an electromagnetic field: "I consider it quite possible that physics cannot be based on the field concept, *i.e.*, on continuous structures [discrete particles]. In that case nothing remains of my castle in the air, gravitation theory included [and of] the rest of modern physics." (*Ideas and Opinions*, 1954). In hindsight, he was correct about the errors of field theory and his general relativity has survived.

21.   The origin of the IN waves and the response of the universe

At first thought, it is puzzling where the in-waves come from. This puzzle is our own fault—a result of looking at the waves of only one particle, and ignoring the waves of all other particles in the universe—gross over simplification. To find reality, we must deal with the real wave-filled universe. When we study this question[5,10] we find a rational origin of the inward waves:

Three hundred years ago Christian Huygens, a Dutch mathematician, found that if a surface containing many separate wave sources was examined at a distance, the combined wavelets appeared as a single wave front having the shape of the surface as shown in Figure 5. This wave front is termed a 'Huygens Combination' of the separate wavelets.

This mechanism is the origin the in-waves, as follows: When an outgoing wave encounters other space resonances (particles), their out-waves are joined with the initial out-wave to form a Huygens Combination wave front. These

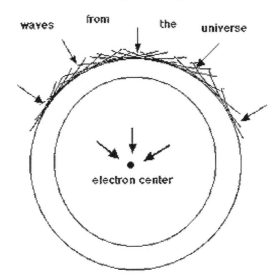

waves **from** the universe

electron center

Figure 6. Formation of in-waves. The out-wave of every particle interacts with other matter in the universe. The response to the outgoing wave is Huygens wavelets from other matter that converge back to the center of the initial out-wave, forming the in-wave. Thus every particle depends on all other particles for its existence. *Every charged particle is a structural part of the universe and the whole universe contributes to each charged particle.*

waves arrive in phase at the initial center forming the in-wave of the initial particle. This occurs throughout the universe so that every particle depends on all others to create its in-wave as in Figure 6. Although particle centers are widely separated, all particles are one unified structure because they share each other's waves.

Part V - The origin of the electron's spin

As an example of the depth of understanding and universality of the Wave Structure of Matter, We describe the origin of the spin of the electron. The physical nature and cause of electron spin has been sought for 75 years ever since Nobel laureate Paul Dirac[4] made a calculation of spin in 1926. His theoretical work predicted the positron, found five years later by C. D. Anderson.

22. Dirac's theory

Dirac was interested in the differences between relativity and quantum theory. Dirac compared the conservation of relativistic energy (E = energy and p = momentum of a particle of mass m) given by

$$E^2 + p^2c^2 = m_0^2c^4 \qquad (6)$$

with Schrödinger's procedure in quantum theory. Schrödinger's procedure was to use an energy statement like Eqn. 6, and change it to a wave equation. He changed the terms for E and p into two wave equation operators, using,

$$E = (ih/2\pi)[\partial\Phi/\partial t], \; p = -(ih/2\pi)[\partial\Phi/\partial r] \qquad (7)$$

Where Φ is the amplitude of the Schrödinger wave function sought. Then the solutions should describe the amplitude of waves of the particle. No one knew why this worked but the results for the H atom are amazingly accurate so it was trusted.

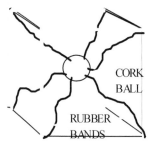

SUPORTING FRAME MADE OF STICKS

A Model of Spherical Rotation

This apparatus is easily made out of a few sticks, a cork, and six rubber bands. The cork can be rotated (taking care not to knot up the rubber bands) continuously without entangling the rubber bands! The cork and bands will return to their initial configuration every two turns.

CORK
BALL
RUBBER
BANDS

This demonstrates a little known variety of rotation. It has application to to particle theory because the spherical rotation does not destroy continuity of space.

Figure 7. Demonstrating spherical rotation.

The procedure was puzzling to Dirac because Eqn. 6 uses squared terms whereas Eqns. 7 cannot be squared! He reasoned that in Eqn. 8 the equivalent of squaring could become part of a matrix algebra. He had a new idea: Try replacing Eqn. 6 with a matrix equation:

$$[Identity]E = [alpha]pc + [beta]m_0c^2 \qquad (8)$$

where the new matrices $[Identity]$, $[alpha]$, and $[beta]$ are 4-operators. This avoided squares of E and p but placed restrictions on the new operators and their solutions.

He found that solutions existed if E and p had fixed values. This theoretical matrix algebra produced correct fixed values of the electron's energy $= mc^2$, and spin $= h/4\pi$ angular momentum units, but it gave no hint of the physical structure of the electron. His new 'Dirac Equation[8] became famous.

Dirac also noticed that only two functions were needed in the electron's Φ solution. So Dirac simplified the algebra by introducing number *pairs*, termed *spinors*, and 2×2 matrices called *spin operators* creating a 2-algebra instead of a 4-algebra. The pair of electron waves, Φ_{in} and Φ_{out} from the WSM are a Dirac spinor, part of the binary universe. A reader-friendly review of Dirac's Equation is found in Eisele[11].

23. The physical mechanism of spin

Spin occurs when the in-waves arrive at the center and change direction and transform into the out-waves. There are strict (boundary) conditions on the transformed amplitudes and polarity of the in- and out-waves: Space cannot be allowed to twist up without limit. The spherical wave amplitudes must continually and smoothly change from being in-waves to out-waves. The in-wave amplitude at the center must be equal and opposite to the out-wave.

It turns out this transformation is possible using a known[12] property of 3D space called *spherical rotation*—a misleading name, there is no true rotation—in which space moves continually around a *point* and returns to its initial state after two turns. In spherical rotation there is no axis like cylindrical rotation of a wheel. Spherical symmetry is preserved because the center of 'rotation' is a point. One direction of rotation produces the electron, the other the positron. This is why every charged particle has an anti-particle.

Batty-Pratt & Racey[13] (1980) analyzed spherical rotation and showed that an exponential oscillator, e^{iwt}, was a spinor. Wolff[8] realized in 1989 that the exponential in-out oscillator waves of the WSM were the *real* physical spinors satisfying the Dirac Equation. It is humbling to realize that only 3D space[12] has this remarkable property shown in Figure 7. If this property of 3D space did not exist, particles and matter could not exist. Life and the universe as we know it could not exist.

24. The equivalence of the WSM and the Dirac equation procedure

It is easy to calculate the rotation rate of an in-wave of frequency (energy) = mc^2/h and wavelength = h/mc. The rotation rate is two turns each cycle. This produces an angular momentum of $\pm h/4\pi$, obtaining Dirac's result simply.

The energy value of electrons is the same for both, but of different signs. The Dirac equation yields $\pm mc^2$ while the WSM says the both electron and positron energy are positive, $+mc^2$. Dirac was forced to interpret the puzzling negative energy as an unseen "sea of negative energy particles." This strange concept has never been observed and has been abandoned.

How can we understand Dirac's negative energy? Look at the electron/positron wave algebra in the Appendix. Write the product of energy and time as negative, that is: $-Et/h = -wt$. But this is the same as switching the in-wave with the out-wave which changes the electron into a positron! Now we see Dirac's result described an anti-particle not a negative energy. He assumed the electron was a discrete particle instead of a wave structure! This mistake has plagued physics for centuries.

25. A model of spherical rotation

Spherical rotation in 3D space can be modeled by a ball held by threads inside a cubical frame shown in Figure 7. The threads represent the coordinates of the space and the rotating ball represents the space at the center of the converging and diverging quantum waves. The ball can be turned about any axis starting from any initial position. If the ball is rotated continuously it returns to its initial configuration after every two rotations. This demonstration appears in the classic book[12] *Gravitation* by Misner, Thorne and Wheeler.

Using the exponential wave solutions for the electron shown in the Appendix, you can *reverse* the spin axis, by reversing time ($t \rightarrow -t$) or by reversing the angular velocity ($w \rightarrow -w$). Both are equivalent to switching the outgoing spherical wave of an electron with the incoming wave. A similar change is an inverted spin state produced by the inversion matrix operating on the spinor. But this does not change the direction of the in/out waves. Thus in these two examples, axis inversion and spin reversal are <u>not</u> the same. But in our human view of cylindrical rotation they are the same. This unique difference is characteristic of the quantum wave electron.

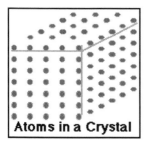

Atoms in a Crystal

Figure 8. Atoms in a crystal. The waves of the electrons in the crystal array produce standing waves along the planes of symmetry. There is no solid structure, substance or material in the crystal. It is the waves that produce the array dimensions, and it is the immense energy density of the space medium that gives it physical strength and rigidity.

26.   Connecting quantum theory and relativity

Before the WSM, there was no physical reason known for the mass increase of relativity. Likewise there were no physical explanations for quantum theory or spin. Were these apparently separate laws connected or not? Indeed, many theorists proclaimed that these phenomena were irreconcilable! Few thought about a connection because most physicists imagined point particles, and were satisfied despite the puzzles. After all, particles had been in the textbooks for years. The connection clue is that both depend on the relative velocity v. Both are Doppler effects due to the relative velocity of two wave structures.

The Appendix shows that the Doppler increase of frequency causes increase of mass (energy or frequency) so that $m = m_0[1 - v^2/c^2]^{-1/2}$ as seen by a moving observer. Rearranged, this equation gives the energy equation used by Dirac:

$$E^2 + p^2c^2 = m_0^2c^4$$

Similarly, the de Broglie wavelength $\lambda = h/p$ is also a Doppler change of wavelength seen by a moving observer (See Appendix) and it leads to the Schrödinger equation.

27.   What is the space medium?

The space medium determines the properties of the waves that propagate in it, as is true for all wave phenomena. Thus space underlies the WSM, the natural laws originating from the WSM, and all the sciences built on the natural laws. But space is not obvious to our bodily senses although we are aware of it, knowing:

A. **The law of inertia** $F = m\mathbf{a}$. For example, our sense of force or energy transfer (to the space medium) when we move or 'heft' a massive object.

B. **The laser gyro.** These instruments now used in most commercial aircraft, contain two laser beams counter-rotating in a quartz prism that measure rotary motion in inertial space. The two beams travel in the space medium, so when the prism rotates, they have opposite frequency shifts and the beat note is proportional to inertial rotation rate.

C. **The e-m constants** μ **and** ε. These are derived from properties of space and determine the indices of refraction in optics.

D. **The rigidity of crystals**. The nuclear wave structures in say a diamond, are held rigidly by standing electron waves forming a lattice in the space medium. This is shown in Figure 8. The geometry of the lattice is determined by the MAP.

E. **The General Theory of Relativity**. The space medium of quantum waves and the space-time of general relativity are one and the same. In both, the density of space is determined by the distribution of matter in the Universe, and, matter properties depend on the space density. This strange feedback between space and matter has been expressed: *Matter tells space what it is and space tells matter how to behave.*

The minimum energy density of space can be approximated using the density of nuclear matter. It is $>10^{46}$ MeV per cubic centimetre, astoundingly large. We have no sensation of its presence despite its existence all around us, because energy transfer only occurs between a resonant source and receiver. Learning more about the space medium is the most exciting and pioneering topic in science today. Prepare yourself for a fascinating adventure.

Part VI - Discussion

We can confidence that the Wave Structure of Matter is the true physical reality of the universe. The logical proof test is *that the experimental measurements of the empirical natural laws must agree with their predictions by the WSM.* They do. In fact, the experimental evidence agrees better with the WSM predictions than with conventional rules. For example, an infinity of charge potential at $r = 0$ is expected from Coulomb's empirical law. It is not found. Instead, the finite experimental value agrees with the WSM. There are more examples: Conventional physics has no explanation for energy exchange, or the Pauli Principle, or spin, or gravity, or charge attraction and repulsion. All of these are predicted correctly by the WSM.

The philosophical conclusions from the connectedness of laws and matter in the universe are thought provoking: *Everything we observe here on Earth— life, mind and matter—depends on the existence of the matter elsewhere in the universe.* Thus if the stars and galaxies were not in the heavens, we could not exist. Matter on Earth and matter in the Universe are necessary to each other.

Einstein was close to the truth. He wrote: *A human being is part of the whole called by us Universe, a part limited in time and space. We experience ourselves, our thoughts and feelings as something separate from the rest, a kind of optical delusion of consciousness. This delusion is a kind of prison for us, restricting us to our personal desire and to affection for a few persons nearest to us. We must free ourselves from the prison by widening our circle of compassion to embrace all living creatures and the whole of nature in its beauty... We shall require a substantially a new manner of thinking if humanity is to survive.*

But the practical value of the WSM theory is the insight it provides allowing scientists to deeply analyze quantum wave structures. In the R&D laboratory, the new insight should advance electronic applications, especially IC and memory devices because their tiny transistor elements use quantum effects to control the flow of currents. The new knowledge will improve bio-technology, communication, and the efficiency of energy transmission. For example, conduction of electric energy along a wire is a quantum energy transfer process. Knowing this, energy losses and costs may be reduced, and transmission distances increased.

Part VII. Mathematical Appendix

28.   Solutions of the wave equation (Principle I)

The wave equation [1], must be written in spherical coordinates because cosmological space has spherical symmetry. Uniform density of the medium (space) is assumed which yields a constant speed c of the waves (and 'light'). There are only two solutions. They are:

$$\text{Outward wave} = \Phi_{out} = (1/r)\Phi_{max}\exp(iwt - ikr) \tag{A}$$

$$\text{Inward wave } = \Phi_{in} = (1/r)\Phi_{max}\exp(iwt + ikr) \tag{B}$$

where Φ = wave amplitude, $k = mc/h$ = wave number, $w = 2\pi f$, r = radius from the wave center, and energy = $E = hf = mc^2$. These two waves are components of charged particles including the electron, positron, proton, and anti-proton. Superposition of the two amplitudes produces a standing wave, and can occur in two ways forming either an electron or a positron. At the center, the inward wave undergoes a rotary-reversal transforming it to the outward wave. This can happen in two ways: CW or CCW. One is the electron, the other the positron, with opposite spins:

$$\text{electron} = \Phi_{in} - \Phi_{out} + \text{CW spin} \tag{C}$$

$$\text{positron} = \Phi_{out} - \Phi_{in} + \text{CCW spin} \tag{D}$$

If you add the electron amplitude to the positron amplitude, the result is zero or *annihilation*, the well-known result, as seen by a check of Equations [A,B,C,D].

You can experiment with particle inversions by changing the (+ or −) signs in the amplitude equations (A and B) of the particles (C and D). To perform a **T**ime inversion, change t to $-t$. To perform a mirror inversion (**P**arity), imagine that the waves are viewed in a mirror. You will see that a positron is a mirror image of the electron. To change a particle to an anti-particle (**C**harge inversion), switch the in-waves and the out-waves, and the spin direction. Thus, successive **C**, **P**, and **T** inversions returns to the initial state which is a proof of the empirical-theoretical **CPT** rule, now seen to be a property of the wave structure.

No time travel. These CPT relations are the physical basis of *Feynman diagrams* that describe the behaviour of electrons and positrons in experimental particle labs. You can now understand Feynman's cryptic statement, "A positron is an electron traveling backward in time." Although this statement led to many sci-fi films about time travel, the fact is the positron does not go backwards. Only its inward and outward waves are opposite to those of the electron. It is still a normal citizen in the particle universe.

29. Origin of special relativity mass increase and the deBroglie wavelength

Write the equation of a SR, as seen by an observer with relative velocity $b = v/c$, as shown in Wolff[8]. Insert relativistic Doppler factors, $g = [1 - v^2/c^2]^{-1/2}$. The amplitudes received by the observer are then,

$$\text{Received amplitude} = \left(1/r\right)\left\{\left(2\Phi_{max}\right)\exp\left[ikg\left(ct+br\right)\right]\sin\left[kg\left(bct + r\right)\right]\right\}$$

This is an *exponential* oscillator modulated by a *sine* factor. The origins of the de Broglie wavelength (QM) and the relativistic energy and momentum (mass increase of SRT) are as follows:

In the *exponetial* factor:

Wavelength = h/mvg = *de Broglie* wavelength with *relativistic* momentum.

Frequency = $kgc/2\pi = gmc^2/h$ = mass frequency with *relativistic* energy.

And in the *sine* factor:

Wavelength = h/mcg = *Compton* wavelength with *relativistic* momentum.

Frequency = $bgmc^2/h = b \times$ (mass frequency) = *relativistic* momentum frequency.

You see that the Doppler factor g causes the correct deBroglie wavelength and SRT mass to appear in the observed waves, as a function of the relative velocity. It is important to note that the effect is symmetrical; it does not depend on whether the relative velocity is $+v$ or $-v$. This symmetry is exactly as observed. Examination of the algebra shows that this is due to the symmetrical presence of *both* the inward and outward waves.

Thus the space resonance *physically* displays all properties of an electron, *viz.*, electric charge, QM, SRT, forces, annihilation, spin, conversion to a positron, and **CPT** relations—all of which were formerly empirical or *theoretical* properties. These physical properties depend on the spherical wave structure and ultimately on the wave medium—space.

To a scientist familiar with wave optics, the truth of the wave structure of an electron seems irrefutable. This is the way waves behave. On the other hand if one has been taught that particles are discrete objects, and if one's career depends on the existence of particulate quarks and gluons and a research contract with a giant accelerator, it is difficult to change belief.

References

1. Clifford, William (1870), "On the Space Theory of Matter," *Cambridge Philosophical Society*, **2**, pp.157-158, Clifford, William (1882), *Mathematical Papers*, Editor: Robert Tucker, Chelsea Publishing, N.Y. 1968.
2. Moore, Walter (1989), *Schrödinger - Life and Thought*, Cambridge U. Press, England, p. 327.

3. Mach, Ernst (1883 - in German), English edition: *The Science of Mechanics*, Open Court, London (1960).

4. Dirac, Paul (1937), "Quantum Electrodynamics" *Nature*, London, **174**, p. 321.

5. Wheeler, J. A. and Feynman, R. (1945), "Interaction with the Absorber...." *Rev. Mod. Phys.*, **17**, p.157.

6. Dirac, Paul (1933*) Proc. Roy. Soc.* London, **A167**, 148.

7. Tetrode, H. (1922), *Zeits fur Physik* **10**, 312.

8. Wolff, Milo (1990), *Exploring the Physics of the Unknown Universe*, ISBN 0-9627787-0-2, Technotran Press, Manhattan Beach, CA.

9. Wolff, Milo (1991), "Microphysics, Fundamental Laws and Cosmology," *Sakharov Memorial Lecture on Physics*, Moscow, May 21-31, 1991, pp. 1131-1150. Nova Scientific Publ., NY (1992).

10. Wolff, Milo (1997), "Exploring the Universe," *Temple University Frontier Perspectives*, **6**, No 2, pp. 44-56.

11. Eisele, John A. (1960), *Modern Quantum Mechanics with Elementary Particle Physics*, John Wiley, NY.

12. Misner, C. W., Thorne, K. and Wheeler, J.A. (1973), *Gravitation*, W.H. Freeman Co. San Francisco, p. 1149.

13. Battey-Pratt, E. and Racey, T. (1980), "Geometric Model of Fundamental Particles," *Intl. J. Theor. Phys.*, **19**, pp. 437-475.

14. Wolff, Milo (1993), "Fundamental Laws, Microphysics and Cosmology," *Physics Essays*, **6**, pp. 181-203.

15. Wolff, Milo (2002), "The Quantum Universe and Electron Spin," pp 517-524, *Gravitation and Cosmology – From the Hubble radius to the Planck Scale*, R. Amoroso *et al.* editors, Kluwer Academic Publishers.

16. Wolff, Milo (1995), "Beyond the Point Particle – A Wave Structure for the Electron," *Galilean Electrodynamics*, Sept-Oct, pp. 83-91.

Light: Neither Particle nor Transverse Wave

William Gaede, Researcher
CorpoAmazonia, Leticia, Colombia
viligaede@yahoo.com

Statistics, variables and equations, and concepts such as energy, force and field tell us nothing about the architectures of light, the atom, and the Universe. The mathematical theories of General and Special Relativity and of Classical and Quantum Mechanics deal exclusively with relations and concepts, and are therefore powerless to describe the shape of physical objects. To discern the architecture of an invisible entity we must merely venture an assumption and logically check whether the proposal explains observation. Pursuant to this method, we show architectures of light, the atom, and the Universe rebuff the abstractions (photon, wave, duality, wave-packet, *etc.*) inferred by Classical and Quantum Mechanics.

Key words: particle, wave, duality, wave-packet, hydrogen atom, static universe, dynamic universe.

The Moon does not move!

For centuries man has debated whether the Universe is static or dynamic. Newton was one who believed in a homogeneous, infinite, and *static* universe, reasoning that only infinite mass and volume could counteract the gravitational attraction of stars and keep them essentially in place. In contrast, relativity champions a *dynamic* universe characterized by self-creation and inflation. However, technically speaking, Newton's universe is not what we really call motionless. Under his peculiar version of *static*, the Moon still orbits a spinning planet and a dog sleeping quietly here on Earth nevertheless moves with respect to Jupiter. If the plain meaning of the word *static* is 'having no motion', the Newtonian universe is ill-conceived. It is imperative, therefore, to distinguish clearly between these mutually exclusive scenarios and reformulate the ageless inquiry using the rigorous definition of static. Is there any movement whatsoever in the Universe? Does the Moon move at all?

In light of this clarification, the reply is now academic. In order to thaw a frozen universe it suffices that a single object move; the remaining ones automatically adjust their distance to it. And since it is a matter of fact that at least your hand moves, we can proceed to synthesize this undeniable reality into a law:

> ***The First Law of Physics or The Dynamic Universe:*** *An object moves with respect to at least one other object in the universe.*

However, before we certify our conclusion, it would seem elementary to begin by defining what we mean by *move*. Newton considered motion to be so self-evident that he thought it trivial to bother with a definition, casually remarking: "I do not define time, space, place and *motion*, as being well known to all." [1] His definitions of absolute and relative motion leave us in even greater doubt as to his understanding: "Absolute motion is the translation of a body from one absolute place to another; and relative motion, the translation from one relative place into another." *Translation* and *motion* are synonyms, and we end up learning nothing from Newton's circular attempt. This is actually quite stunning considering that Newton is credited with having discovered the three laws of *motion*. Can theorists assert that a particle *collided* in an accelerator or that light *travels* at 300,000 km/sec if they haven't formally defined the word motion? Can Quantum Mechanics (QM) regard indeterminacy to be a 'principle' if advocates haven't first distinguished between *position* and *momentum*? Can we take for granted that our universe is dynamic if we have but vague notions of the terms *location* and *translate*? It is inconceivable to answer such questions unless we firmly anchor certain foundations:

distance: Linear space, gap, or separation between the surfaces of two objects. [2]

position: An imaginary volume of space occupied by an object; the object itself. [3]

location: The set of distances of an object from the remaining matter in the universe.

In a universe consisting of a single shape, the object merely has position: the object itself. For location to acquire meaning, two or more objects must inhabit the frame or field of view. Motion requires more than one location, and time, that we compare two locations or motions. Whereas distance, position, and location are attributes of a static universe, motion and time conceptually belong to a dynamic world, and whereas time necessarily involves observers, motion is contextual. [4]

motion: Two or more locations of an object.

time: A qualitative relation between two locations of an object or between the motions of two objects (*e.g.*, before, after). A *second*, in contrast, is a specific quantitative relation between the movement of one object *and* a reference. (*e.g.*, the orbit of Earth versus the trajectory of a caesium wave). [5]

For example, energy is not an object, but a concept. [6] Energy lacks the one attribute that would allow this term to be classified as 'physical': shape. [2] Likewise, boundaries are implicit requisites for *distance* and *location*, a fact that by extension denies *concepts* such as love or energy the ability to move. In Physics, motion is a property circumscribed to *objects*. For instance, the Moon is an object, a photograph—a single frame of the film. The orbit of our satellite is a video, the set of locations that comprises one revolution. However, the Moon can either be here or there. Like any object it enjoys the *static* attribute of being at a single location. The Moon does not *live* stretched out across an infinitesimal interval of time (*i.e.*, between two marks on the time axis [$t > 0$]). The Moon *exists* in a cross-section of time (*i.e.*, conceptually a single cut through the time axis [$t = 0$]). The reason conscious observers perceive motion is that they remember the Moon's previous location, the next-to-the-last frame on the film. In the absence of this now vanished reference the Moon and every atom comprising the Moon only have location. Each and every atom exists at the cutting edge of universal events, at the limit of motion, in a cross-section of time: the atom itself. Without memory, the Moon does not move! It merely has location. These definitions now give us justification to amend our First Law:

The First Law of Physics or The Static Universe: *An object has location.*

It is important to note that Mathematics plays no role in a static universe. There is not a single variable, function, or equation that can depict a physical object. Without prior experience, we cannot derive the shape of a sphere from the expression [x^3] or from an equation such as [$v = (4\pi r^3)/3$]. Mathematical symbols represent *relations* inferred after an object is observed. Hence, Mathematics has no authority to tell us what things look like; it is a language restricted solely to characterizing the dynamic universe (*i.e.*, relations). Variables, functions, and equations depict motion and are limited to communicating *how* objects or behaviours *compare* with standards and references. Therefore, if Mathematics underlies General and Special Relativity and Classical and Quantum Mechanics, these theories are wholly unsuited for the task of illustrating our static universe. What does light look like when it is standing still? What would the photograph reveal if Atom Man took a snapshot of hydrogen? What is the nature of the Universe before we introduce an observer? We cannot hope to answer these questions with statistics, functions, or motion-embodying concepts such as energy [6], force [7], or field [8]. What we can do is make an assumption—propose an object—and logically and empirically check whether the model successfully simulates what we observe.

Why light is not a particle

Consistent with the foregoing methodology, Newton [9] proposed a hypothesis. He assumed that light is comprised of discrete corpuscles that together form a ray or beam. He argued that a stream of bullets could account for: reflection, refraction, and rectilinear propagation. His theory predicted that light should travel faster through denser media. However, it either has been or can be shown that each of his arguments is without merit.

a. *Reflection.* A simple experiment shows that your hand interferes with the free passage of light to generate a shadow. If, as Newton argued, light consists of corpuscles, the implication is that each corpuscle is three-dimensional (3D), possesses a surface, and strikes your palm. Bohr's [10] still accepted 'quantum jump' theory holds that the atom emits corpuscles (light) when the electron falls to a lower energy orbit, state, or band. [11],[12] However, we just settled that energy is a concept. Therefore, if light is comprised of particles, the atom in effect becomes a mechanism that somehow converts a concept (energy) into a physical object (a particle). Should this assessment be incorrect, the onus shifts to advocates to explain how a corpuscle of light is manufactured inside the atom. How does something that has no shape or physical dimensions (energy) acquire such attributes?

Another simple experiment demonstrates that two beams intersecting perpendicularly do not interact. In order to avoid billiard ball style collisions, particle surfaces must be regarded as being physically transparent to each other. This saddles advocates with the burden of explaining how a 3D mirror manages to turn back these ethereal particles. Monte Carlo simulations reduce corpuscular reflection and refraction to a hit or miss phenomenon. [13] While ingenious, this statistical gimmick falls short of answering why a particle of light travels roundtrip through several feet of transparent pool water without being deflected by countless protons and electrons whereas the same particle finds it immeasurably harder to cross a 1 mm sheet of opaque paper. Clearly, if a mirror reflects light and glass allows it through, neither thickness nor molecular makeup of the material is the discriminating factor. [14]

b. *Refraction.* Foucault [15] proved Newton's 'refrangibility' prediction to be false.

c. *Rectilinear Propagation.* A specific particle cannot mimic the verified wave behaviour of light (oscillate about a fixed axis) and be said to be moving rectilinearly in the direction of the ray.

These objections show that none of Newton's behavioural justifications for the corpuscle passes a more rigorous examination. We add to this list that, in a laboratory setting, marbles would fail to simulate the stochastic explanation offered for interference fringes. [16] Where such straight-forward prediction is refuted by experiment, the scientific method demands that advocates resolve the discrepancy, namely, the subatomic world's refusal to follow rational Newtonian mechanics.

A more pertinent objection to the corpuscular hypothesis comes from a little regarded, but well-established principle of optics: ray reversibility. [17] Whether reflecting or refracting, a beam of light inexplicably retraces its path. Neither particle nor wave—both of which are outgoing, one-way mechanisms—can simulate re-tracking, especially if source and mirror are light years apart.

As shown by the following arguments, discrete particles are also notoriously incompatible with waves.

a.   Let us assume that a plane transverse wave is comprised of discrete corpuscles. Under this scheme, frequency is the number of cycles completed by a given particle in an arbitrary period of time, and wavelength, the separation between two adjacent corpuscles of the crest or trough. Therefore, a particle does not vibrate up and down while the 'disturbance' passes through, for this contradicts the notion of frequency. Rather, the particle completes a sinusoidal trajectory (*i.e.*, up, down, *and forward*). However, this behaviour presents three insurmountable problems:

1. An oscillating particle must travel faster than c and faster yet during amplitude increases.
2. Newton's 1st and 3rd Laws require that an extrinsic object account for the sinusoidal cycle of each discrete particle.
3. The 2D wave is comprised of 3D particles.

Any one of these observations defeats the initial assumption that a plane transverse wave could consist of particles.

b.   Faraday and Maxwell Laws require mutual induction between magnetic and electric fields. The discrete particle hypothesis is incompatible with continuity and orthogonality.

c.   Electric lines of force begin and end on charges whereas magnetic lines of force form closed loops. Discrete particles are at odds with the continuity required to generate these architectures.

The scientific method requires that we discard a hypothesis if a single observation contradicts our prediction. [18] Here we have listed several well-documented objections to the corpuscle. Hence, this reasonable rule of thumb renders the particle hypothesis as well as the mathematics developed for it null and void. By extension, all explanations that assume light to be a corpuscle (*e.g.*, Eddington's alleged confirmation of Einstein's Theory of General Relativity [19]) are also declared moot.

Why light is not a transverse wave

In contrast to the particle, Huygens [20] proposed that light rays consist of longitudinal waves propagating in accordance with the principle that carries his name. Years later, Fresnel [21] introduced plane transverse waves to model polarization. Newton objected to waves, among other reasons, because proponents failed to specify the nature of the intermediary. When Michelson [22] finally demolished the 200-year-old aether, proponents, undaunted, replaced it with 'nothing'. Today, wave theorists continue to insist that waves are the undulation of nothing! [23]

This ethereal hypothesis, however, bypasses established rules of science. The scientific method requires that the prosecutors begin by defining 'nothing'. [2] Afterwards, they must conceptualize how nothing (or the flat ribbon they usually use to model the wave) interacts with a 3D hand to generate a shadow.

If the allegation is that the wave converts to particle upon contact, the scientific method mandates in addition that this mystical 0D or 2D to 3D metamorphosis be accounted for in the theory as well.

A more poignant criticism of transverse waves arises from the bizarre standing wave simulation. The reason fluids and gases cannot support transverse waves is that they are comprised of *discrete* molecules. For the standing wave to oscillate about an imaginary axis connecting Sun and Earth, it must necessarily be *continuous* (made of a single piece) and be attached at both ends. Only the force of *pull* can generate straightness in Physics. Photons, particles, waves, and wave-packets are out-going, *push* mechanisms incapable of modeling properties such as straightness or ray reversibility under dynamic conditions.

Wave theory has also been shown to be inconsistent with the photoelectric [24] and Compton [25] effects, and has yet to justify why electric lines of force begin and end on charges while magnetic lines of force loop around. If we further factor in that a wave is what something *does* as opposed to what something *is* [26], the transverse wave hypothesis crumbles like a house of cards. Again, these arguments constitute sufficient reasons to reject the transverse plane wave as a structural object and viable candidate for light. Any equation or explanation ever devised around transverse waves is hereby rendered without effect (*e.g.*, Young's slit experiment [27]).

Why light is not a wave-packet

By the end of the first quarter of the 20th Century it became apparent that light could be neither particle nor wave, a predicament that warranted brainstorming new architectures. Instead, theorists chose a regrettable unscientific path. The still-accepted Copenhagen Interpretation (CI) that resulted from the 5th Solvay Conference of 1927 blends Heisenberg's Uncertainty Principle [28], Born's probability wave [29], Bohr's Principle of Complementarity [30], and personal opinions of Heisenberg [31] into an unfathomable compromise known as *duality*. With regard to light, the CI states essentially that a photon 'behaves' both as a particle and as a wave, and that the nature of the experiment determines which of these aspects light will exhibit. Therefore, the claim that the CI is a 'physical' interpretation of the quantum formalism is quite misleading. The CI addresses location, behaviour, and perceptions; it says nothing about structure.

Moreover, the scientific method requires that the relevant definitions and objects be established *a priori* and remain unalterable throughout the subsequent theory. The CI flows in reverse. It retroactively infers the assumptions from experiment and tailors architectures to suit its arguments. If the results of the slit experiment are explainable in terms of waves, the photon retroactively converts to a wave, and if the results of the photoelectric effect are explainable in terms of particles, the wave now becomes a particle. It is this 'tenderness' of the two strategic hypotheses of QM—the photon and the electron—that leads Baierlein [32] to conclude that light *is* simultaneously a 2D wave *and* a 3D particle. But here again Quantum Mechanics steamrolls over science. The scientific method compels proponents to incorporate within their theory the process by which an object intermittently loses and regains a physical dimension such as width to support such assertions.

Actually, these issues may no longer be relevant. Duality has made a more radical transition by transmuting from tentative principle to permanent object. Penrose [16] and Ridley [33] not only discuss, but also illustrate the peculiar entity known as a *wave-packet*, an unfathomable cross between a quantum of energy and the ripples of nothing. Indeed, if as the establishment suggests, light consists of discrete packages of energy, proponents have no further excuse not to illustrate this admittedly finite structure. [34] QM has, in effect, transformed a formless concept into a solid!

Thus, the wave-packet divides the mainstream into two camps. Those who still yearn to visualize the architecture of light have yet to come to terms with the fact that whatever shape they concoct for the photon, wave-packet, wave-function, probability wave, or state vector is guaranteed to be irreconcilable with certified behaviours of light. Rather than incorporate the favourable aspects of waves and particles as proponents intend, the wave-packet embodies the weaknesses of both. The objections made earlier to waves and corpuscles have not disappeared with this forceful integration, and we are nevertheless left with the uneasy feeling that the wave-packet is a wave made of particles, in turn made of waves, and so on *ad infinitum*. On the other hand, those who have surrendered to abstraction are assuming that they actually have a physical structure before them during the thought experiment. This is not Physics, but semblance of Physics. Duality is an *ad hoc* concept, and the wave-packet, an extremely misleading term that stealthily attempts to pass a mathematical equation for a physical object. The wave-packet is the grandest monument, the most amusing symbol of the establishment's impotence to elucidate the correct architecture of light through the language of Mathematics.

The shape of light

If all experimental results can be expressed in terms of frequency and wavelength, light is obviously some sort of wave, but what kind of wave is it? The first clue arises from a basic observation. If your hand intercepts light, this wave is not longitudinal or transverse but 3D! If, in addition, common sense leads us to suspect that there is a Grand Unified Theory (GUT) that marries light (an outward force) with gravity (an inward force), we must simply brainstorm all the rational mechanisms that can simultaneously generate push and pull. [7]

One candidate that incorporates these features is a rope. You torque a taut rope and the sinusoidal signal instantly makes its presence felt at the opposite end. Coincidentally, the object attached there immediately feels tugged in your direction. This twined, DNA-like structure consists of anti-parallel, orthogonal electric and magnetic *threads*, physical entities quite unlike the ethereal *field* proposed by Classical and Quantum Mechanics. [8] Although it may be argued that the plane transverse wave is but a lengthwise cross-section of the rope, this observation actually underscores the motion-architecture dichotomy inherent in these irreconcilable systems. The 2D transverse wave is *dynamically* undulant whereas the 3D rope is *structurally* wavy. In a universe devoid of motion, the 2D transverse wave becomes what wave theorists allege today: absolutely nothing. The rope, instead, need not be torqued to acquire a rippled shape because this is its natural static state. The rope is also unique in that it is already attached at and tugs from both ends.

The rope configuration makes it apparent why light consists of both object (particle) and motion (wave). In a motionless state, the rope contains no information; it is actually an unremarkable object. We note, however, that consistent with wave theory, the shorter the links (wavelength), the greater the number of links that fit in a given length (frequency), making the rope the only physical configuration that can justify the constancy of the expression ($c = f * \lambda$). If we twist the rope the signal blitzes out at lightning speed, a phenomenon that could easily confuse the keenest observer. The signal travels rectilinearly along the taut intermediary while the rope moves in space, thus embodying *straightness* and *curving* (of the signal) in a single mechanism. More telling is the subtle bidirectional flow. When observing the middle of a spinning rope, the signal travels in diametrical directions. Suddenly, ray reversibility is also demystified. [17] If we torque the rope from both ends, the links get longer the farther we are from origin. Now the results of the Harvard Tower experiment make sense. [35] A rope twirls simultaneously clockwise (CW) and counterclockwise (CCW) depending on which end we face, and when compelled to reverse rotation, the opposite end has no alternative but to follow through. Now, EPR is also descrambled: it is no longer necessary to invoke mathematical wizardry or quantum magic to explain this phenomenon. [36] The electric and magnetic strands may spin counter to each other, so a cross-section of the rope would show 'photons' with both CW and CCW spin states. [37] As a bonus we have simplicity and perfect symmetry, features generally favoured by intuition. By making the assumption that light is a torque wave generated by a two-strand rope, we have tentatively been able to account for a series of well-established structural and dynamic aspects of light:

Static or structural aspects of light

1. Duality: why light behaves both as a particle (object) and as a wave (motion)
2. Maxwell and Faraday Laws: why the electric and magnetic 'fields' run 90° to each other and appear to induce each other into being
3. Why light is sinusoidal
4. Why the expression $c = f * \lambda$ is a constant
5. The physical interpretation of amplitude
6. Why wavelength increases away from the center of the Earth [35]

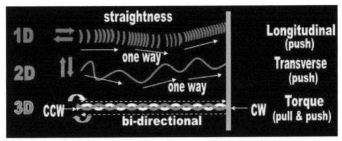

Fig. 1 How longitudinal (spring), transverse (ribbon), and torque (rope) waves measure up.
a. *Straightness*. Neither back and forth (long.) nor up and down (trans.) generate straightness. In Physics only the force of pull generates straightness. As the taut rope rotates in space it give the impression that light curves.
b. *Speed and Direction*. The twist of a taut rope sends the signal almost instantaneously in both directions (i.e., ray reversibility). The rope configuration embodies push and pull.
c. *EPR*. The rope twirls CW from one end and CCW from the other.
d. *Amplitude*. Only the rope achieves consistent amplitude.
e. *Orthogonality*. Only the rope configuration can justify why the electric 'field' runs at 90° to the magnetic 'field'.

Dynamic or motion-related aspects of light

1. Why light travels so fast.
2. Why light travels rectilinearly even during refraction and diffraction
3. Why each 'field' oscillates around an imaginary axis
4. The Principle of Ray Reversibility: retracing of the optical path [17]
5. EPR: instantaneous mutual influence of diametrically directed photons [36]
6. Simultaneous CW and CCW spin of a 'photon' [37]

The crucial attribute, however, is that we have integrated the two opposing forces of the universe—push and pull—into a single mechanism. Hence, in principle, the rope has the potential to unify light with gravity. Let's now factor in the hydrogen atom.

Why the electron is not a discrete bead

Thomson [38], Millikan [39], Rutherford [14], and Bohr [10], pioneered the discrete, planetary model of the hydrogen atom. This prototype consists of a positively charged bowling ball circled by a much lighter, negatively charged bead. Lewis [40] developed the shell model, and de Broglie [41] presented an electron that extends in an integral number of waves around Rutherford's proton sphere. Schrödinger [42] and Born [29] gave yet another physical interpretation by conceptualizing the electron as a cloud. The question actually begs a much simpler answer. The scientific method demands that proponents decide in advance whether the single, S-orbital electron of hydrogen is a cloud, a shell, a ring, or a bead and then to use this hypothesis consistently throughout their dissertation. And the fact remains that the prosecutors have overwhelmingly voted for the bead. Despite wholesale denials and disclaimers the architecture used in Quantum Mechanics is still Bohr's debunked planetary model. The

American Heritage Dictionary (*AHD*) defines an *ion* as an atom that has gained or lost one or more discrete electron beads. Ridley [43] defines *electric current* as the flow of discrete electron particles. Davies [44] portrays scattering as the exchange of supernatural virtual photons between two discrete electron marbles. And Ebbing [45] depicts covalent bonding as the inter-atomic sharing of discrete electron golf balls. Indeed, discrete electrons underlie Lewis's shell theory. Schrödinger lent credibility to de Broglie's Saturnian hypothesis, but the matter-wave equation relies on discrete quantities for electron mass and charge, implying that a finite object underlies it nevertheless. And Born's [46] electron cloud is really a cloud of probability, the region around the nucleus where a discrete electron bead is likely to be found. Therefore, the hydrogen atom in use today continues to be the Ptolemaic anachronism conjured by the Fathers of Quantum. The integral wave, the shell, and the cloud are not architectural models of a physical electron, but *regions* occupied by electron beads.

However, the discrete bead model of the electron runs into insurmountable obstacles and must be discarded once and for all by serious scientists for a host of reasons, among them:

a. Quantum has yet to offer a physical interpretation of *positive* and *negative*.
b. Newton's 1st and 3rd Laws require contact between two surfaces for one of them to change course. The quantum mechanical hydrogen atom tacitly has two interfaces: proton-field and field-electron. The electron bead is physically bound to and orbits the nucleus thanks to this intermediary known as *field*. However, field is not a physical object, but a concept. [8] Hence, QM implicitly has a *concept* binding the physical bead to the bowling ball. (Fig. 2 - Hydrogen)
c. QM has electron beads occupying energy levels. [10] However, Feynman [6] candidly confesses that physicists have no idea what energy *is*. Feynman has the scientific method backwards. The scientific method requires Feynman to tell the jury what X is for, or else the jury cannot understand what he is talking about. If Feynman doesn't know what X is, then he should not be allowed to use X in his dissertation.
d. Valence Band (VB) and Molecular Orbital (MO) theory hold not that the beads themselves, but rather that their trajectories (*i.e.*, orbitals) interact! [45]
e. VB and MO theories self-servingly avoid quantum paradox by alleging not that the negative bead, but that its charge-indifferent P-orbital (the *region* where an electron bead is likely to be found) runs unprejudiced through a barrier of positive protons in a nucleus and out the other side (Fig. 2 - Neon).

Until these objections are addressed, the scientific method requires us to reject that the electron can possibly be a discrete bead. Meanwhile, any theory and mathematics ever developed around the orbiting bead model is hereby rendered moot (*e.g.*, VB and MO theories).

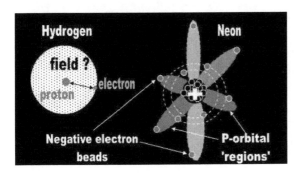

Fig. 2 The hydrogen and neon atoms of Quantum Mechanics
Hydrogen. QM has 'something' known as a 'field' physically binding the electron bead to the proton bowling ball. However, a *field* is not a physical object. It is defined as a 'region' of space (i.e., a concept).
Neon. QM self-servingly avoids paradox by running the charge-neutral p-orbital 'region' (a concept) rather than the negative electron 'bead' (an object) through the positive nucleus.

The shape of the hydrogen atom

If QM's cartoonesque atom is not viable, is there a rational model we can replace it with? Let's invoke EM threads again and suppose that the verb of MO theory is really a noun. Thus, the quantum *orbital* becomes what MO theory has assumed it to be for decades: a balloon. Electric and magnetic threads converge on the hydrogen atom from the remaining atoms in the Universe. The electric component of the rope continues straight into the center of the atom, where, at a radius of 10^{-15} m, the proton behaves as the impenetrable sphere Rutherford [14] experienced. The proton is a tiny dandelion, an intersection of electric threads extending outwards towards every atom in existence. [47] The magnetic thread forks out from the electric thread at the boundary of the electron shell and begins to circumvent the proton in a de Broglie-like [41] undulating pattern. Countless such threads approaching from near and far knit a wavy ball-of-yarn surface that encapsulates the dandelion in a Lewis-like [40] membrane known as the electron (Fig. 3). Under these assumptions all atoms, detected and undetected, are bound to each other via EM threads. The Universe is criss-crossed and threaded throughout by the medium through which the torque signal (light) travels. [48] Indeed, the underlying proposal is that matter consists of a single, closed-loop thread in the whole of space. This thread converts to tiny spherical knots known as hydrogen atoms, which then, consistent with current theory, fuse to form heavier atoms or join to form molecules, which in turn serve as building blocks for macro objects like human beings. Under the rope hypothesis, what prevents our planet from escaping the solar system is that every atom on Earth is physically connected *via* EM threads to every atom that comprises the Sun.

We now integrate the electron balloon, the proton dandelion, and the electromagnetic rope that interconnects two such structures into a simple, symmetric system consisting of two hydrogen atoms bound by a two-strand rope (Fig.

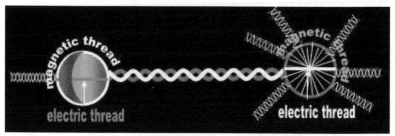

Fig. 3 Cross-section of two hydrogen atoms interconnected by an EM rope

3). We rely on this simple system to explain the most fundamental attributes of the hydrogen atom from a strictly physical perspective.

Once we integrate the rope, the electron shell, and the proton dandelion, it becomes apparent why two vibrating atoms act like they were connected by a *stiff spring* [49]—because indeed they are. For the rope to remain straight, any brusque motion of one atom necessarily shakes the one at the opposite end. But light is exchanged independently of the vibrating motion of the atom. The torque signal is the result of pumping. Consistent with Bohr's theory [10], when the electron balloon spontaneously contracts to a smaller radius (*i.e.*, makes a 'quantum jump' to a lower energy state), the rope picks up the slack. The hydrogen atom is said to have 'emitted' a photon, a 'quantum of energy', which predictably is measured as an integral number of 'wavelengths' (*i.e.*, rope links). This 'packet' does not travel towards infinity, but along the taut rope to the atom connected at the other end. When the balloon expands, it does so at the expense of EM rope, which again is drawn in an integral number of wavelengths. By its very nature, this model disallows intermediate 'orbitals' or energy bands. Hence, every rope converging on our atom is constantly torqued from both ends. The atom is a tiny heart perpetually pumping torque waves in and out of the system. The friction generated at every point along the electron surface where electric and magnetic threads fork out is a composite Millikan [39] measured as *charge*, a physical interaction that misleads Baeirlein [32] to conclude that a signal struck a wall and morphed into a particle. This model enables us to explain observations heretofore considered supernatural by QM advocates.

Static or structural aspects of the hydrogen atom

    a. Why the electron has wave, particle, shell, and cloud-like properties
    b. Structural stability: why the electron does not spiral into the nucleus
    c. The absence of intermediate 'energy bands'
    d. Why a P-'orbital' runs through the 'positive' nucleus
    e. de Broglie integral waves
    f. Why electric 'lines of force' begin and end on 'charges'.

Dynamic or motion-related aspects of the hydrogen atom

    a. Bohr's mystical quantum jump

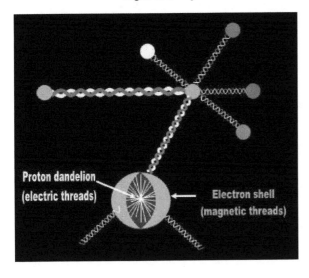

Fig. 4 The static universe

b. Why the atom emits and absorbs light in discrete quantum packets
c. Why light has constant velocity irrespective of the motion of the source
d. The Mössbauer Effect: recoilless emission [50]
e. Why light 'travels' as a wave and departs and arrives as a 'particle'
f. The physical interpretation of *charge*

From a universal point of view, the rope hypothesis also explains:

a. the physical meaning of c^2 (light travels in both directions from one atom to all others and *vice versa*)
b. Mach's Principle [51]
c. the inverse square rule of gravity (or why the Earth orbits the Sun)

We have now finished our introduction to Thread Theory (TT). We have arrived at the end of our quest: an illustration of the static universe we inhabit. The words spoken by Democritus 2500 years ago serve as appropriate closing remarks. He purportedly said '*The only existing things are atoms and empty space; the rest is mere opinion.*' We rephrase his materialistic vision in order to incorporate light and state that: '*The only existing things are atoms interconnected by ropes, both drifting in empty space; the rest is mere opinion.*'

Conclusion

For the past 400 years, Physics has entertained but two alternatives for light: wave and corpuscle; for the last 200 years theorists have used loosely defined concepts such as points, energy, field, and force to describe structure; and in the last 100 years Quantum Mechanics has treated photons and electrons as particles. Indeed, the apologetic CI is a result of QM's stubborn insistence on a corpuscular universe. Most of the equations of Quantum Mechanics and the work of hundreds of 'particle' physicists revolve around discrete particles.

However, without exception, every subatomic particle ever identified lacks the one attribute that would qualify it as such: shape. And nevertheless, even if mathematicians were successful in illustrating a single subatomic particle ever 'invented' to explain observation, QM would still fail. The reason theoretical physicists have yet to find gravity is that discrete corpuscles have no physical way of generating the force of pull, the only 'force' that QM has yet to explain rationally.

The purpose of this article was to underscore known weaknesses in the quantum/classical, wave/particle models and to introduce Thread Theory. The hypotheses just formulated are the pillars of a quite complex theory of light and gravity that arises as an alternative to purely mathematical endeavours. Thread Theory seeks to demystify longstanding physical paradoxes, among which we include the most notorious:

a. How do intersecting beams manage to elude interaction? (Under TT the question becomes: how do threads avoid tangling?) This is actually the only question Theoretical Physics needs to answer today!

b. EPR, tunnelling, ray reversibility, Mössbauer effect, polarization, *etc.*

c. How does a single entity generate both fringes in the double slit experiment and instantaneous electric current during the photoelectric effect?

d. What physical mechanism embodies the two forces of the universe: gravity and electromagnetism, respectively pull and push?

e. How do magnets physically manage to attract and repel (i.e., the physical meaning of positive and negative)?

f. How does an electric current physically propagate and generate a field?

Development of these topics are well beyond the scope of the instant paper. Meanwhile, the open-minded reader may want to ponder the words of Dr. Penrose [16], "Somehow, Nature contrives to build a consistent world in which particles and field-oscillations are the same thing! Or, rather, her world consists of some more subtle ingredient, the words 'particle' and 'wave' conveying but partially appropriate pictures." The rope hypothesis may very well be the subtle ingredient that has eluded detection since man began to tinker with light.

Notes and References

1. Newton, I., *The Mathematical Principles of Natural Philosophy*, Pemberton, 3rd Edition (1726), Translated by Andrew Motte 1729.

2. An object is 'that which has shape'. Gaede, W., *What* is an object!, *Apeiron* 10, No. 1 (Jan 2003). http://redshift.vif.com/JournalFiles/V10NO1PDF/V10N1gae.PDF

3. From a conceptual point of view, position is the volume of space displaced or occupied by an object. However, we argued that space is a place as opposed to an object (See ref. # 2). Then, no such displacement or occupation occurs. The object merely has a location with respect to the remaining objects in the universe. Under this assumption there is no meaning or purpose for the word *position* in Physics.

4. An object can have only one location. Hence, if an object successively occupies more than one location, it moves by definition and not because an observer is present to record the incident. An undetectable star moves or not, irrespective of observers. However, without memory, without a comparison between the previous location and the current one, we cannot talk about motion. Therefore, motion is contextual: we must settle in advance whether an observer is required or not.

5. Note that without a comparison all we have is motion! (*i.e.*, 30 km for Earth and either the length of a wave or the trajectory of a particle for the emission of caesium.)

6. Energy is defined as 'capacity...', an attribute of an object as opposed to an object in its own right. *The American Heritage Dictionary* (*AHD*), 4th ed, 2000. http://dictionary.reference.com. Capacity is a latent ability and conceptually unrelated to shape. Mathematically, energy = mass * (*velocity of light*)^2. These conflicting notions necessarily conduce to logical inconsistencies: *a.* Capacity (concept) and velocity (motion) are irreconcilable. *b.* 'Transfer of energy' implies the movement of velocity or of capacity. *c.* Quantum holds that light is comprised of discrete packets of energy, tacitly implying that energy has a perimeter or surface (*i.e.*, shape). The scientific experiment to test this last assertion is simple: the proponent merely needs to illustrate a packet. Only then can we subsequently visualize two locations (motion) of energy. But then again, the notion that energy is a physical object is irreconcilable with either velocity (two or more locations) or capacity (an attribute of an object). To complicate matters further, Glashow alleges that space is 'seething energy'. Glashow, S., *The Charm of Physics*, American Institute of Physics, 1991. See also Villard, R., Lloyd, R., Astrophysics challenged by dark energy finding, SPACE.com, April 10, 2001. http://www.space.com/scienceastronomy/ generalscience/darkenergy_folo_010410.html. Therefore, advocates that assume that light is the undulation of *nothing* are tacitly equating energy with nothing. With so many irreconcilable notions of the word energy, it is not surprising that Feynman wrote that physicists have no idea what energy *is*. Feynman, R., *Six Easy Pieces*, Perseus (1996), pp. 69-72.

7. Sheldon Glashow laments that "[Einstein] never cared to learn about the two other forces that exist in the universe, beyond gravity and electromagnetism." Glashow, S., *Interactions*, Warner Books (1989). Perhaps in Quantum Mechanics there are four forces, but in Physics there are only two. Force can be succinctly defined as 'cause of resistance or acceleration'. *AHD*. In Physics, there are only two forms of contact we can imagine that may change the state of rest of an object: push and pull. The weak interaction is conceptually a force of push and the strong interaction is conceptually a force of pull. The burden shifts to QM advocates to show how the atomic force that exiles particles differs from push and how the force responsible for the integrity of the nucleus differs from pull. Meanwhile, Einstein's intuition is vindicated.

8. *Field* is defined as 'a region of space...', a *region* is 'a portion of space', and *space* is 'the infinite extension of *field*. *AHD*. Classical and Quantum Mechanics propose that light is comprised of electric and magnetic *fields*, and QM alleges that the electron is bound to the proton via an electrostatic field. Therefore, aside from relying on circular definitions, advocates of Classical and Quantum Mechanics have summarily converted a concept into a substance.

9. Newton, I., *Opticks*, based on the 4th ed. London 1730, Dover, 1952.

10. Bohr, N., *Phil Mag* **26**, No 151 (1913), 1-25, 476-502, 857-875.

11. Eschner, J., *et al.*, Light interference from single atoms and their mirror images, *Nature* **413**, 495-498 (2001). http://www.nature.com/cgi-taf/DynaPage.taf?file=/nature/journal/v413/n6855/abs/413495a0_fs.html.

12. Nogues, G., *et al.*, Seeing a single photon without destroying it, *Nature* **400**, 239-242 (1999). http://www.nature.com/cgi-taf/DynaPage.taf?file = /nature/journal/ v400/n6741/abs/400239a0_fs.html

13. de Mul, F., *et al.*, Laser Doppler Velocimetry and Monte Carlo simulations on models for blood perfusion in tissue, *Applied Optics* 34, 1995, p. 6595—6611. Göbel, G., *et al.*, Monte Carlo simulations of light scattering in inhomogeneous spheres, M. Pinar Menguc, Ed., *Radiative transfer— II*, Proceedings of the Second International Symposium on Radiation Transfer. Kusadasi, Turkey,

July 1997. Begell House, New York 1998. http://diogenes.iwt.uni-bremen.de/vt/laser/papers/ kusadasi.pdf.

14. Rutherford showed that atoms are mostly empty space. Rutherford, E., *Phil Mag.* **21**, 669 (1911).

15. Foucault, L., *Compt. Rend. Acad. Sci.* **30** (1850) 551. We note that Foucault as well as Michelson *calculated* the speed of light and concluded that it travels slower through denser media. Their versions fail to explain what mechanism induces a wave or particle to regain speed after it returns to the less dense medium.

16. Penrose, R., *The Emperor's New Mind*, (Oxford University Press, U.K., 1999) Ch. 6.

17. Although scarcely alluded to in theoretical physics, the Principle of Ray Reversibility (PRR) is widely used in optics. See, for example, S. Klein and D. Garcia, "Line of sight and alternative representations of aberrations of the eye," *JRefract Surg* **16**, Sep/Oct 2000. http://www.cs.berkeley.edu/~ddgarcia/papers/JRefractSurg2000.pdf. Derived from Fermat's Principle, the PRR states that, whether reflecting or refracting, a ray of light retraces its path. The theoretical physicist should now stop to ponder what this entails. In the context of a two body system separated by several light years, the PRR implies that a discrete 'photon' retraces its exact path during the return trip! If the PRR is correct, General Relativity and Quantum Mechanics are entirely divorced from reality. A second interpretation of the PRR is that light travels in diametrical directions simultaneously. Again, this explanation excludes one-way waves and photons.

18. Popper, K., *The Logic of Scientific Discovery*, p. 30, Basic Books, New York, 1959. Hawking, S., *A Brief History of Time*, Bantam, (1998), p. 13.

19. Dyson, F., Eddington, A., and Davidson, C., A determination of the deflection of light by the sun's gravitational field made during the total eclipse of May 29, 1919, *Mem. Royal Astronomical Society* **62**, (1920) 291.

20. Huygens, C., *Traité de la Lumière*, Leiden: Van der Aa, 1690.

21. Fresnel, M., *Oeuvres complètes d'Augustin Fresnel: tome deuxième.* Paris, Imprimerie Impériale. (1868).

22. Michelson, A., *Am. J. of Science* **22** (1881) 120-129.

23. "Electromagnetic waves differ from other transverse and longitudinal waves in that they do not need a medium ...to travel through." Gunderson, P., *The Handy Physics Answer Book*, (Visible Ink Press, Detroit, 1999).

24. Einstein, A., *Ann. d. Physik* **17** (1905) 132.

25. Compton, A., *Philos. Mag* **45**, (1923) 1121.

26. A wave is a 'disturbance'. *AHD*

27. Young, T., *Phil. Trans. Roy. Soc.*, London xcii (1802) 12, 387.

28. Heisenberg mathematically demonstrated that we cannot know both the position and momentum of a particle to arbitrary accuracy. Heisenberg, W., *Zeitschrift fur Physik* **43**, 172 (1927). The Uncertainty Principle (UP) actually has a more down-to-earth explanation independent of observers or Mathematics. Heisenberg discovered that when the ball moves, it doesn't stand still, and when the ball stands still, it doesn't move. The UP is the result of not following the scientific method. The scientific method requires Heisenberg to begin by defining the words position, location, and motion. These fundamental qualitative concepts belong exclusively to Physics and precede any mathematical relation or standard. They cannot be defined in terms of coordinates, number lines, or mathematical symbols. They must be defined with grammar.

29. Born interpreted Schrödinger's wave function as the 'probability' of finding a particle within a given volume of space. Born, M., *Zeitschrift für Physik* **37**, 863 (1926a); **38**, 803 (1926b). Heisenberg reinterprets that the wave function does not have a physical location, but is actually an encryption of the observer's 'knowledge' of the particle's whereabouts. Heisenberg, W., *The Physical Principles of the Quantum Theory*, (Dover, NY, 1930) Trans. Carl Eckhart and F.C. Hoyt. Hence, neither wave-packet nor duality is related in any way to the structure of the particle itself.

This also follows from the dynamic nature of the components of the state vector (*i.e.*, energy, momentum, time, spin, *etc.*). The wave-function is all motion and no substance.

30. Complementarity primarily underscores the dual wave/particle behaviour of subatomic particles such as photons and electrons. N. Bohr, "Can quantum-mechanical description of physical reality be considered complete?" *Phys. Rev.* **48** (1935) 696. It says nothing of what light looks like in the 'behaviour-less' static universe.

31. Heisenberg makes the observer the central figure of any experiment without which there is no independent reality. He emphasizes measurement and knowledge limitations, and advises against intruding with opinions about reality and meaning. Heisenberg, W., *The Physical Principles of the Quantum Theory*, (Dover, NY, 1930) Trans. Carl Eckhart and F.C. Hoyt. Hence, none of Heisenberg's philosophical arguments have anything to do with the structural nature of light.

32. 'Light travels as a wave but departs and arrives as a particle'. Baierlein, R., *Newton to Einstein*, (Cambridge University Press, UK, 1992), p. 170.

33. Ridley, B., *Time, Space and Things*, (Cambridge University Press, U. K., 1984) pp. 12-13.

34. A photon is defined as a discrete particle (*AHD*) or as an indivisible elementary *particle (WordNet 1.6)*, Princeton University (1997). See also Buks, E., *et al.*, Dephasing in electron interference by a 'which-path' detector, Nature 391, (1998) 871-874. http://www.nature.com/cgi-taf/DynaPage.taf?file=/nature/journal/v391/n6670/abs/391871a0_fs.html

35. Pound, R, Rebka, G., *Phys. Rev. Lett.* 4 337 (1960); 4 275 (1960).

36. Einstein, A., Podolsky, B., Rosen, N., Can Quantum-Mechanical Description of Physical reality be considered complete? *Phys. Rev.* **47**, p. 777 (1935); Mair, A., *et al.*, Entanglement of the orbital angular momentum states of photons, *Nature* **412**, 313-316(2001). http://www.nature.com/cgi-taf/DynaPage.taf?file=/nature/journal/v412/n6844/abs/412313a0_fs.html

37. Waldman laments that in quantum linear polarization is a superposition of two circular polarizations where each photon must be thought of as spinning CW and CCW simultaneously! Waldman, G., *Introduction to Light*, (Prentice-Hall, Englewood Cliffs, NJ, 1983), p. 80.

38. Thomson, J., *Philos. Mag.* **44,** 293 (1897).

39. Millikan, R., *Philos. Rev.* **32**, 349 (1911).

40. Lewis, G., *J. Am. Chem. Soc.*, **38,** 762 (1916).

41. de Broglie, L., *Phil Mag* **47**, No 278 (1924).

42. Schrödinger, E., *Ann. der Physik* **79**, 361, (1926).

43. Ridley, pp. 18-19.

44. Davies, P., *The Forces of Nature*, (Cambridge University Press, U.K., 1986) Ch. 4.

45. Ebbing, D., Gammon, S., *General Chemistry*, Houghton Mifflin College; 6th Ed (Jan 1999).

46. Boslough, J., Worlds within the atom, *Nat. Geo.*, (May 1985) pp. 634-663.

47. "The gravitational attraction relative to the electrical repulsion between two electrons is 1 divided by 4.17×10^{42}! The question is, where does such a large number come from?" Feynman, R. *Six Easy Pieces*, Perseus (1996).

48. "But the electromagnetic theory is not based on action at a distance as was Newton's. Space is thought of as being threaded throughout with electrical and magnetic tensions." Disney, M., *The Hidden Universe*, MacMillan; (1985).

49. "A chemical bond acts like a stiff spring connecting nuclei. As a result, the nuclei in a molecule vibrate, rather than maintaining fixed positions relative to each other." Ebbing, p. 278.

50. Mössbauer, R., Nuclear resonance absorption of gamma rays in Ir^{191}, *Naturwissenschaften* **45**, 538 (1958).

51. Mach, E., *The Science of Mechanics: A Critical and Historical Account of Its Development*. La-Salle: Open Court (1960).

A Matter of Life... or Hypothesis on the Role of Electron Waves in Creating "Order out of Chaos"

Françoise Tibika-Apfelbaum
Department of Inorganic Chemistry
The Hebrew University, Jerusalem
Email: francoise@huji.ac.il

The intention of the present paper is to underline the continuity between matter and life, and to advance our investigation of matter.

Early in the last century, Oparin, a pioneer in studies of the origin of life, reached the following conclusion: "...the simplest living organisms originated gradually by a long evolutionary process from organic substances..." and "...the numerous attempts to discover some specific "vital energies" resident only in organisms invariably ended in total failure..."(1)

Today, this approach is central to most theories on the origin of life (2).

However, given the very nature of life, Oparin's statement contained a major difficulty: life was characterized primarily by its irreversibility, while matter was then considered essentially reversible.

Indeed, the irreversibility of matter, as expressed by the second law of thermodynamics, according to which the disorder of the world increases constantly, was thought and taught to be a consequence of our human limitation rather than an intrinsic property of matter. The limitation is that we do not live long enough to observe all the possible states a system can adopt. Our short stay on earth allows us to witness only those states with a high probability.

From this perspective, matter could not "give birth" to life since both realms were totally alien to each other, each "living" a different time. It would have been like watching a video film of a pregnant woman giving birth and suddenly finding the live baby in our living room.

The question that seemed unavoidable in Oparin's time was: How, in the long evolutionary process mentioned above, did we pass from a reversible world to an irreversible one?

Strangely enough, in spite of the importance of the question, the answer was found much later, in Prigogine's work.

Indeed, in his study on non-equilibrium thermodynamics (for which he won the Nobel Prize in chemistry in 1977), Prigogine showed that matter irreversibility is part of the very nature of matter. It is not an illusion, but the ineluctable result of the dynamics of large populations of particles constantly colliding. He showed that these collisions lead to resonances between the degrees of freedom of the particles. and it is these resonances which are responsible for time-symmetry breaking (3).

In other words, it is not because we are not able to live billions of years that we will never see a gas mixture spontaneously separating, but because irreversibility is an inherent quality of matter, as it is for life. Matter is irreversible because it is made of a multitude of particles. This may, by the way, lead to a satisfying answer to Schrödinger's queries: Why are atoms so small compared to our own dimensions? Why is the number of atoms so large? (4)

Thanks to Prigogine, in the passage from matter to life, there was no question of jumping from one reality to another. Continuity between inert matter and the living realm was established, Darwin's evolution theory could be "stretched" to Mendeleev's table, and Oparin's statement returned to its proper place.

It is not the purpose of this article to dwell on the mathematical development which led Prigogine to his results, but to focus on another area of Prigogine's work in an attempt to further transform our perception of inert matter.

Prigogine studied systems far from equilibrium. This domain of thermodynamics is the domain of the living. We constantly consume energy to keep us away from equilibrium. Indeed, "running away from equilibrium" is at the basis of our life; it corresponds to our instinct of life; it is at the root of the teleonomic character defined by Monod (5). In this context, metabolism and replication are seen as sophisticated strategies used by an organism to resist equilibrium. For all of us, equilibrium is synonymous with death: it is when it losses its ability to "resist equilibrium" that an organism disintegrates and dies.

On the other hand, in the world of matter, the "instinct" of chemical systems is to rush to equilibrium, spontaneously, at the first opportunity, as iron attracted by a magnetic field.

It is known that a spaceship can be projected out of the field of earth's gravitational attraction. Similarly, one might ask: Is it possible to remove a chemical system from the field of equilibrium attraction?

As long as energy is provided a chemical system can be kept far from equilibrium. The shock was to see that, under specific conditions, far from equilibrium, certain chemical systems, the oscillating reactions, spontaneously generate organized structures. These spectacular and unexpected structures are known today as dissipative structures.

The phenomenon was observed as early as 1921, but for decades it was rejected as an artifact by most chemists because it seemed to violate the second principle of thermodynamics. (6) Tremendous effort was made to fight this rejection and convince a reluctant community to open their eyes and minds to unexpected but real facts. Nearly half a century later, in 1968, Prigogine and his co-workers presented a mathematical model for those reactions, showing full compliance with the second principle of thermodynamics. (Yet, as late as 1972, some scepticism remained. (6))

What we want to remember about these systems is this sentence from Prigogine, in which he refers to their spontaneous auto-organization, or to what he later called "order out of chaos":

> Such a degree of order stemming from the activity of billions of molecules seems incredible.... To change the color all at once, molecules must have a way to "communicate." The system has to act "as a whole. (7, p.148)

Though this conclusion leaves us quite perplexed, coming from Prigogine, it is worthy of consideration.

In 1924, Louis de Broglie, considering the dual nature of light as suggested by Einstein (1905), thought of the possible dual nature of matter—corpuscular and wave-like. De Broglie suggested the following mathematical relation, which assigns a specific wavelength λ to any particle of momentum p, $\lambda = h / p$ (h = Planck's constant).

Since then, this relation has been confirmed with great success throughout experiments on elementary particles, not only electrons (8), but also atoms (9) and neutrons. (10) The correspondence between experiment and theory is impressive.

Two years later in 1926, Schrödinger found, for bound particles such as electrons, the function that describes this wave, the wave function ψ that revolutionized physics.

Consequently, the building blocks of our universe became hybrids of two components, one exhibiting properties taken from the physics of waves, and the other exhibiting properties taken from the physics of rigid bodies. However, even today, we have no precise idea what this hybrid really is. This is what the Heisenberg Uncertainty Principle stipulates: if one focuses on one part of the hybrid, the other part becomes hazy. Moreover, when scaling up the phenomenon and dealing with large populations of atoms and their electrons, the wave component of this hybrid seems to be "lost" and one is left with a rigid body. Indeed, the wavelength associated with any macroscopic object is smaller than the dimensions of any physical system, and no wave phenomena, such as interference, can be observed, not even for dust particles.

Therefore, except for limited and specific cases (like those mentioned above), we have dropped the idea of a tangible picture of the wave component of that hybrid and turned to a mathematical interpretation, given by Max Born, who showed that $|\psi(x)|^2$ gives the probability of finding a particle at point x.

However we should not forget that at the atomic scale, there is no such thing as a rigid body and that what we call matter is an aggregate of these strange hybrids, an "intensive state" (by analogy to intensive properties) of something we still cannot grasp.

De Broglie, who died in 1987, witnessed the tremendous success of quantum theory, but did not agree with the abstract interpretation given to "his" wave. He wrote: "...By which strange coincidence could a representation of probabilities propagate in space through time like a physical wave able to be reflected, refracted and diffracted?" (11)

Similar reservations were expressed by Schrödinger: "...The mathematical representations used by theoreticians must be only a manner of describing with precision the nature of the considered phenomena, and must not be reduced to a simple intellectual gymnastics," (12)—as well as by Einstein and Planck. Einstein refused to accept that all we could observe was a probability. He wrote to his friend Born (13): "I am not satisfied with the idea that we possess a machinery that enables us to prophesize but to which we are not able to give a clear sense." For Einstein, quantum mechanics was not wrong but incomplete, unfinished.

Since the interpretation given to $|\psi(x)|^2$ has lead to exact predictions and, as such, forms the cornerstone of the highly successful quantum theory, attempts to give a tangible description of ψ, the wave function of the electron, have declined. Indeed, it is so hard to envisage what this wave function is that we cannot help but wonder if solving the wave-particle duality enigma might not be like transcending body-mind duality.

However, after this digression, we return to dissipative structures and the communication between molecules, mentioned by Prigogine.

During an oscillating reaction there is a reversal of the spontaneous motion of electrons from maximum disorder, (i.e., the most probable electron arrangement) to organized structures. Prigogine said that this organization is the result of communication between molecules. Our hypothesis is that, if communication between molecules indeed exists, it is accomplished through the wave associated with their electrons. Even though these waves are not directly observable, they may induce observable effects. It seems plausible that information about these structures may be transmitted through coherence between these electron waves.

Acknowledgments

I dedicate this paper to the memory of Ilya Prigogine, and thank Danielle Storper-Perez, without whom this article could not have been written.

References

1.   Oparin A.I. *The Origin of Life*, Dover Publications Inc, New York, 2nd Ed, p.60 and p.246, 1953.

2.   a) Lifson S. On the Crucial Stages in the Origin of Animated Matter, *J Mol Evol*, 44:1-8, 1997. b) Elitszur AC. *Time and Consciousness*, Ministry of Defence publishing, Israel, 1994.

3.   Prigogine I. *The End of Certainty*, The Free Press, New-York, p 67, 1996.

4.   Schrödinger E. *What is life?*, Cambridge University Press, 1945.

5.   Monod J. *Chance and Necessity*, Vintage Books Edition, N-Y, p.13, 1972.

6.   Epstein I.R and Pojman J.A. *An Introduction to Nonlinear Chemical Dynamics*, Oxford University Press, New York, Oxford, 1998.

7.   Prigogine I and Stengers I. *Order out of Chaos*, Bantam New Age Books, p.148, 1984.

8.   Davisson C and Germer L.H. Diffraction of Electrons by a Crystal of Nickel, *Phys Rev* 30:705-740, 1927.

9.   Carnal O and Mlynek J. Young's double-slit experiment with atoms: A simple atom interferometer, *Phys Rev Lett*, 66:2689-2692 ,1991.

10. Gähler R and Zeilinger A. Wave-optical Experiments with Very Cold Neutrons, *Am J Phys*, 59(4): 316-324, 1991.

11. de Broglie L. *Certitudes et Incertitudes de la Science*, Eds Albin Michel, Paris, 1966.

12. Lochak G. Convergence and Divergence between the Ideas of de Broglie and Schrödinger in Wave Mechanics, *Foundations of Physics*, 17(12): 1189-1203,1987.

13. Einstein A and Born M. *Le Grand Débat de la Mécanique Quantique*, La Recherche, 3:137-144, 1972.

The Electron as an Extended Structure in Cosmonic Gas

Adolphe Martin
10299 avenue Bois de Boulogne, #7014
Montreal, QC H4N 2W4

The ether (vacuum, zero-point energy, spacetime, *etc.*) is modeled as a gas composed of elementary particles called cosmons. Physical fields are defined by mechanical properties of a cosmonic gas. These gas properties are necessary and sufficient to derive the structure of the electron as an extended fundamental particle, based on known (measured) properties, such as mass, isospin, hypercharge, and two concentric charges of opposite sign equivalent to the single electron charge *e*.

Introduction

Previous studies of relativity showed that Einstein-Minkowski space-time is an isomorphism of Galilean space and time [Martin 1994b, 1998], thereby unifying electromagnetism with classical mechanics, and making it possible to use classical mechanics to describe the nature of physical phenomena. A new model is proposed to explain known physical phenomena [Martin 1994a]. The material substratum is modeled as a gas composed of elementary particles called cosmons which are agitated in all directions. This "cosmonic gas" is assumed to pervade the Universe, even the space between and within fundamental particles.

It will be shown that the well-known mechanical gas properties [Loeb, 1961] are sufficient to account for the electrical, magnetic and gravitational fields and other physical phenomena. Arguments will then be presented to show that fundamental particles are spatially extended configurations of cosmonic gas. Lastly, the results of calculations to determine electron structure will be presented.

Cosmons

Cosmons have no moving parts, and hence no internal energy and, according to Einstein, no rest, inertial or gravitational mass. The space between cosmons is absolute void. Cosmons are thus alone in the Universe, and there are no fields or forces at this level. The sole property of a cosmon, its diameter, determines a certain volume of space that is forbidden to other cosmons. The cosmonic gas is the substance of which the universe is made.

Interaction with other cosmons occurs only during encounters, where there is an exchange of velocity components along the line of centres, with a quantum space jump of one cosmon diameter. The velocity components normal to the line of centres remain with each cosmon. Between encounters, cosmons move at constant velocity (speed and direction).

Laws of physics from cosmonic gas equations

In accordance with kinetic gas theory, cosmon velocity varies from zero to indefinitely large values in a Maxwell distribution, the mean speed of agitation C, with $5C^2 = 2c^2$, c being the speed of light. Because they have zero spin, cosmons are bosons, and their energy varies according to Planck's distribution law. Classical concepts such as mass, charge and magnetism, are properties of the gas at this level.

In the vacuum, which is space devoid of fields and fundamental particles, cosmonic gas properties are spatially uniform by definition. Local vacuum properties (denoted by the subscript 0), are taken as the zero level of all physical measurements.

Fields are produced by local gradients or variations of gas properties due to the presence of cosmons. In the cosmonic gas, the total pressure (the sum of kinetic energy density and static pressure) defines the quantity electric charge density × electric potential. The total pressure gradient parallel to streamline accelerates the gas element, giving it velocity, thus producing electric current. The gradient of total pressure normal to streamline results from vector product of gas flow × vorticity (twice the rotational velocity of the gas element). The induced velocity due to a vortex line element at any point around it is proportional to this pressure gradient normal to streamline and inversely proportional to the cube of distance from vortex element to the point considered. This induced velocity is normal to both the pressure gradient and the vorticity. "This is exactly the Biot-Savard Law in electrodynamics, from which the magnetic field in the neighbourhood of a current carrying wire can be calculated." (Prandtl and Tietjens, 1957: p. 206) Magnetic lines of force are simply vortex lines in the cosmonic gas. These are the necessary conditions for the existence of electromagnetic fields.

Constant total pressure corresponds to irrotational flow of the cosmonic gas, which is the signature of purely mechanical phenomena, as in acceleration of neutral matter. Static pressure and kinetic energy density then vary in complementary fashion.

There are no forces between cosmons. Therefore, the term of the Van der Waals equation which accounts for potential between molecules of gas has no role in the cosmonic gas. In the resulting Clausius equation, $P(1 - b/V) = NkT$, gravitation is due to the volume term. When this term is negligible, the equation accounts for the phenomena of QED. When this term is not negligible, it assumes the form $-2\pi N/(N_{max} - N)$, where N_{max} is the numeric density (concentration) of cosmons at maximum compaction (mean free path being 0). This dimensionless factor multiplies the pressure p in the cosmonic gas equation. This negative pressure accounts for the gravitational term. It thus reproduces, at a different scale, all the effects of the electromagnetic and weak interactions, which then become the strong and colour forces of QCD.

A black hole in cosmonic gas theory is defined as a configuration where the gravitational term of the Clausius equation is greater than the pressure term, which produces gravitational collapse, and cosmon number density reaches its maximum value, N_{max}. There is no singularity.

The main equations of quantum physics are derived from the viscosity formula for gases. The Planck constant thus corresponds to the minimum quantum of action in the cosmonic gas. It can be defined as the cosmonic gas viscosity coefficient per cosmon, or μcL, where μ is mean cosmon mass and L is mean free path. The Einstein ($E = mc^2$) and Planck ($E = h\nu$) energy formulae follow automatically. Highly transient quantum mechanical phenomena are due to cosmonic gas viscosity effects. Quasi-permanent quantum mechanical phenomena are due to mechanical resonance (quantum conditions), which neutralizes viscosity effects, giving the cosmonic gas superfluid properties.

Gas vortex in fundamental particles

Fundamental particles are thus seen as spinning concentrations of gas (vortices) exhibiting gradients of pressure, density, and temperature. As a result, a moving fundamental particle will automatically possess wave properties. The wave-particle duality is thereby explained.

According to Maxwell, wherever the electric field has a divergence (sum of partial differentials), there is an electric charge density. Consequently, fields—both electromagnetic and gravitational—which extend indefinitely far in space (as do their charge, mass and energy), are part of the fundamental particles, which in turn must also extend indefinitely far in space. The general applicability of wave mechanics, the apparent nonlocality of fundamental particles, and diffraction phenomena are thus explained.

When the Coulomb law is modified by replacing the invariant e with a charge e_r which varies from 0 at the centre to e at $r \to \infty$, the problem of infinities is removed, and with it the need for renormalization.

In gas dynamics there are two very stable velocity configurations in three dimensions. Both have toroidal geometry.

Circular vortex

The circular vortex is analogous to a circular smoke ring. Series of tori are centred on a common axis, their meridians, defined by planes passing through the axis, being circles whose centre describes a circle in the equatorial plane normal to the axis. These torus meridians are streamlines of the gas. This pattern is rotated about the axis, making it spherical.

Planes normal to the axis cut the torus surfaces in circular lines, all centred on the common
axis. These circular lines are vortex lines in the gas.
Hence the name "circular vortex."

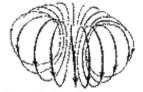

Spin vortex

Figure 1 - Circular vortex.
Streamlines follow the torus
surfaces.

The geometrical arrangement of the spin vortex
(also called "spherical") is the same as the circular
vortex. However, the flow and vortex lines are interchanged. The streamlines
are circles concentric with the axis, while the vortex lines are the tori meridians.

This pattern exists in every *spinning* concentration. The closer the merid-
ian circles of the tori, the higher the rotational velocity. In purely mechanical
configurations (irrotational flow), the rotational velocity becomes mechanical
spin, and neutral currents follow the streamlines. In electrical concentrations
(rotational flow), the pattern is equivalent to iso-spin, while electrical currents
follow the streamlines, and the vortex lines are the magnetic lines of force of
the magnetic field.

Moving particle

The circular vortex pattern, combined with a uniform flow pattern, produces
the flow around a sphere. There is no flow across the sphere, so the gas inside
the sphere moves with it. The gas outside the sphere, however, does flow past
the sphere. This combined velocity pattern accompanies every concentration
moving in the gas, including the photon. (The Compton effect is a vectorial exchange of energy and momentum components of the circular vortices.) The
higher the velocity of the concentration, the higher the velocity along the
streamlines and the greater the vorticity on the vortex lines. At the two inter-
sections of the axis of motion with the sphere, the velocity falls to zero relative
to the concentration, and the pressure goes to a maximum. The distance between these two points is the deBroglie wavelength.

The sphere is infinitely large at zero velocity of the concentration, and decreases in size as the velocity increases. It shrinks to zero at infinite Galilean
velocity V or when Einstein velocity $v = c$. The relation between these two velocities is $v/c = \tanh(V/c)$.

Similar purely mechanical patterns at constant
total pressure account for the kinetic energy of neutral concentrations. The flow pattern (circular vortex) due to the concentration's velocity, combined
with the iso-spin pattern (spherical vortex), when
viewed in the direction of motion, produces a precession of the spin axis to the left, normal to the

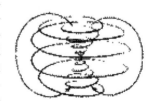

Figure 2 – Spin vortex.
Streamlines are circles centred
on the vertical axis.

**Table 1. Fundamental particle spectrum from Q = (t + Y) – Y/2
(values in units of e/6)**

Y \ Q	-4/3	-1	-2/3	-1/3	0	1/3	2/3	1	4/3	
						t = -1			t = 0	
							t = -½			t = ½
2					6-6 Z		12-6 W+Z			
5/3						7-5 Y		13-5 X		
4/3				2-4*			8-4		t = 1	
1					3-3 n ν		9-3 e p			
2/3			-2-2 2d			4-2		10-2 2u		
1/3				-1-1 d	vacuum		5-1 u			
0		-6+0 W π			0-0 γ q**		6-0 W π			
-1/3			-5+1 u			1+1 d				
-2/3	-10+2 2u				-4+2		2+2 2d			
-1		-9+3 e p			-3+3 n ν					
-4/3			-8+4			-2+4*				
-5/3	-13+5 X			-7+5 Y						
-2		-12+6 W+Z			-6+6 Z					
						X & Y (theor.)				

* impossible
** gluon

velocity, of all ½-spin matter particles. Anti-particles precess to the right.

For integral iso-spins, –1, 0, +1, the particle spin is to the right or left. Because the spin axis is parallel to the velocity in these particles, the streamlines of the spin vortex are parallel to the vortex lines of the circular vortex, and similarly, vortex lines of the spin vortex are parallel to the streamlines of the circular vortex. There is no induction and no precession of the spin axis.

Fundamental particle spectrum

When fundamental particles are classified by their charge Q (abscissa) and hypercharge Y (ordinate), the above table results, with each row representing iso-spin (–1, –½, 0, +½, +1), according to the Gell-Mann formula $Q = t + Y/2$.

In leptons, however, where gravity is negligible, equilibrium can be obtained only with two concentric charges of opposite sign, and the formula must be modified to $Q = (t + Y) - Y/2$, which yields the same charge as the classic

Gell-Mann formula when the particle is viewed from a distance. The $(t + Y)$ term includes the additive or subtractive spin effect on the central charge where it has its greatest effect. This component, which is normal to the toroidal vortex line, combines with the centrifugal force due to spin. This resultant adds to the radial total pressure gradient, thereby contributing to the charge of the particle. The $Y/2$ term is a charge, of opposite sign and concentric to the $(t + Y)$ term, that is produced by vacuum polarization. It extends indefinitely from the central charge to the surrounding space. Neutral particles are mass concentrations with charges of equal and opposite sign.

The values of these terms are multiplied by 6 to give integral values in the chart. They form a continuous quantized series from −13 to +13, including hypothetical unobserved X and Y particles. The odd values are distributed on half-integer iso-spin lines $t = \pm\frac{1}{2}$, and correspond to left-handed fermion particles. Even values are concentrated on integral spin lines corresponding to bosons.

The values $(Y + t) = 0$, with $Y/2 = 0$, correspond to the vacuum level. It can be seen that the values of particles and anti-particles of each species are symmetrical (opposite sign) about the vacuum values. This condition is required for photon energy to transform into a particle/anti-particle pair from the vacuum, together with the inverse (photon pair production).

Calculation of electron structure

Due to the particle/anti-particle symmetry of opposite charges, it is assumed that the properties at the centre of the electron have the same values as the vacuum. This symmetry requires that the functions of gas properties also be symmetrical about the vacuum values. Since we assume a gas medium, the Lane-Emden function for the equilibrium of a polytropic gas sphere (Chandrasekhar, 1939) meets all the requirements for stationary fundamental particles. This function is also a requirement for the equilibrium of any spherically symmetric configuration (Kompaneyets, 1961).

A polytropic gas sphere with index $n = 5$ is the only one that extends to infinity and yet has finite mass and charge. Electrical concentrations are obtained by setting $G = 1$, replacing M by e and the pressure term p by total pressure P, thus making allowance for spin kinetic energy and iso-spin charge in a spherical geometry. For a circular vortex, the relations between vorticity (2 × velocity), orbital velocity and velocity potential are nearly the same as for the polytropic gas sphere properties between charge density, field, and electric potential with the same functions along the radius (Prandtl and Tietjens, 1957). This explains why isospin acts as part of the total charge. In our calculations, the same value of isospin ratio $t = \frac{1}{2}$, is part of each constituent charge, as in the total charge.

The calculations have been performed for the electron concentration with the Lane-Emden function applied to each of two superimposed charges e_{c1} and e_{c2} of opposite sign. The functions, derived directly from the mass, charge, iso-

Table 2 – Calculated values for electron structure

A	B	C	D	E	F	G	H	I	J
Electron		mass m	9.10E-28	$e_1/e = 3/2$	$e_2/e = ½$	$k(mc^2)$	t	9.10E-28	mass m_e
CONSTANTS			-7.2E-13	2		0.470588	0.5	4.80286	e E-10
a_1/a_2	$1-a_1/a_2^2$	$(B4^{3/5})$	$a_1/a_2^{6/5}$	e_{c1}/e	e_{c2}/e	e_1/e	$-e_2/e$	2.81785	r_e E-13
0.247511	0.938738	0.962779	0.187203	2	1	1.5	0.5	2.41E-07	$3\pi e^2/32r_e$
a_1/a_2	a_2/a_1	e_1/e	(G4–C6)	a/a_1	a/a_2	q/e	r_e/a	0.625864	$51\pi/256$
0.247511	4.040219	1.5	-6.6E-16	0.266489	0.065959	1	1.59779	5.12E+27	$3e/4\pi r_e 3$
c_o	h	K_1	K_2	ρ_{e1}/ρ_e	$-\rho_{e2}/\rho_e$	$3e/4\pi a^3$	$\sin^2\alpha_{o1}$	1.59779	$256/51\pi$
2.99E+10	6.62E-27	0.001009	0.000331	1.007639	0.007639	2.09E+28	0.914127	0.470588	$k(mc^2)$
a_1/a_2	$1-a_1/a_2$	e_{c1}/e	e_{c2}/e	r_e/a_1	r_e/a_2	ρ_e	ψ_c	P_c	$\Delta\Omega$
0.247511	0.752488	2	1	0.425794	0.105389	7.85E+26	211.9759	1.66E+29	-7.2E-13
T_o	r_o/a_1	e_{c1} E-10	e_{c2} E-10	a_1 E-13	a_2 E-13	r_o/r_e	r_o E-13	mc^2 E-07	$8/17mc^2$
2.736	3.262691	9.60572	4.80286	6.617856	26.73759	7.662588	21.59202	8.186193	3.852326

spin, and hypercharge, bear a strong resemblance to the Schrödinger wave forms for a free electron. The general equations are for concentrations with two charges, given known electron values, and values of particle charges e_1 and e_2, from the particle spectrum chart. The resulting parameter values represent stability conditions for the electron structure.

$$-\Omega = \frac{2}{3}mc^2\left[\frac{5}{3} - \frac{t^2}{(t+Y/2)^2}\right]^{-1} = \frac{8}{17}mc^2 = \frac{3\pi}{32}\frac{e^2}{a}$$

$$a = \frac{51\pi e^2}{256mc^2}; \quad \frac{e^2}{mc^2} = r_e \text{ (classical electron radius)}$$

$$\frac{e_{c1}}{e} = \left[\left(\frac{e_1}{e}\right)^{2/5}\cdot\left\{1-\left(\frac{a_1}{a_2}\right)^2\right\}^{3/5} + \left(\frac{e_{c1}}{c}-1\right)^{2/5}\left(\frac{a_1}{a_2}\right)^{6/5}\right]^{5/2}$$

and with $e = e_1 - e_2 = e_{c1} - e_{c2}$

$$\frac{e_{c1}^2}{a_1} - \frac{e_{c2}^2}{a_2} = \frac{256}{51\pi}mc^2$$

$$\left(\frac{e_{c1}}{e}\right)^2\left(\frac{a_2}{a_1}\right) - \left(\frac{e_{c2}}{e}\right)^2 = a_2\frac{256mc^2}{51\pi e^2} = \frac{a_2}{a}; \quad \frac{a_1}{a} = \frac{a_2}{a}\left(\frac{a_1}{a_2}\right)$$

where e_1 is inner charge for $0 \le r \le r_0$, e_2 is outer charge from $r_0 \le r \le \infty$, while e_{c1} is major polytropic charge, e_{c2} is minor polytropic charge for both $0 \le r \le \infty$, and r_0 is radius of inner charge e_1, and a_1 is the radius of a sphere at constant density ρ_{c1} with charge e_{c1}; a_2 is the radius of a sphere at constant density ρ_{c2} with charge e_{c2}, as isospin ratio of $t = ½$ is applied to each charge.

The calculations for the resolution of these equations were made in a computer spreadsheet (Table 2). Values of a_1/a_2 and ec_1/e are iterated until value of e_1/e calculated from the inverse of the equation for ec_1/e reached the

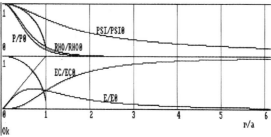

Figure 3 – Polytropic values against r/a for a single charge. This configuration is physically unstable, since the distributed charge would explode by self-repulsion.

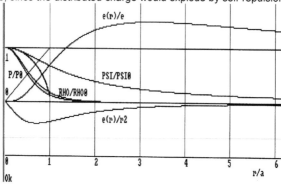

Figure 4 – Polytropic values for two polytropic charges of opposite sign in equilibrium.

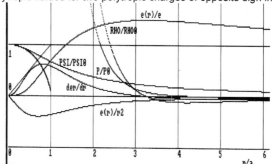

Figure 5 – Same as graph 2, but with P and ρ values ×100. This shows that ρ passed at
zero values at r_0/a_1 = 3.26 as calculated, determining the maximum of e_r/e.

required value of $e_1/e = 1.5$. The remaining values relevant to electron structure
are given in Table 2.

These lead to values of charge density ρ_c, electrical potential ψ_c, and total
cosmonic gas pressure P_c, as follows:

$$\rho_c = \frac{3}{4\pi} \sum \frac{e_c}{a^3} = 7.85 \times 10^{26} \text{ esu cm}^{-3}$$

$$\psi_c = \sum \frac{e_c}{6a} = 2.1198 \times 10^2 \text{ erg esu}^{-1}$$

$$P_c = \rho_c \psi_c = 1.66 \times 10^{29} \text{ erg cm}^{-3}$$

With polytropic index 5, ψ varies as $\cos\alpha$, ρ varies as $\cos^5\alpha$, and P varies as $\cos^6\alpha$ ($0 \leq \alpha \leq \pi/2$). The ratio of varying electron charge to constant Coulomb charge is defined as e_r/e shown in the graph in Figure 4. The electron's electrical field $E = e_r/r^2 = e/r^2(e_r/e)$. The ratio of the electron's field to the Coulomb field is e_r/e. Thus e_r/e is the ratio of the electron charge and field to Coulomb values.

The results of these calculations are amenable to experimental verification, probably in electron-electron collisions.

References

Chandrasekhar, S. (1939). *An introduction to the study of stellar structure*, Dover, New York.

Kompaneyets, A.S. (1961). *Theoretical physics*, Dover, New York.

Loeb, L.B. (1961). *The kinetic theory of gases*, 3rd edition, Dover, New York.

Martin, A. and Keys. C.R. (1994a). "The ether revisited," in: Barone, M. and Selleri, F. (eds) *Frontiers of fundamental physics*, Plenum, New York.

Martin, A. (1994b). Light Signals in Galilean relativity, *Apeiron* Nr. 18, pp. 20-25.

Martin, A. (1998). "Reception of light signals in Galilean space-time," in: *Open questions in relativistic physics*, Selleri, F. (ed.), Apeiron, Montreal.

Prandtl, L. and Tietjens O.G. (1957). *Fundamentals of hydro- and aero-mechanics*. Dover, New York.

Made in the USA
San Bernardino, CA
28 January 2017